SD Card Projects Using the PIC Microcontroller

SD Card Projects Using the PIC Microcontroller

Dogan Ibrahim

AMSTERDAM • BOSTON • HEIDELBERG • LONDON
NEW YORK • OXFORD • PARIS • SAN DIEGO
SAN FRANCISCO • SINGAPORE • SYDNEY • TOKYO
Newnes is an imprint of Elsevier

Newnes is an imprint of Elsevier
30 Corporate Drive, Suite 400, Burlington, MA 01803, USA
The Boulevard, Langford Lane, Kidlington, Oxford, OX5 1GB, UK

Library of Congress Cataloging-in-Publication Data
Ibrahim, Dogan.
SD card projects using the PIC microcontroller / Dogan Ibrahim.
p. cm.
Includes bibliographical references and index.
ISBN 978-1-85617-719-1 (alk. paper)
1. Microcontrollers—Programming. 2. Programmable controllers. 3. Computer storage devices. I. Title.
TJ223.P76.I275 2010
004.16—dc22

2009041498

British Library Cataloguing-in-Publication Data
A catalogue record for this book is available from the British Library.

For information on all Newnes publications,
visit our Web site, www.elsevierdirect.com

Printed in the United States of America

10 11 12 9 8 7 6 5 4 3 2 1

Typeset by: diacriTech, Chennai, India

Copyright Exceptions

Thanks is due to Microchip Ltd for their technical support and permission to include MPLAB IDE, MDD library, and Student Version of the MPLAB C18 compiler on the Web site that accompanies this book.

Contents

Preface

A microcontroller is a single-chip microprocessor system that contains data and program memory, serial and parallel input–output, timers, and external and internal interrupts, all integrated into a single chip that can be purchased for as little as $2.00. Approximately 40% of microcontroller applications are in office automation, such as PCs, laser printers, fax machines, intelligent telephones, and so forth. Approximately one-third of microcontrollers are found in consumer electronic goods. Products like CD players, hi-fi equipment, video games, washing machines, cookers, etc., fall into this category. The communications market, automotive market, and military share the rest of the application areas.

Flash memory cards are high-capacity nonvolatile read-write type semiconductor memories used in many domestic, commercial, and industrial applications. For example, portable electronic devices like digital cameras, video recorders, MP3 players, GPS receivers, laptop computers, and many more domestic and office products use some form of flash memory cards. Currently, there are many types of flash memory cards. Some of the popular cards are secure digital (SD) card, compact flash card, memory stick card, smart media card, and so on.

This book is about SD memory cards; it gives the basic working theory of the cards and describes how they can be used in PIC microcontroller-based electronic projects. Eighteen fully tested and working projects are given in the book to show how SD cards can be used for storing large amounts of data.

This book has been written with the assumption that the reader has taken a course on digital logic design and has been exposed to writing programs using at least one high-level programming language. Knowledge of the C programming language will be useful. In addition, familiarity with at least one member of the PIC16F series of microcontrollers will be an advantage. Knowledge of assembly language programming is not required because all the projects in the book are based on C language.

Chapter 1 presents the basic features of microcontroller systems. It also introduces the important topic of number systems and describes how to convert a given number from one base into another base.

Chapter 2 provides a review of the PIC18F series of microcontrollers. The various features of these microcontrollers are described in detail.

Chapter 3 provides brief details about commonly used memory cards. SD cards are currently the most widely used memory cards. The technical details and communication methods of these cards are described in the chapter.

Chapter 4 begins with a short tutorial on C language and then examines the features of the MPLAB C18 compiler used in all of the projects in this book. A fully working student version of the compiler is also given on the Web site that accompanies this book.

Chapter 4 also covers the advanced features of the MPLAB C18 language. Topics like built-in functions, simulators, and libraries are discussed, along with working examples.

Chapter 5 explores the various software and hardware development tools for the PIC18 series of microcontrollers and gives examples of various commercially available development kits. In addition, development tools like simulators, emulators, and in-circuit debuggers are described, with examples.

Chapter 6 provides some simple projects using the PIC18 series of microcontrollers and the MPLAB C18 language compiler. All the projects in the chapter are based on the PIC18F series of microcontrollers, and all the projects have been tested and are working. The chapter should be useful for those who are new to PIC microcontrollers and for those who want to extend their knowledge of programming the PIC18F series of microcontrollers using the MPLAB C18 compiler.

Chapter 7 is about the PIC microcontroller SPI bus interface. SD cards are usually used in SPI bus mode, and this chapter should provide an invaluable introduction to the SPI bus and its programming using the MPLAB C18 compiler.

In this book, the Microchip SD card function library, known as the memory disk drive (MDD) library, is used in all SD card–based projects. Chapter 8 gives the details of the MDD functions and describes how they can be used in projects to create files on the SD card and how to read and write these files.

Chapter 9 provides 18 working and fully tested SD card–based microcontroller projects. The block diagram, circuit diagram, full program listing, and description of each program are given for each project. The projects include simple topics like creating files on an SD card, formatting a card, and reading and writing to the card. In addition, SD card–based complex data-logging projects are given, where ambient temperature and pressure are read and stored on the SD card with real-time stamping. The data can then be exported into a spreadsheet program, such as Excel, and the change in the temperature or pressure can be analyzed statistically or plotted against time.

The Web site accompanying this book contains all the program source files and HEX files of the projects described in the book. In addition, a copy of the student version of MPLAB C18 compiler is included on the Web site.

Prof. Dr. Dogan Ibrahim
September, 2009

About the Web Site

The Web site accompanying this book contains the following folders and files:

MPLAB IDE:	MPLAB IDE software package
C18:	Student version of the MPLAB C18 compiler
MDD:	Microchip MDD File I/O System Library
FIGURES:	Figures used in this book (.TIFF and .JPG)
FIGURES-BMP:	Figures used in this book (.BMP)
TABLES:	Tables used in this book
PROGRAMS:	A list of programs used in this book (.C and .HEX)
DRAWINGS:	Circuit diagrams used in this book (.DSN)

CHAPTER 1

Microcontroller Systems

1.1 Introduction

The term *microcontroller* or *microcomputer* is used to describe a system that includes a minimum of a microprocessor, program memory, data memory, and input–output (I/O). Some microcontroller systems include additional components, such as timers, counters, analog-to-digital (A/D) converters, and so on. Thus, a microcontroller system can be anything from a large computer having hard disks, floppy disks, and printers to a single-chip embedded controller.

In this book, we are going to consider only the type of microcontrollers that consist of a single silicon chip. Such microcontroller systems are also known as *embedded controllers*, and they are used in office equipment like PCs, printers, scanners, copy machines, digital telephones, fax machines, and sound recorders. Microcontrollers are also used in household goods, such as microwave ovens, TV remote control units, cookers, hi-fi equipment, CD players, personal computers, and fridges. Many microcontrollers are available in the market. In this book, we shall look at programming and system design using the programmable interface controller (PIC) series of microcontrollers manufactured by Microchip Technology Inc.

1.2 Microcontroller Systems

A microcontroller is a single-chip computer. *Micro* suggests that the device is small and *controller* suggests that the device can be used in control applications. Another term used for microcontrollers is *embedded controller*, because most of the microcontrollers are built into (or embedded in) the devices they control. For example, microcontrollers with dedicated programs are used in washing machines to control the washing cycles.

A microprocessor differs from a microcontroller in many ways. The main difference is that a microprocessor requires several other external components for its operation, such as program memory and data memory, I/O devices, and an external clock circuit. In general, a microprocessor-based system usually consists of several supporting chips interconnected and operating together. The power consumption and the cost of a microprocessor-based system are, thus, usually high. A microcontroller on the other hand has all the support chips incorporated inside the same chip. All microcontrollers operate on a set of instructions (or the user

D.O.I.: 10.1016/B978-1-85617-719-1.00005-1

program) stored in their memory. A microcontroller fetches the instructions from its program memory one by one, decodes these instructions, and then carries out the required operations.

Microcontrollers have traditionally been programmed using the assembly language of the target device. Although assembly language is fast, it has several disadvantages. An assembly program consists of mnemonics, and it is difficult to learn and maintain a program written using assembly language. Also, microcontrollers manufactured by different firms have different assembly languages, and the user is required to learn a new language every time a new microcontroller is to be used. Microcontrollers can also be programmed using one of the traditional high-level languages, such as Basic, Pascal, or C. The advantage of high-level language is that it is much easier to learn than an assembler. Also, very large and complex programs can easily be developed using a high-level language. For example, it is rather a complex task to multiply two floating point numbers using assembly language. The similar operation, however, is much easier and consists of a single statement in a high-level language. In this book, we shall be learning the programming of PIC microcontrollers using the popular C18 high-level C programming language developed by Microchip Inc.

In general, a single chip is all that is required to have a running microcontroller system. In practical applications, additional components may be required to allow a microcomputer to interface to its environment. With the advent of the PIC family of microcontrollers, the development time of a complex electronic project has been reduced from many days to several hours.

Basically, a microcomputer executes a user program that is loaded in its program memory. Under the control of this program, data is received from external devices (inputs), manipulated, and then sent to external devices (outputs). For example, in a simple microcontroller-based temperature data logging system, the temperature is read by the microcomputer using a temperature sensor. The microcomputer then saves the temperature data on an SD card at predefined intervals. Figure 1.1 shows the block diagram of our simple temperature data logging system.

The system shown in Figure 1.1 is a very simplified temperature data logger system. In a more sophisticated system, we may have a keypad to set the logging interval and an

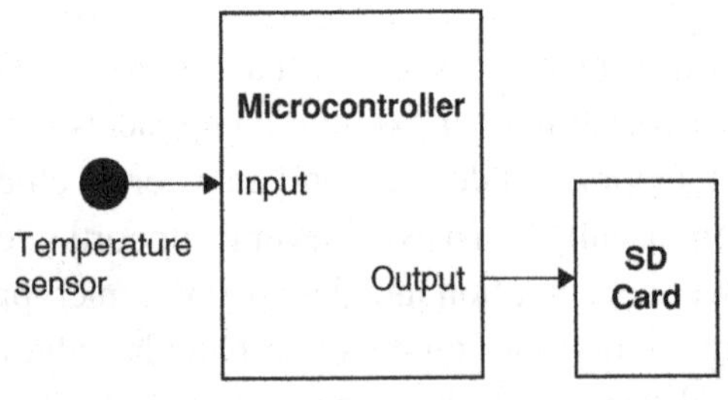

Figure 1.1: Microcontroller-Based Temperature Data Logger System

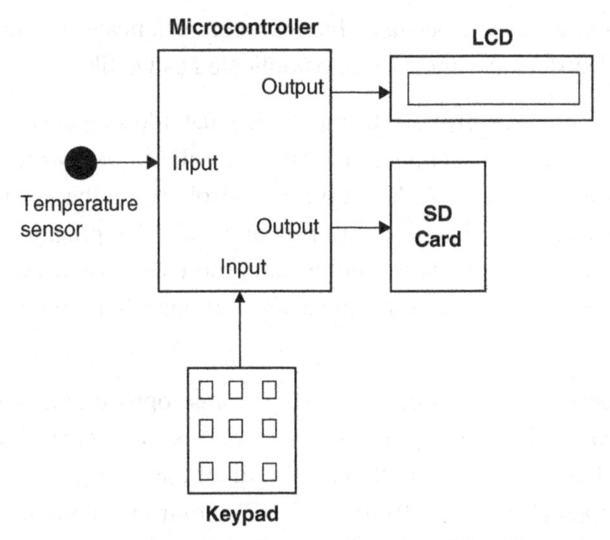

Figure 1.2: Temperature Data Logger System with a Keypad and LCD

LCD to display the current temperature. Figure 1.2 shows the block diagram of this more sophisticated temperature data logger system.

We can make our design even more sophisticated (see Figure 1.3) by adding a real-time clock chip (RTC) to provide the absolute date and time information so that the data can be saved with date and time stamping. Also, the temperature readings can be sent to a PC every second for archiving and further processing. For example, a graph of the temperature change can be

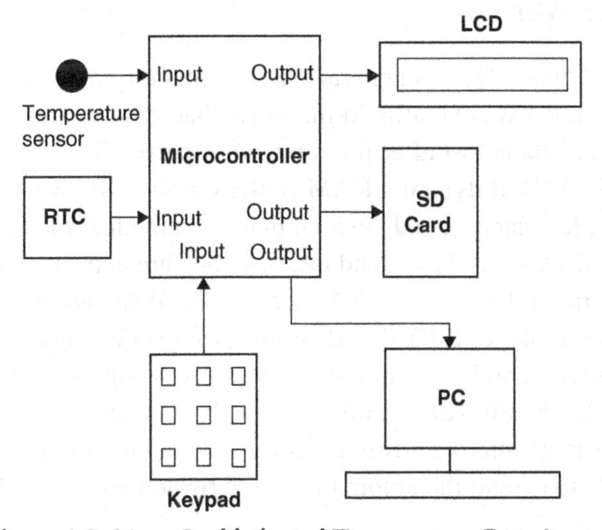

Figure 1.3: More Sophisticated Temperature Data Logger

plotted on the PC. As you can see, because the microcontrollers are programmable, it is very easy to make the final system as simple or as complicated as we like.

A microcontroller is a very powerful electronic device that allows a designer to create sophisticated I/O data manipulation under program control. Microcontrollers are classified by the number of bits they process. Eight-bit microcontrollers are the most popular ones and are used in most microcontroller-based monitoring and control applications. Microcontrollers of 16 and 32 bits are much more powerful but usually more expensive and not required in many small-to-medium-size, general-purpose applications where microcontrollers are generally used.

The simplest microcontroller architecture consists of a microprocessor, program and data memory, and I/O circuitry. The microprocessor itself consists of a central processing unit (CPU) and the control unit (CU). The CPU is the brain of the microprocessor, where all the arithmetic and logic operations are performed. The CU controls the internal operations of the microprocessor and sends out control signals to other parts of the microprocessor to carry out the required instructions.

Memory is an important part of a microcontroller system. Depending upon the type used, we can classify memory into two groups: program memory and data memory. Program memory stores the application program written by the programmer and is usually nonvolatile; i.e., data is not lost after the removal of power. Data memory is where the temporary data used in a program is stored and is usually volatile; i.e., data is lost after the removal of power.

There are basically six types of memory, as summarized below.

1.2.1 Random Access Memory

Random access memory (RAM) is a general-purpose memory that usually stores the user data in a program. RAM is volatile in the sense that it cannot retain data in the absence of power; i.e., data is lost after the removal of power. The RAM in a system is either static RAM (SRAM) or dynamic RAM (DRAM). The SRAMs are fast, with access time in the range of a few nanoseconds, which makes them ideal memory chips in computer applications. DRAMs are slower and because they are capacitor based they require refreshing every several milliseconds. DRAMs have the advantage that their power consumption is less than that of SRAMs. Most microcontrollers have some amount of internal RAM, commonly 256 bytes, although some microcontrollers have more and some have less. For example, the PIC18F452 microcontroller has 1536 bytes of RAM, which should be enough for most microcontroller-based applications. In most microcontroller systems, it is possible to extend the amount of RAM by adding external memory chips if desired.

1.2.2 Read Only Memory

Read only memory (ROM) is a type of memory that usually holds the application program or fixed user data. ROM is nonvolatile. If power is removed from ROM and then reapplied, the original data will still be there. ROMs are programmed at the factory during the manufacturing process and their content cannot be changed by the user. ROMs are only useful if you have developed a microcontroller-based application and wish to order several thousand microcontroller chips preprogrammed with this program.

1.2.3 Programmable Read Only Memory

Programmable read only memory (PROM) is a type of ROM that can be programmed in the field, often by the end user, using a device called a PROM programmer. PROM is used to store an application program or constant data. Once a PROM has been programmed, its contents cannot be changed again. PROMs are usually used in low production applications where only several such memories are required.

1.2.4 Erasable Programmable Read Only Memory

Erasable programmable read only memory (EPROM) is similar to ROM, but the EPROM can be programmed using a suitable programming device. EPROMs have a small clear glass window on top of the chip where the data can be erased under strong ultraviolet light. Once the memory is programmed, the window should be covered with dark tape to prevent accidental erasure of the data. An EPROM must be erased before it can be reprogrammed. Many development versions of microcontrollers are manufactured with EPROMs where the user program can be stored. These memories are erased and reprogrammed until the user is satisfied with the program. Some versions of EPROMs, known as one time programmable (OTP) EPROMs, can be programmed using a suitable programmer device, but these memories cannot be erased. OTP memories cost much less than EPROMs. OTP is useful after a project has been developed completely, and it is required to make many copies of the final program memory.

1.2.5 Electrically Erasable Programmable Read Only Memory

Electrically erasable programmable read only memory (EEPROM) is a nonvolatile memory. These memories can be erased and can also be reprogrammed using suitable programming devices. EEPROMs are used to save constant data, such as configuration information, maximum and minimum values of a measurement, and identification data. Some microcontrollers have built-in EEPROMs. For example, PIC18F452 contains a 256-byte EEPROM where each byte can be programmed and erased directly by applications software. EEPROMs are usually very slow. The cost of an EEPROM chip is much higher than that of an EPROM chip.

1.2.6 Flash EEPROM

Flash EEPROM is another version of EEPROM type memory. This memory has become popular in microcontroller applications and is used to store the user program. Flash EEPROM is nonvolatile and is usually very fast. The data can be erased and then reprogrammed using a suitable programming device. Some microcontrollers have only 1K of flash EEPROM, while some others have 32 K or more. The PIC18F452 microcontroller has 32 KB of flash memory.

1.3 Microcontroller Features

Microcontrollers from different manufacturers have different architectures and different capabilities. Some may suit a particular application while others may be totally unsuitable for the same application. Some of the hardware features of microcontrollers in general are described in this section.

1.3.1 Buses

The connections between various blocks of a computer system are called buses. A bus is a common set of wires that carry a specific type of information. In general, every computer system has three buses: address bus, data bus, and control bus.

An address bus carries the address information in a computer system. It is a unidirectional bus having 16 bits in small computer systems and 32 or more bits in larger systems. An address bus usually carries the memory addresses from the CPU to the memory chips. This bus is also used to carry the I/O addresses in many computer systems.

A data bus carries the data in a computer system. It is a bidirectional bus having 8 bits in small systems and 14, 16, 32, or even more bits in larger systems. A data bus carries the memory data from the CPU to the memory chips. In addition, data is carried to other parts of a computer via the data bus.

The control bus is usually a smaller bus and is used to provide control signals to most parts of a computer system. For example, memory read and write control signals are carried by the control bus.

1.3.2 Supply Voltage

Most microcontrollers operate with the standard logic voltage of +5 V. Some microcontrollers can operate at as low as +2.7 V and some will tolerate +6 V without any problems. You should check the manufacturers' data sheets about the allowed limits of the power supply voltage. For example, PIC18F452 microcontrollers can operate with a power supply +2 to +5.5 V.

A voltage regulator circuit is usually used to obtain the required power supply voltage when the device is to be operated from a mains adaptor or batteries. For example, a 5-V regulator is required if the microcontroller is to be operated using a 9-V battery.

1.3.3 The Clock

All microcontrollers require a clock (or an oscillator) to operate. The clock is usually provided by connecting external timing devices to the microcontroller. Most microcontrollers will generate clock signals when a crystal and two small capacitors are connected. Some will operate with resonators or external resistor-capacitor pair. Some microcontrollers have built-in timing circuits and they do not require any external timing components. If the application is not time sensitive, then external or internal (if available) resistor-capacitor timing components should be used to lower the costs.

An instruction is executed by fetching it from the memory and then decoding it. This usually takes several clock cycles and is known as the *instruction cycle*. In PIC microcontrollers, an instruction cycle takes four clock periods. Thus, the microcontroller is actually operated at a clock rate, which is a quarter of the actual oscillator frequency. For example, in a PIC microcontroller operating at 4-MHz clock, the instruction cycle time is only 1 μs (frequency of 1 MHz). The PIC18F series of microcontrollers can operate with clock frequencies up to 40 MHz.

1.3.4 Timers

Timers are important parts of any microcontroller. A timer is basically a counter, which is driven either by an external clock pulse or by the internal oscillator of the microcontroller. A timer can be 8 or 16 bits wide. Data can be loaded into a timer under program control and the timer can be stopped or started by program control. Most timers can be configured to generate an interrupt when they reach a certain count (usually when they overflow). The interrupt can be used by the user program to carry out accurate timing-related operations inside the microcontroller. The PIC18F series of microcontrollers have at least three timers. For example, the PIC18F452 microcontroller has three built-in timers.

Some microcontrollers offer capture and compare facilities where a timer value can be read when an external event occurs or the timer value can be compared to a preset value and an interrupt generated when this value is reached. Most PIC18F microcontrollers have at least two capture and compare modules.

1.3.5 Watchdog

Most microcontrollers have at least one watchdog facility. The watchdog is basically a timer that is normally refreshed by the user program, and a reset occurs if the program fails to refresh the watchdog. The watchdog timer is used to detect serious problems in programs, such as the program being in an endless loop. A watchdog is a safety feature that prevents runaway software and stops the microcontroller from executing meaningless and unwanted code. Watchdog facilities are commonly used in real-time systems where it is required to regularly check the successful termination of one or more activities.

1.3.6 Reset Input

A reset input is used to reset a microcontroller externally. Resetting puts the microcontroller into a known state such that the program execution starts usually from address 0 of the program memory. An external reset action is usually achieved by connecting a push-button switch to the reset input such that the microcontroller can be reset when the switch is pressed.

1.3.7 Interrupts

Interrupts are very important concepts in microcontrollers. An interrupt causes the microcontroller to respond to external and internal (e.g., a timer) events very quickly. When an interrupt occurs, the microcontroller leaves its normal flow of program execution and jumps to a special part of the program known as the interrupt service routine (ISR). The program code inside the ISR is executed and upon return from the ISR the program resumes its normal flow of execution.

The ISR starts from a fixed address of the program memory. This address is known as the *interrupt vector address*. Some microcontrollers with multi-interrupt features have just one interrupt vector address, while some others have unique interrupt vector addresses, one for each interrupt source. Interrupts can be nested such that a new interrupt can suspend the execution of another interrupt. Another important feature of a microcontroller with multi-interrupt capability is that different interrupt sources can be given different levels of priority. For example, the PIC18F series of microcontrollers have low-priority and high-priority interrupt levels.

1.3.8 Brown-Out Detector

Brown-out detectors are also common in many microcontrollers, and they reset a microcontroller if the supply voltage falls below a nominal value. Brown-out detectors are safety features, and they can be employed to prevent unpredictable operation at low voltages, especially to protect the contents of EEPROM type memories if the supply voltage falls.

1.3.9 A/D Converter

An A/D converter is used to convert an analog signal like voltage to digital form so that it can be read and processed by a microcontroller. Some microcontrollers have built-in A/D converters. It is also possible to connect an external A/D converter to any type of microcontroller. A/D converters are usually 8–10 bits having 256–1024 quantization levels. Most PIC microcontrollers with A/D features have multiplexed A/D converters where more than one analog input channel is provided. For example, the PIC18F452 microcontroller has 10-bit, 8-channel A/D converters.

The A/D conversion process must be started by the user program and it may take several hundreds of microseconds for a conversion to complete. A/D converters usually generate interrupts when a conversion is complete so that the user program can read the converted data quickly.

A/D converters are very useful in control and monitoring applications because most sensors (e.g., temperature sensor, pressure sensor, and force sensor) produce analog output voltages that cannot be read by a microcontroller without an A/D converter.

1.3.10 Serial I/O

Serial communication (also called RS232 communication) enables a microcontroller to communicate with other devices using the serial RS232 communication protocol. For example, a microcontroller can be connected to another microcontroller or to a PC and exchange data using the serial communication protocol. Some microcontrollers have built-in hardware called universal synchronous-asynchronous receiver-transmitter (USART) to implement a serial communication interface. The baud rate and the data format can usually be selected by the user program. If serial I/O hardware is not provided, it is easy to develop software to implement the serial data communication using any I/O pin of a microcontroller. The PIC18F series of microcontrollers have built-in USART modules.

Some microcontrollers (e.g., PIC18F series) incorporate a serial peripheral interface (SPI) or an integrated interconnect (I^2C) hardware bus interface. These enable a microcontroller to interface to other compatible devices easily.

1.3.11 EEPROM Data Memory

EEPROM type data memory is also very common in many microcontrollers. The advantage of an EEPROM is that the programmer can store nonvolatile data in such a memory and can also change this data whenever required. For example, in a temperature monitoring application, the maximum and the minimum temperature readings can be stored in an EEPROM. Then, if the power supply is removed for whatever reason, the values of the latest readings will still be available in the EEPROM. The PIC18F452 microcontroller has 256 bytes of EEPROM. Some other members of the family have more (e.g., PIC18F6680 has 1024 bytes) EEPROMs.

1.3.12 LCD Drivers

LCD drivers enable a microcontroller to be connected to an external LCD display directly. These drivers are not common because most of the functions they provide can be implemented in the software. For example, the PIC18F6490 microcontroller has a built-in LCD driver module.

1.3.13 Analog Comparator

Analog comparators are used where it is required to compare two analog voltages. Although these circuits are implemented in most high-end PIC microcontrollers, they are not common in other microcontrollers. The PIC18F series of microcontrollers have built-in analog comparator modules.

1.3.14 Real-Time Clock

Real-time clock (RTC) enables a microcontroller to have absolute date and time information continuously. Built-in real-time clocks are not common in most microcontrollers because they can easily be implemented by either using a dedicated RTC or by writing a program.

1.3.15 Sleep Mode

Some microcontrollers (e.g., PIC) offer built-in sleep modes where executing this instruction puts the microcontroller into a mode where the internal oscillator is stopped and the power consumption is reduced to an extremely low level. The main reason for using the sleep mode is to conserve the battery power when the microcontroller is not doing anything useful. The microcontroller usually wakes up from the sleep mode by external reset or by a watchdog time-out.

1.3.16 Power-on Reset

Some microcontrollers (e.g., PIC) have built-in power-on reset circuits, which keep the microcontroller in reset state until all the internal circuitry has been initialized. This feature is very useful as it starts the microcontroller from a known state on power-up. An external reset can also be provided where the microcontroller can be reset when an external button is pressed.

1.3.17 Low-Power Operation

Low-power operation is especially important in portable applications where the microcontroller-based equipment is operated from batteries. Some microcontrollers (e.g., PIC) can operate with less than 2 mA at a 5-V supply and approximately 15 μA at a 3-V supply. Some other microcontrollers, especially microprocessor-based systems where there could be several chips, may consume several hundred milliamperes or even more.

1.3.18 Current Sink/Source Capability

This is important if the microcontroller is to be connected to an external device that may draw large current for its operation. PIC microcontrollers can source and sink 25 mA of current from each output port pin. This current is usually sufficient to drive light-emitting diodes (LEDs), small lamps, buzzers, small relays, etc. The current capability can be increased by connecting external transistor switching circuits or relays to the output port pins.

1.3.19 USB Interface

USB is currently a very popular computer interface specification used to connect various peripheral devices to computers and microcontrollers. Some PIC microcontrollers provide built-in USB modules. For example, PIC18F2X50 has built-in USB interface capabilities.

1.3.20 Motor Control Interface

Some PIC microcontrollers (e.g., PIC18F2X31) provide motor control interface.

1.3.21 Controller Area Network Interface

Controller area network (CAN) bus is a very popular bus system used mainly in automation applications. Some PIC18F series of microcontrollers (e.g., PIC18F4680) provide CAN interface capabilities.

1.3.22 Ethernet Interface

Some PIC microcontrollers (e.g., PIC18F97J60) provide Ethernet interface capabilities. Such microcontrollers can easily be used in network-based applications.

1.3.23 ZigBee Interface

ZigBee is an interface similar to Bluetooth and is used in low-cost wireless home automation applications. Some PIC18F series of microcontrollers provide ZigBee interface capabilities, making the design of such wireless systems very easy.

1.4 Microcontroller Architectures

Usually two types of architecture are used in microcontrollers (see Figure 1.4): *Von Neumann architecture* and *Harvard architecture*. Von Neumann architecture is used by a large percentage of microcontrollers, where all memory space is on the same bus and instruction and data use the same bus. In the Harvard architecture (used by the PIC microcontrollers),

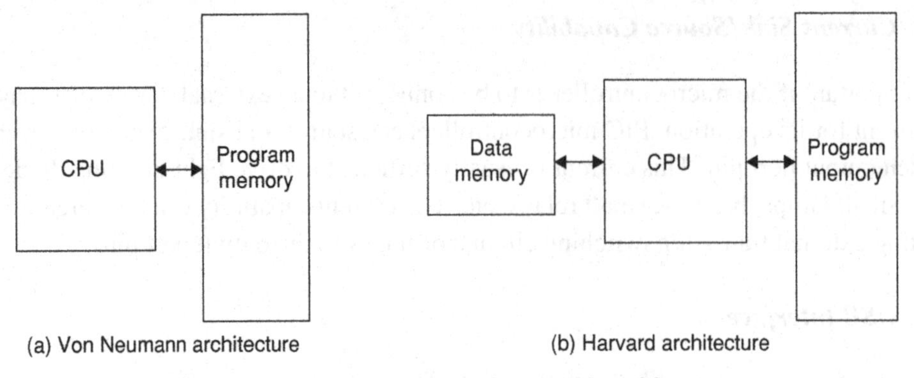

Figure 1.4: Von Neumann and Harvard Architectures

code and data are on separate buses, and this allows the code and data to be fetched simultaneously, resulting in an improved performance.

1.4.1 Reduced Instruction Set Computer and Complex Instruction Set Computer

Reduced instruction set computer (RISC) and complex instruction set computer (CISC) refer to the instruction set of a microcontroller. In an 8-bit RISC microcontroller, data is 8 bits wide but the instruction words are more than 8 bits wide (usually 12, 14, or 16 bits), and the instructions occupy one word in the program memory. Thus, the instructions are fetched and executed in one cycle, resulting in an improved performance.

In a CISC microcontroller, both data and instructions are 8 bits wide. CISC microcontrollers usually have over 200 instructions. Data and code are on the same bus and cannot be fetched simultaneously.

1.5 Choosing a PIC Microcontroller

Choosing a microcontroller for an application requires taking into account the following factors:

- Microcontroller speed
- The number of I/O pins required
- The peripheral devices required (e.g., USART and A/D converter)
- The memory size (RAM, flash, EEPROM, etc.)
- Power consumption
- Physical size

1.6 Number Systems

The efficient use of a microprocessor or a microcontroller requires a working knowledge of binary, decimal, and hexadecimal numbering systems. This section provides a background for those who are unfamiliar with these numbering systems and who do not know how to convert from one number system to another one.

Number systems are classified according to their bases. The numbering system used in everyday life is base 10 or the decimal number system. The most commonly used numbering system in microprocessor and microcontroller applications is base 16 or hexadecimal. In addition, base 2 (binary) or base 8 (octal) number systems are also used.

1.6.1 Decimal Number System

As you all know, the numbers in this system are 0, 1, 2, 3, 4, 5, 6, 7, 8, and 9. We can use the subscript 10 to indicate that a number is in decimal format. For example, we can show the decimal number 235 as 235_{10}.

In general, a decimal number is represented as follows:

$$a_n \times 10^n + a_{n-1} \times 10^{n-1} + a_{n-2} \times 10^{n-2} + \cdots + a_0 \times 10^0$$

For example, decimal number 825_{10} can be shown as follows:

$$825_{10} = 8 \times 10^2 + 2 \times 10^1 + 5 \times 10^0$$

Similarly, decimal number 26_{10} can be shown as follows:

$$26_{10} = 2 \times 10^1 + 6 \times 10^0$$

or

$$3359_{10} = 3 \times 10^3 + 3 \times 10^2 + 5 \times 10^1 + 9 \times 10^0$$

1.6.2 Binary Number System

In the binary number system, there are two numbers: 0 and 1. We can use the subscript 2 to indicate that a number is in binary format. For example, we can show binary number 1011 as 1011_2.

In general, a binary number is represented as follows:

$$a_n \times 2^n + a_{n-1} \times 2^{n-1} + a_{n-2} \times 2^{n-2} + \cdots + a_0 \times 2^0$$

For example, binary number 1110_2 can be shown as follows:

$$1110_2 = 1 \times 2^3 + 1 \times 2^2 + 1 \times 2^1 + 0 \times 2^0$$

Similarly, binary number 10001110_2 can be shown as follows:

$$10001110_2 = 1 \times 2^7 + 0 \times 2^6 + 0 \times 2^5 + 0 \times 2^4 + 1 \times 2^3 + 1 \times 2^2 + 1 \times 2^1 + 0 \times 2^0$$

1.6.3 Octal Number System

In the octal number system, the valid numbers are 0, 1, 2, 3, 4, 5, 6, and 7. We can use the subscript 8 to indicate that a number is in octal format. For example, we can show octal number 23 as 23_8.

In general, an octal number is represented as follows:

$$a_n \times 8^n + a_{n-1} \times 8^{n-1} + a_{n-2} \times 8^{n-2} + \cdots + a_0 \times 8^0$$

For example, octal number 237_8 can be shown as follows:

$$237_8 = 2 \times 8^2 + 3 \times 8^1 + 7 \times 8^0$$

Similarly, octal number 1777_8 can be shown as follows:

$$1777_8 = 1 \times 8^3 + 7 \times 8^2 + 7 \times 8^1 + 7 \times 8^0$$

1.6.4 Hexadecimal Number System

In the hexadecimal number system, the valid numbers are 0, 1, 2, 3, 4, 5, 6, 7, 8, 9, A, B, C, D, E, and F. We can use the subscript 16 or H to indicate that a number is in hexadecimal format. For example, we can show hexadecimal number 1F as $1F_{16}$ or as $1F_H$.

In general, a hexadecimal number is represented as follows:

$$a_n \times 16^n + a_{n-1} \times 16^{n-1} + a_{n-2} \times 16^{n-2} + \cdots + a_0 \times 16^0$$

For example, hexadecimal number $2AC_{16}$ can be shown as follows:

$$2AC_{16} = 2 \times 16^2 + 10 \times 16^1 + 12 \times 16^0$$

Similarly, hexadecimal number $3FFE_{16}$ can be shown as follows:

$$3FFE_{16} = 3 \times 16^3 + 15 \times 16^2 + 15 \times 16^1 + 14 \times 16^0$$

1.7 Converting Binary Numbers into Decimal

To convert a binary number into decimal, write the number as the sum of the powers of 2.

Example 1.1

Convert binary number 1011_2 into decimal.

Solution

Write the number as the sum of the powers of 2:

$$
\begin{aligned}
1011_2 &= 1 \times 2^3 + 0 \times 2^2 + 1 \times 2^1 + 1 \times 2^0 \\
&= 8 + 0 + 2 + 1 \\
&= 11
\end{aligned}
$$

or $1011_2 = 11_{10}$.

Example 1.2

Convert binary number 11001110_2 into decimal.

Solution

Write the number as the sum of the powers of 2:

$$
\begin{aligned}
11001110_2 &= 1 \times 2^7 + 1 \times 2^6 + 0 \times 2^5 + 0 \times 2^4 + 1 \times 2^3 + 1 \times 2^2 + 1 \times 2^1 + 0 \times 2^0 \\
&= 128 + 64 + 0 + 0 + 8 + 4 + 2 + 0 \\
&= 206
\end{aligned}
$$

or $11001110_2 = 206_{10}$.

Table 1.1 shows the binary equivalent of decimal numbers from 0 to 31.

Table 1.1: Binary Equivalent of Decimal Numbers

Decimal	Binary	Decimal	Binary
0	00000000	16	00010000
1	00000001	17	00010001
2	00000010	18	00010010
3	00000011	19	00010011
4	00000100	20	00010100

—cont'd

Table 1.1: Binary Equivalent of Decimal Numbers —cont'd

Decimal	Binary	Decimal	Binary
5	00000101	21	00010101
6	00000110	22	00010110
7	00000111	23	00010111
8	00001000	24	00011000
9	00001001	25	00011001
10	00001010	26	00011010
11	00001011	27	00011011
12	00001100	28	00011100
13	00001101	29	00011101
14	00001110	30	00011110
15	00001111	31	00011111

1.8 Converting Decimal Numbers into Binary

To convert a decimal number into binary, divide the number repeatedly by 2 and take the remainders. The first remainder is the least significant digit (LSD) and the last remainder is the most significant digit (MSD).

Example 1.3

Convert decimal number 28_{10} into binary.

Solution

Divide the number by 2 repeatedly and take the remainders:

28/2 → 14 Remainder 0 (LSD)

14/2 → 7 Remainder 0

7/2 → 3 Remainder 1

3/2 → 1 Remainder 1

1/2 → 0 Remeinder 1 (MSD)

The required binary number is 11100_2.

Example 1.4

Convert decimal number 65_{10} into binary.

Solution

Divide the number by 2 repeatedly and take the remainders:

$$
\begin{array}{lll}
65/2 \rightarrow 32 & \text{Remainder } 1 & \text{(LSD)} \\
32/2 \rightarrow 16 & \text{Remainder } 0 & \\
16/2 \rightarrow 8 & \text{Remainder } 0 & \\
8/2 \rightarrow 4 & \text{Remainder } 0 & \\
4/2 \rightarrow 2 & \text{Remainder } 0 & \\
2/2 \rightarrow 1 & \text{Remainder } 0 & \\
1/2 \rightarrow 0 & \text{Remainder } 1 & \text{(MSD)}
\end{array}
$$

The required binary number is 1000001_2. ■

Example 1.5

Convert decimal number 122_{10} into binary.

Solution

Divide the number by 2 repeatedly and take the remainders:

$$
\begin{array}{lll}
122/2 \rightarrow 61 & \text{Remainder } 0 & \text{(LSD)} \\
61/2 \rightarrow 30 & \text{Remainder } 1 & \\
30/2 \rightarrow 15 & \text{Remainder } 0 & \\
15/2 \rightarrow 7 & \text{Remainder } 1 & \\
7/2 \rightarrow 3 & \text{Remainder } 1 & \\
3/2 \rightarrow 1 & \text{Remainder } 1 & \\
1/2 \rightarrow 0 & \text{Remainder } 1 & \text{(MSD)}
\end{array}
$$

The required binary number is 1111010_2. ■

1.9 Converting Binary Numbers into Hexadecimal

To convert a binary number into hexadecimal, arrange the number in groups of four and find the hexadecimal equivalent of each group. If the number cannot be divided exactly into groups of four, insert zeroes to the left-hand side of the number.

Example 1.6

Convert binary number 10011111_2 into hexadecimal.

Solution

First, divide the number into groups of four and then find the hexadecimal equivalent of each group:

$$10011111 = \underset{9}{1001}\ \underset{F}{1111}$$

The required hexadecimal number is $9F_{16}$.

Example 1.7

Convert binary number 1110111100001110_2 into hexadecimal.

Solution

First, divide the number into groups of four and then find the equivalent of each group:

$$1110111100001110 = \underset{E}{1110}\ \underset{F}{1111}\ \underset{0}{0000}\ \underset{E}{1110}$$

The required hexadecimal number is $EF0E_{16}$.

Example 1.8

Convert binary number 111110_2 into hexadecimal.

Solution

Because the number cannot be divided exactly into groups of four, we have to insert zeroes to the left of the number:

$$111110 = \underset{3}{0011}\ \underset{E}{1110}$$

The required hexadecimal number is $3E_{16}$.

Table 1.2 shows the hexadecimal equivalent of decimal numbers 0 to 31.

Table 1.2: Hexadecimal Equivalent of Decimal Numbers

Decimal	Hexadecimal	Decimal	Hexadecimal
0	0	16	10
1	1	17	11
2	2	18	12
3	3	19	13
4	4	20	14
5	5	21	15
6	6	22	16
7	7	23	17
8	8	24	18
9	9	25	19
10	A	26	1A
11	B	27	1B
12	C	28	1C
13	D	29	1D
14	E	30	1E
15	F	31	1F

■

1.10 Converting Hexadecimal Numbers into Binary

To convert a hexadecimal number into binary, write the 4-bit binary equivalent of each hexadecimal digit.

■ Example 1.9

Convert hexadecimal number $A9_{16}$ into binary.

Solution

Writing the binary equivalent of each hexadecimal digit

$$A = 1010_2 \quad 9 = 1001_2$$

The required binary number is 10101001_2.

■

Example 1.10

Convert hexadecimal number $FE3C_{16}$ into binary.

Solution

Writing the binary equivalent of each hexadecimal digit

$$F = 1111_2 \quad E = 1110_2 \quad 3 = 0011_2 \quad C = 1100_2$$

The required binary number is 1111111000111100_2.

■

1.11 Converting Hexadecimal Numbers into Decimal

To convert a hexadecimal number into decimal, we have to calculate the sum of the powers of 16 of the number.

Example 1.11

Convert hexadecimal number $2AC_{16}$ into decimal.

Solution

Calculating the sum of the powers of 16 of the number:

$$2AC_{16} = 2 \times 16^2 + 10 \times 16^1 + 12 \times 16^0$$

$$= 512 + 160 + 12$$

$$= 684$$

The required decimal number is 684_{10}.

■

Example 1.12

Convert hexadecimal number EE_{16} into decimal.

Solution

Calculating the sum of the powers of 16 of the number

$$EE_{16} = 14 \times 16^1 + 14 \times 16^0$$

$$= 224 + 14$$

$$= 238$$

The required decimal number is 238_{10}.

■

1.12 Converting Decimal Numbers into Hexadecimal

To convert a decimal number into hexadecimal, divide the number repeatedly by 16 and take the remainders. The first remainder is the LSD and the last remainder is the MSD.

■ Example 1.13

Convert decimal number 238_{10} into hexadecimal.

Solution

Dividing the number repeatedly by 16

$238/16 \rightarrow 14$ Remainder 14 (E)   (LSD)

$14/16 \rightarrow 0$ Remainder 14 (E)   (MSD)

The required hexadecimal number is EE_{16}.

■

■ Example 1.14

Convert decimal number 684_{10} into hexadecimal.

Solution

Dividing the number repeatedly by 16

$684/16 \rightarrow 42$ Remainder 12 (C)   (LSD)

$42/16 \rightarrow 2$ Remainder 10 (A)

$2/16 \rightarrow 0$ Remainder 2   (MSD)

The required hexadecimal number is $2AC_{16}$.

■

1.13 Converting Octal Numbers into Decimal

To convert an octal number into decimal, calculate the sum of the powers of 8 of the number.

Example 1.15

Convert octal number 15_8 into decimal.

Solution

Calculating the sum of the powers of 8 of the number

$$
\begin{aligned}
15_8 &= 1 \times 8^1 + 5 \times 8^0 \\
&= 8 + 5 \\
&= 13
\end{aligned}
$$

The required decimal number is 13_{10}.

Example 1.16

Convert octal number 237_8 into decimal.

Solution

Calculating the sum of the powers of 8 of the number

$$
\begin{aligned}
237_8 &= 2 \times 8^2 + 3 \times 8^1 + 7 \times 8^0 \\
&= 128 + 24 + 7 \\
&= 159
\end{aligned}
$$

The required decimal number is 159_{10}.

1.14 Converting Decimal Numbers into Octal

To convert a decimal number into octal, divide the number repeatedly by 8 and take the remainders. The first remainder is the LSD and the last remainder is the MSD.

Example 1.17

Convert decimal number 159_{10} into octal.

Solution

Dividing the number repeatedly by 8

$$159/8 \rightarrow 19 \quad \text{Remainder } 7 \quad \text{(LSD)}$$

$$19/8 \rightarrow 2 \quad \text{Remainder } 3$$

$$2/8 \rightarrow 0 \quad \text{Remainder } 2 \quad \text{(MSD)}$$

The required octal number is 237_8.

Example 1.18

Convert decimal number 460_{10} into octal.

Solution

Dividing the number repeatedly by 8

$$460/8 \rightarrow 57 \quad \text{Remainder } 4 \quad \text{(LSD)}$$

$$57/8 \rightarrow 7 \quad \text{Remainder } 1$$

$$7/8 \rightarrow 0 \quad \text{Remainder } 7 \quad \text{(MSD)}$$

The required octal number is 714_8.

Table 1.3 shows the octal equivalent of decimal numbers 0–31.

Table 1.3: Octal Equivalent of Decimal Numbers

Decimal	Octal	Decimal	Octal
0	0	16	20
1	1	17	21
2	2	18	22
3	3	19	23
4	4	20	24
5	5	21	25
6	6	22	26
7	7	23	27
8	10	24	30
9	11	25	31
10	12	26	32
11	13	27	33
12	14	28	34
13	15	29	35
14	16	30	36
15	17	31	37

1.15 Converting Octal Numbers into Binary

To convert an octal number into binary, write the 3-bit binary equivalent of each octal digit.

Example 1.19

Convert octal number 177_8 into binary.

Solution

Write the binary equivalent of each octal digit:

$$1 = 001_2 \quad 7 = 111_2 \quad 7 = 111_2$$

The required binary number is 001111111_2.

Example 1.20

Convert octal number 75_8 into binary.

Solution

Write the binary equivalent of each octal digit:

$$7 = 111_2 \quad 5 = 101_2$$

The required binary number is 111101_2.

1.16 Converting Binary Numbers into Octal

To convert a binary number into octal, arrange the number in groups of three and write the octal equivalent of each digit.

Example 1.21

Convert binary number 110111001_2 into octal.

Solution

Arranging in groups of three

$$110111001 = \underset{6}{110} \; \underset{7}{111} \; \underset{1}{001}$$

The required octal number is 671_8.

1.17 Negative Numbers

The most significant bit of a binary number is usually used as the sign bit. By convention, for positive numbers this bit is 0 and for negative numbers this bit is 1. Table 1.4 shows the 4-bit positive and negative numbers. The largest positive and negative numbers are +7 and –8, respectively.

To convert a positive number into negative, take the complement of the number and add 1. This process is also called the 2's complement of the number.

Example 1.22

Write decimal number –6 as a 4-bit number.

Solution

First, write the number as a positive number, then find the complement and add 1:

```
0110  +6
1001  complement
   1  add 1
----
1010  which is –6
```

Table 1.4: Four-Bit Positive and Negative Numbers

Binary Numbers	Decimal Equivalent
0111	+7
0110	+6
0101	+5
0100	+4
0011	+3
0010	+2
0001	+1
0000	0
1111	–1
1110	–2
1101	–3
1100	–4
1011	–5
1010	–6
1001	–7
1000	–8

Example 1.23

Write decimal number −25 as an 8-bit number.

Solution

First, write the number as a positive number, then find the complement and add 1:

00011001	+25
11100110	complement
1	add 1
11100111	which is −25

1.18 Adding Binary Numbers

The addition of binary numbers is similar to the addition of decimal numbers. Numbers in each column are added together with a possible carry from a previous column. The primitive addition operations are as follows:

$$0 + 0 = 0$$

$$0 + 1 = 1$$

$$1 + 0 = 1$$

$$1 + 1 = 10 \quad \text{generate a carry bit}$$

$$1 + 1 + 1 = 11 \quad \text{generate a carry bit}$$

Some examples are given below.

Example 1.24

Find the sum of binary numbers 011 and 110.

Solution

We can add these numbers as in the addition of decimal numbers:

011	First column:	1 + 0 = 1
+ 110	Second column:	1 + 1 = 0 and a carry bit
1001	Third column:	1 + 1 = 10

Example 1.25

Find the sum of binary numbers 01000011 and 00100010.

Solution

We can add these numbers as in the addition of decimal numbers:

```
  01000011   First column:    1 + 0 = 1
+ 00100010   Second column:   1 + 1 = 10
----------
  01100101   Third column:    0 + carry = 1
             Fourth column:   0 + 0 = 0
             Fifth column:    0 + 0 = 0
             Sixth column:    0 + 1 = 1
             Seventh column:  1 + 0 = 1
             Eighth column:   0 + 0 = 0
```

■

1.19 Subtracting Binary Numbers

To subtract two binary numbers, convert the number to be subtracted into negative and then add the two numbers.

Example 1.26

Subtract binary number 0010 from 0110.

Solution

First, let's convert the number to be subtracted into negative:

```
0010   number to be subtracted
1101   complement
   1   add 1
----
1110
```

Now, add the two numbers:

$$\begin{array}{r} 0110 \\ +\ 1110 \\ \hline 0100 \end{array}$$

Because we are using 4 bits only, we cannot show the carry bit.

■

1.20 Multiplication of Binary Numbers

Multiplication of two binary numbers is same as the multiplication of decimal numbers. The four possibilities are as follows:

$$0 \times 0 = 0$$

$$0 \times 1 = 0$$

$$1 \times 0 = 0$$

$$1 \times 1 = 1$$

Some examples are given below.

■ Example 1.27

Multiply the two binary numbers 0110 and 0010.

Solution

Multiplying the numbers

$$\begin{array}{r} 0110 \\ 0010 \\ \hline 0000 \\ 0110 \\ 0000 \\ 0000 \\ \hline 001100 \end{array} \text{ or } 1100$$

In this example, 4 bits are needed to show the final result.

■

Example 1.28

Multiply binary numbers 1001 and 1010.

Solution

Multiplying the numbers

```
      1001
      1010
      ----
      0000
     1001
    0000
   1001
   -------
   1011010
```

In this example, 7 bits are required to show the final result. ■

1.21 Division of Binary Numbers

The division of binary numbers is similar to the division of decimal numbers. An example is given below.

Example 1.29

Divide binary number 1110 by binary number 10.

Solution

Dividing the numbers

```
     111
  10)1110
     10
     --
      11
      10
      --
       10
       10
       --
       00
```

gives the result 111_2. ■

1.22 Floating Point Numbers

Floating point numbers are used to represent noninteger fractional numbers and are used in most engineering and technical calculations, for example, 3.256, 2.1, and 0.0036. The most commonly used floating point standard is the IEEE standard. According to this standard, floating point numbers are represented with 32 bits (single precision) or 64 bits (double precision).

In this section, we will look at the format of 32-bit floating point numbers only and see how mathematical operations can be performed with such numbers.

According to the IEEE standard, 32-bit floating point numbers are represented as follows:

```
31  30        23  22                          0
X   XXXXXXXX  XXXXXXXXXXXXXXXXXXXXXXX
↑       ↑                       ↑
sign    exponent                mantissa
```

The most significant bit indicates sign of the number, where 0 indicates positive and 1 indicates negative.

The 8-bit exponent shows the power of the number. To make the calculations easy, the sign of the exponent is not shown, but instead excess 128 numbering system is used. Thus, to find the real exponent, we have to subtract 127 from the given exponent. For example, if the mantissa is "10000000," the real value of the mantissa is 128 – 127 = 1.

The mantissa is 23 bits wide and represents the increasing negative powers of 2. For example, if we assume that the mantissa is "11100000000000000000000," the value of this mantissa is calculated as follows: $2^{-1} + 2^{-2} + 2^{-3} = 7/8$.

The decimal equivalent of a floating point number can be calculated using the following formula:

$$\text{Number} = (-1)^s \, 2^{e-127} \, 1 \cdot f,$$

where $s = 0$ for positive numbers, 1 for negative numbers,
e = exponent (between 0 and 255), and
f = mantissa.

As shown in the above formula, there is a hidden "1" before the mantissa; i.e., mantissa is shown as "1 · *f*."

The largest and the smallest numbers in 32-bit floating point format are as follows:

The largest number

0 11111110 11111111111111111111111

This number is $(2 - 2^{-23})\ 2^{127}$ or decimal 3.403×10^{38}. The numbers keep their precision up to six digits after the decimal point.

The smallest number

0 00000001 00000000000000000000000

This number is 2^{-126} or decimal 1.175×10^{-38}.

1.23 Converting a Floating Point Number into Decimal

To convert a given floating point number into decimal, we have to find the mantissa and the exponent of the number and then convert into decimal as shown above.

Some examples are given here.

Example 1.30

Find the decimal equivalent of the floating point number given below:

0 10000001 10000000000000000000000

Solution

Here,

$$\text{sign} = \text{positive}$$

$$\text{exponent} = 129 - 127 = 2$$

$$\text{mantissa} = 2^{-1} = 0.5$$

The decimal equivalent of this number is $+1.5 \times 2^2 = +6.0$.

■

Example 1.31

Find the decimal equivalent of the floating point number given below:

0 10000010 11000000000000000000000

Solution

In this example,

$$\text{sign} = \text{positive}$$

$$\text{exponent} = 130 - 127 = 3$$

$$\text{mantissa} = 2^{-1} + 2^{-2} = 0.75$$

The decimal equivalent of the number is $+1.75 \times 2^3 = 14.0$. ■

1.23.1 Normalizing the Floating Point Numbers

Floating point numbers are usually shown in normalized form. A normalized number has only one digit before the decimal point (a hidden number 1 is assumed before the decimal point).

To normalize a given floating point number, we have to move the decimal point repetitively one digit to the left and then increase the exponent after each move.

Some examples are given below.

Example 1.32

Normalize the floating point number 123.56.

Solution

If we write the number with a single digit before the decimal point, we get

$$1.2356 \times 10^2$$

■

Example 1.33

Normalize the binary number 1011.1_2.

Solution

If we write the number with a single digit before the decimal point, we get

$$1.0111 \times 2^3$$

■

1.23.2 Converting a Decimal Number into Floating Point

To convert a given decimal number into floating point, we have to carry out the following steps:

- Write the number in binary
- Normalize the number
- Find the mantissa and the exponent
- Write the number as a floating point number

Some examples are given below.

Example 1.34

Convert decimal number 2.25_{10} into floating point.

Solution

Writing the number in binary

$$2.25_{10} = 10.01_2$$

Normalizing the number,

$$10.01_2 = 1.001_2 \times 2^1$$

Here, $s = 0$, $e - 127 = 1$ or $e = 128$, and $f = 00100000000000000000000$.

(Remember that a number 1 is assumed on the left-hand side, even though it is not shown in the calculation.) We can now write the required floating point number as follows:

s	e	f
0	10000000	(1)001 0000 0000 0000 0000 0000

or the required 32-bit floating point number is

$$01000000000100000000000000000000$$

■

Example 1.35

Convert the decimal number 134.0625_{10} into floating point.

Solution

Writing the number in binary

$$134.0625_{10} = 10000110.0001$$

Normalizing the number

$$10000110.0001 = 1.00001100001 \times 2^7$$

Here, $s = 0$, $e - 127 = 7$ or $e = 134$, and $f = 00001100001000000000000$

We can now write the required floating point number as follows:

s	e	f
0	10000110	(1)00001100001000000000000

or the required 32-bit floating point number is

01000011000001100001000000000000

1.23.3 Multiplication and Division of Floating Point Numbers

The multiplication and division of floating point numbers is rather easy and the steps are given below:

- Add (or subtract) the exponents of the numbers
- Multiply (or divide) the mantissa of the numbers
- Correct the exponent
- Normalize the number

The sign of the result is the EXOR of the signs of the two numbers.

Because the exponent is processed twice in the calculations, we have to subtract 127 from the exponent.

An example is given below to show the multiplication of two floating point numbers.

Example 1.36

Show the decimal numbers 0.5_{10} and 0.75_{10} in floating point and then calculate the multiplication of these numbers.

Solution

We can convert the numbers into floating point as follows:

$$0.5_{10} = 1.0000 \times 2^{-1}$$

Here, $s = 0$, $e - 127 = -1$ or $e = 126$ and $f = 0000$

or

$$0.5_{10} = 0\ 01110110\ (1)000\ 0000\ 0000\ 0000\ 0000\ 0000$$

Similarly,

$$0.75_{10} = 1.1000 \times 2^{-1}$$

Here, $s = 0$, e = 126, and $f = 1000$

or

$$0.75_{10} = 0\ 01110110\ (1)100\ 0000\ 0000\ 0000\ 0000\ 0000$$

Multiplying the mantissas, we get "(1)100 0000 0000 0000 0000 0000." The sum of the exponents is 126 + 126 = 252. Subtracting 127 from the mantissa, we obtain 252 − 127 = 125. The EXOR of the signs of the numbers is 0. Thus, the result can be shown in floating point as follows:

$$0\quad 01111101\quad (1)100\ 0000\ 0000\ 0000\ 0000\ 0000$$

The above number is equivalent to decimal 0.375 (0.5 × 0.75 = 0.375), which is the correct result.

■

1.23.4 Addition and Subtraction of Floating Point Numbers

The exponents of floating point numbers must be the same before they can be added or subtracted. The steps to add or subtract floating point numbers is as follows:

- Shift the smaller number to the right until the exponents of both numbers are the same. Increment the exponent of the smaller number after each shift.
- Add (or subtract) the mantissa of each number as an integer calculation, without considering the decimal points.
- Normalize the obtained result.

An example is given below.

■ Example 1.37

Show decimal numbers 0.5_{10} and 0.75_{10} in floating point and then calculate the sum of these numbers.

Solution

As shown in Example 1.36, we can convert the numbers into floating point as follows:

$$0.5_{10} = 0\ 01110110\quad (1)000\ 0000\ 0000\ 0000\ 0000\ 0000$$

Similarly,

$$0.75_{10} = 0\ 01110110\ (1)100\ 0000\ 0000\ 0000\ 0000\ 0000$$

Because the exponents of both numbers are the same, there is no need to shift the smaller number. If we add the mantissa of the numbers without considering the decimal points, we get

```
 (1)000 0000 0000 0000 0000 0000
 (1)100 0000 0000 0000 0000 0000
 -------------------------------- +
(10)100 0000 0000 0000 0000 0000
```

To normalize the number, we can shift it right by one digit and then increment its exponent. The resulting number is

```
0 01111111 (1)010 0000 0000 0000 0000 0000
```

The above floating point number is equal to decimal number 1.25, which is the sum of decimal numbers 0.5 and 0.75.

To convert floating point numbers into decimal and decimal numbers into floating point, the freely available program given in the following Web site can be used:

http://www.babbage.cs.qc.edu/IEEE-754/Decimal.html

■

1.24 Binary-Coded Decimal Numbers

Binary-coded decimal (BCD) numbers are usually used in display systems like LCDs and seven-segment displays to show numeric values. BCD data is stored in either packed or unpacked forms. Packed BCD data is stored as two digits per byte and unpacked BCD data is stored as one digit per byte. Unpacked BCD data is usually returned from a keypad or a keyboard. The packed BCD is more frequently used, and this is the format described in the remainder of this section.

In packed BCD, each digit is a 4-bit number from 0 to 9. As an example, Table 1.5 shows the packed BCD numbers between 0 and 20.

Table 1.5: Packed BCD Numbers Between 0 and 20

Decimal	BCD	Binary
0	0000	0000
1	0001	0001
2	0010	0010
3	0011	0011
4	0100	0100
5	0101	0101

Table 1.5: Packed BCD Numbers Between 0 and 20 —cont'd

Decimal	BCD	Binary
6	0110	0110
7	0111	0111
8	1000	1000
9	1001	1001
10	0001 0000	1010
11	0001 0001	1011
12	0001 0010	100
13	0001 0011	1101
14	0001 0100	1110
15	0001 0101	1111
16	0001 0110	1 0000
17	0001 0111	1 0001
18	0001 1000	1 0010
19	0001 1001	1 0011
20	0010 0000	1 0100

Example 1.38

Write the decimal number 295 as a packed BCD number.

Solution

Writing the 4-bit binary equivalent of each digit

$$2 = 0010_2 \qquad 9 = 1001_2 \qquad 5 = 0101_2$$

The required packed BCD number is $0010\ 1001\ 0101_2$.

Example 1.39

Write the decimal equivalent of packed BCD number $1001\ 1001\ 0110\ 0001_2$.

Solution

Writing the decimal equivalent of each 4-bit group, we get the decimal number 9961.

1.25 Summary

This chapter has given an introduction to the microprocessor and microcontroller systems. The basic building blocks of microcontrollers have been described briefly. The chapter has also provided an introduction to various number systems, and has described how to convert a given number from one base into another base. The important topics of floating point numbers and floating point arithmetic have also been described with examples.

1.26 Exercises

1. What is a microcontroller? What is a microprocessor? Explain the main differences between a microprocessor and a microcontroller.
2. Give some example applications of microcontrollers around you.
3. Where would you use an EPROM?
4. Where would you use a RAM?
5. Explain what types of memory are usually used in microcontrollers.
6. What is an I/O port?
7. What is an A/D converter? Give an example of use of this converter.
8. Explain why a watchdog timer could be useful in a real-time system.
9. What is serial I/O? Where would you use serial communication?
10. Why is the current sinking/sourcing important in the specification of an output port pin?
11. What is an interrupt? Explain what happens when an interrupt is recognized by a microcontroller.
12. Why is brown-out detection important in real-time systems?
13. Explain the differences between a RISC-based microcontroller and a CISC-based microcontroller. What type of microcontroller is PIC?
14. Convert the following decimal numbers into binary:

    a) 23 b) 128 c) 255 d) 1023
    e) 120 f) 32000 g) 160 h) 250

15. Convert the following binary numbers into decimal:

    a) 1111 b) 0110 c) 11110000
    d) 00001111 e) 10101010 f) 10000000

16. Convert the following octal numbers into decimal:

a) 177	b) 762	c) 777	d) 123
e) 1777	f) 655	g) 177777	h) 207

17. Convert the following decimal numbers into octal:

a) 255	b) 1024	c) 129	d) 2450
e) 4096	f) 256	g) 180	h) 4096

18. Convert the following hexadecimal numbers into decimal:

a) AA	b) EF	c) 1FF	d) FFFF
e) 1AA	f) FEF	g) F0	h) CC

19. Convert the following binary numbers into hexadecimal:

a) 0101	b) 11111111	c) 1111	d) 1010
e) 1110	f) 10011111	g) 1001	h) 1100

20. Convert the following binary numbers into octal:

a) 111000	b) 000111	c) 1111111	d) 010111
e) 110001	f) 11111111	g) 1000001	h) 110000

21. Convert the following octal numbers into binary:

a) 177	b) 7777	c) 555	d) 111
e) 1777777	f) 55571	g) 171	h) 1777

22. Convert the following hexadecimal numbers into octal:

a) AA	b) FF	c) FFFF	d) 1AC
e) CC	f) EE	g) EEFF	h) AB

23. Convert the following octal numbers into hexadecimal:

a) 177	b) 777	c) 123	d) 23
e) 1111	f) 17777777	g) 349	h) 17

24. Convert the following decimal numbers into floating point:

a) 23.45	b) 1.25	c) 45.86	d) 0.56

25. Convert the following decimal numbers into floating point and then calculate their sum:
    0.255 and 1.75

26. Convert the following decimal numbers into floating point and then calculate their product:
    2.125 and 3.75

27. Convert the following decimal numbers into packed BCD:

a) 128	b) 970	c) 900	d) 125

28. Convert the following decimal numbers into unpacked BCD:
    a) 128 b) 970 c) 900 d) 125

29. Convert the following packed BCD numbers into decimal:
    a) 0110 0011 b) 0111 0100 c) 0001 0111

CHAPTER 2

PIC18F Microcontroller Series

The PIC16 series of microcontrollers have been around for many years. Although they are excellent general-purpose microcontrollers, they have certain limitations. For example, the program and data memory capacities are limited, the stack is small, and the interrupt structure is primitive – all interrupt sources share the same interrupt vector. The PIC16 series of microcontrollers also do not provide direct support for advanced peripheral interfaces, such as USB and CAN bus, and it is rather complex to interface to such devices easily. The instruction set of these microcontrollers is also limited. For example, there are no instructions for multiplication or division and branching is rather simple and is made out of a combination of *skip* and *goto* instructions. Microchip Inc. has developed the PIC18 series of microcontrollers for high-pin-count, high-density, and complex applications.

Figure 2.1 shows the current PIC microcontroller family of products. At the lowest end of the family, we have the PIC10 microcontrollers, operating at approximately 5 MIPS and with small form factors, less memory, and a low cost. Then we have the PIC12 and PIC16 series of microcontrollers with midrange architectures, 5–8 MIPS operating performance, and reasonable size of memory. The microcontrollers of the PIC18 family are advanced high-performance devices, with 10–16 MIPS, and offer a large amount of memory with various on-chip peripheral support modules. As shown in Figure 2.1, the higher end of the family consists of 16-bit devices, such as the PIC24 and the dsPIC, and 32-bit devices, such as the dsPIC33 and the PIC32 series.

The PIC18 microcontroller family consists of three architectures: the standard PIC18F series, the PIC18J series, and the PIC18K series. PIC18J series are 10–12 MIPS, low-voltage, high-performance microcontrollers with integrated USB, Ethernet, or LCD. PIC18K series are 16 MIPS, high-performance, and low-power devices.

In this book, we shall be using the standard PIC18F series of microcontrollers. The PIC18F microcontrollers can be used in cost-efficient solutions for general-purpose applications written in C, using a real-time operating system (RTOS), and require complex communication protocol stack, such as TCP/IP, CAN, USB, or ZigBee. PIC18F devices provide flash program memory in sizes from 8 to 128 KB and data memory from 256 to 4 KB, operating at 2.0–5.0 V at speeds from DC to 40 MHz.

The basic features of the PIC18F series of microcontrollers are as follows:

- 77 instructions
- PIC16 source code compatible

D.O.I.: 10.1016/B978-1-85617-719-1.00006-3

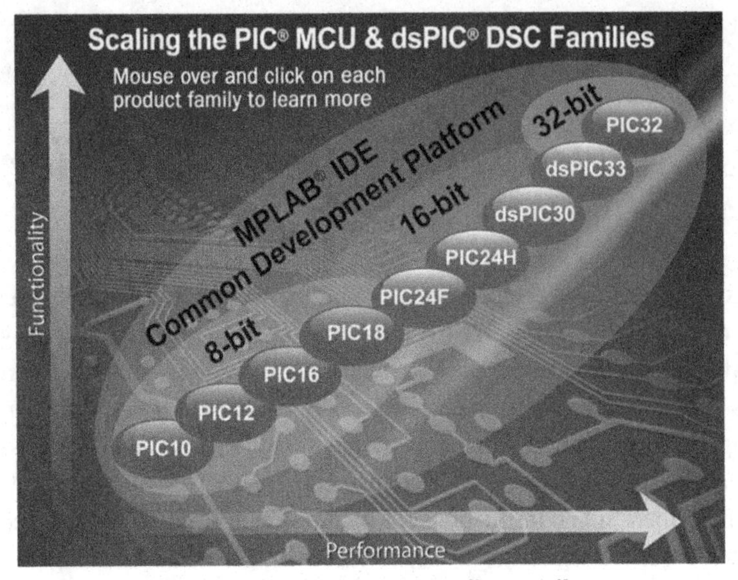

Figure 2.1: PIC Microcontroller Family

- Program memory addressing up to 2 MB
- Data memory addressing up to 4 KB
- DC to 40-MHz operation
- 8 × 8 hardware multiplier
- Interrupt priority levels
- 16-bit wide instructions, 8-bit wide data path
- Up to two 8-bit timer/counters
- Up to three 16-bit timer/counters
- Up to four external interrupts
- High-current (25 mA) sink/source capability
- Up to five capture/compare/pulse width modulation (PWM) modules
- Master synchronous serial port module (serial peripheral interface [SPI] and I²C modes)
- Up to two universal synchronous-asynchronous receiver-transmitter (USART) modules
- Parallel slave port (PSP)
- Fast 10-bit analog-to-digital (A/D) converter

- Programmable low-voltage detection (LVD) module
- Power-on reset (POR), power-up timer (PWRT), and oscillator start-up timer (OST)
- Watchdog timer (WDT) with on-chip RC oscillator
- In-circuit programming

In addition, some microcontrollers in the family offer the following special features:

- Direct CAN 2.0B bus interface
- Direct USB 2.0 bus interface
- Direct LCD control interface
- TCP/IP interface
- ZigBee interface
- Direct motor control interface

There are many devices in the PIC18F family, and most of them are source compatible with each other. Table 2.1 gives the characteristics of some of the popular devices in this family. In this chapter, the PIC18FXX2 microcontrollers are chosen for detailed study. Most of the other microcontrollers in the family have similar architectures.

Table 2.1: The 18FXX2 Microcontroller Family

Feature	PIC18F242	PIC18F252	PIC18F442	PIC18F452
Program memory (bytes)	16K	32K	16K	32K
Data memory (bytes)	768	1536	768	1536
EEPROM (bytes)	256	256	256	256
I/O Ports	A,B,C	A,B,C	A,B,C,D,E	A,B,C,D,E
Timers	4	4	4	4
Interrupt sources	17	17	18	18
Capture/compare/PWM	2	2	2	2
Serial communication	MSSP USART	MSSP USART	MSSP USART	MSSP USART
A/D converter (10 bits)	5 channels	5 channels	8 channels	8 channels
Low-voltage detect	Yes	Yes	Yes	Yes
Brown-out reset	Yes	Yes	Yes	Yes
Packages	28-pin DIP 28-pin SOIC	28-pin DIP 28-pin SOIC	40-pin DIP 44-pin PLCC 44-pin TQFP	40-pin DIP 44-pin PLCC 44-pin TQFP

Most readers may be familiar with the programming and applications of the PIC16F series. Before going into detailed information on the PIC18F series, it is worthwhile to look at the similarities of PIC16F and PIC18F and the new features of the PIC18F series.

Similarities of PIC16F and PIC18F are as follows:

- Similar packages and pinouts
- Similar names and functions of special function registers (SFRs)
- Similar peripheral devices
- Subset of PIC18F instruction set
- Similar development tools

New features of the PIC18F series are as follows:

- Number of instructions doubled
- 16-bit instruction word
- Hardware 8 × 8 multiplier
- More external interrupts
- Priority-based interrupts
- Enhanced status register
- Increased program and data memory size
- Bigger stack
- Phase-locked loop (PLL) clock generator
- Enhanced input–output (I/O) port architecture
- Set of configuration registers
- Higher speed of operation
- Lower-power operation

2.1 PIC18FXX2 Architecture

As shown in Table 2.1, the PIC18FXX2 series consists of four devices. PIC18F2X2 micro-controllers are 28-pin devices, and PIC18F4X2 microcontrollers are 40-pin devices. The architectures of both groups are almost identical except that the larger devices have more I/O ports and more A/D converter channels. In this section, we shall be looking at the architecture of the PIC18F452 microcontrollers in detail. The architectures of other standard PIC18F series microcontrollers are very similar, and the knowledge gained in this section should be enough to understand the operation of other PIC18F series microcontrollers.

The pin configuration of the PIC18F452 microcontroller (DIP package) is shown in Figure 2.2. This is a 40-pin microcontroller housed in a DIL package and has a pin configuration similar to that of the popular PIC16F877.

Figure 2.3 shows the internal block diagram of the PIC18F452 microcontrollers. The CPU is at the center of the diagram and consists of an 8-bit ALU, an 8-bit working accumulator register (WREG), and an 8×8 hardware multiplier. The higher byte and the lower byte of a multiplication are stored in two 8-bit registers called PRODH and PRODL, respectively.

The program counter and the program memory are shown at the top left corner of the diagram. Program memory addresses consist of 21 bits and are capable of accessing 2 MB of program memory locations. PIC18F452 has only 32 KB of program memory, which requires only 15 bits; thus, the remaining six address bits are redundant and not used. A table pointer provides access to tables and to the data stored in the program memory. The program memory contains a 31-level stack, which is normally used to store the interrupt and subroutine return addresses.

The data memory can be seen at the top central part of the diagram. The data memory address bus is 12 bits wide and is capable of accessing 4 KB of data memory locations. As we shall study later, the data memory consists of the SFR and the general-purpose registers (GPR), all organized in banks.

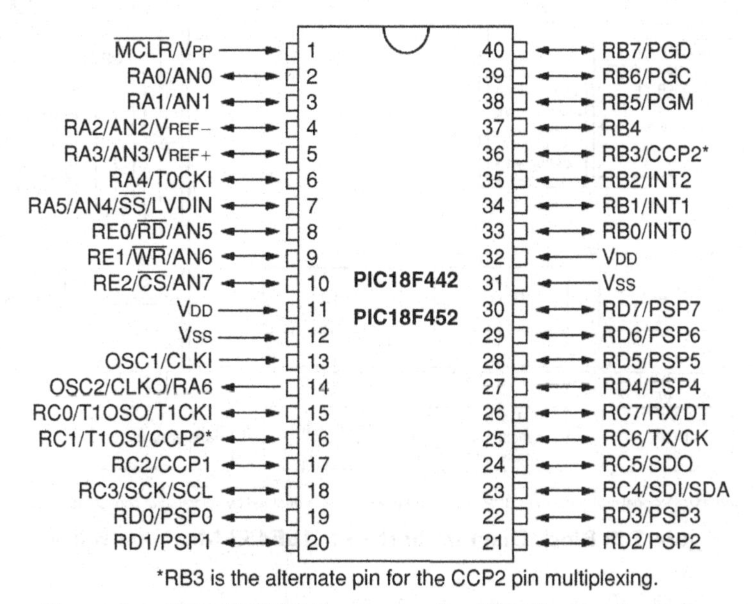

Figure 2.2: PIC18F452 Microcontroller DIP Pin Configuration

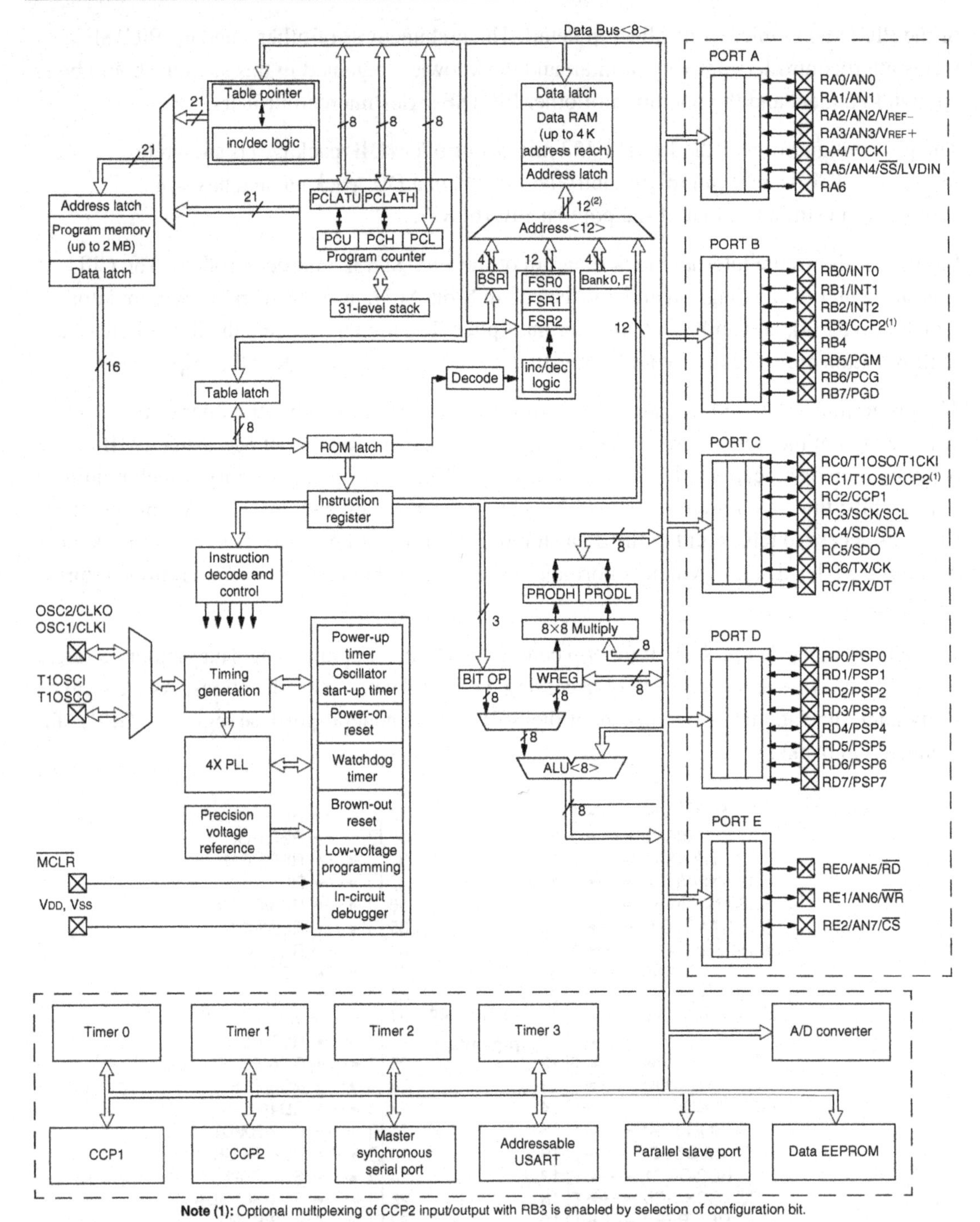

Figure 2.3: Block Diagram of the PIC18F452 Microcontroller

The bottom part of the diagram shows the timers/counters, capture/compare/PWM registers, USART, A/D converter, and the EEPROM data memory. PIC18F452 consists of

- Four counters/timers
- Two capture/compare/PWM modules
- Two serial communication modules
- Eight 10-bit A/D converter channels
- 256-byte EEPROM

The oscillator circuit is located at the left-hand side of the diagram. This circuit consists of

- PWRT
- OST
- POR
- WDT
- Brown-out reset (BOR)
- Low-voltage programming
- In-circuit debugger (ICD)
- PLL circuit
- Timing generation circuit

The PLL is new to the PIC18F series, and it provides the option of multiplying the oscillator frequency to speed up the overall operation. The WDT can be used to force a restart of the microcontroller in the event of a program crash. The ICD is useful during program development, and it can be used to return diagnostic data, including the register values, as the microcontroller is executing a program.

The I/O ports are located at the right-hand side of the diagram. PIC18F452 consists of five parallel ports named PORTA, PORTB, PORTC, PORTD, and PORTE. Most port pins have multiple functions. For example, PORTA pins can be used as either parallel I/O or analog inputs. PORTB pins can be used as either parallel I/O or interrupt inputs.

2.1.1 Program Memory Organization

The program memory map is shown in Figure 2.4. Each PIC18F member has a 21-bit program counter and hence is capable of addressing 2 MB of memory space. User memory space on the PIC18F452 microcontroller is 00000H to 7FFFH. Accessing a nonexistent

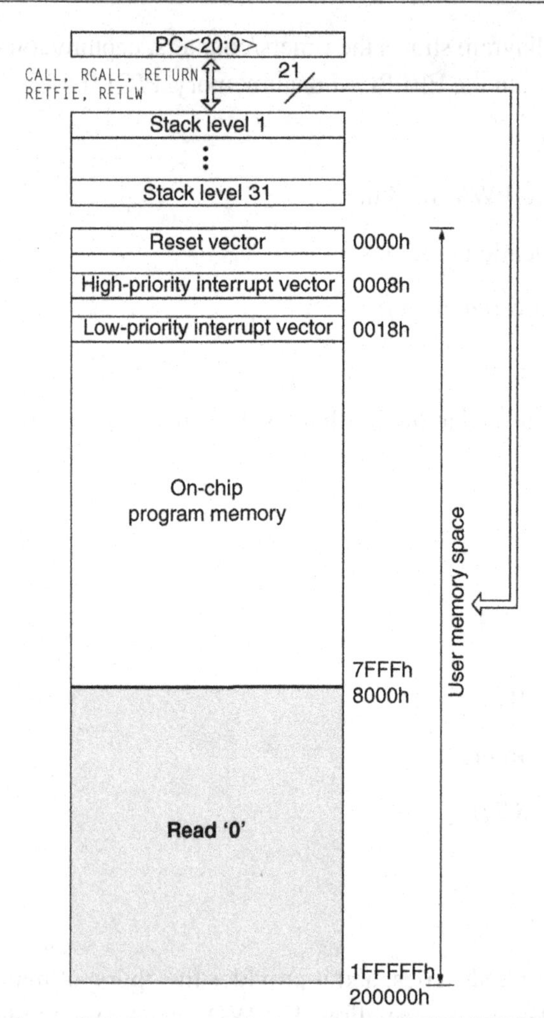

Figure 2.4: Program Memory Map of PIC18F452

memory location (8000H to 1FFFFFH) will cause a read of all 0s. The reset vector where the program starts after a reset is at address 0000H. Addresses 0008H and 0018H are reserved for the vectors of high-priority and low-priority interrupts, respectively, and interrupt service routines must be written to start at one of these locations.

The PIC18F microcontroller has a 31-entry stack that is used to hold the return addresses for subroutine calls and interrupt processing. The stack is not a part of the program or a data memory space. The stack is controlled by a 5-bit stack pointer, which is initialized to 00000 after a reset. During a subroutine call (or interrupt), the stack pointer is first incremented, and the memory location pointed to by the stack pointer is written using the contents of the program counter. During a return from a subroutine call (or interrupt), the memory location pointed to by the stack pointer is decremented. The projects in this book are based on C language. Subroutine and interrupt call/return operations are handled automatically by the C language compiler, and thus their operation is not described here in detail.

Program memory is addressed in bytes and instructions are stored as 2 or 4 bytes in program memory. The least significant byte of an instruction word is always stored in an even address of the program memory.

An instruction cycle consists of four cycles: A fetch cycle begins with the program counter incrementing in Q1. In the execution cycle, the fetched instruction is latched into the instruction register in cycle Q1. This instruction is then decoded and executed during the Q2, Q3, and Q4 cycles. A data memory location is read during Q2 and written during Q4.

2.1.2 Data Memory Organization

The data memory map of the 18F452 microcontroller is shown in Figure 2.5. The data memory address bus is 12 bits, with the capability of addressing up to 4 MB. The memory in general consists of 16 banks, each of 256 bytes. PIC18F452 has 1536 bytes of data memory (6 banks × 256 bytes each) occupying the lower end of the data memory. Bank switching is done automatically when using a high-level language compiler, and thus the user need not worry about selecting memory banks during programming.

The special function register (SFR) occupies the upper half of the top memory bank. SFR contains registers that control the operations of the microcontroller, such as the peripheral devices, timers/counters, A/D converter, interrupts, USART, and so on. Figure 2.6 shows the SFR registers of the PIC18F452 microcontroller.

2.1.3 The Configuration Registers

The PIC18F452 microcontrollers have a set of configuration registers (PIC16 series had only one configuration register). Configuration registers are programmed during the programming of the flash program memory by the programming device. These registers are shown in Table 2.2. The descriptions of these registers are given in Table 2.3. Some of the important configuration registers are described in this section in detail.

CONFIG1H

This configuration register is at address 300001H and is used to select the microcontroller clock sources. The bit patterns are shown in Figure 2.7.

CONFIG2L

This configuration register is at address 300002H and is used to select the brown-out voltage bits. The bit patterns are shown in Figure 2.8.

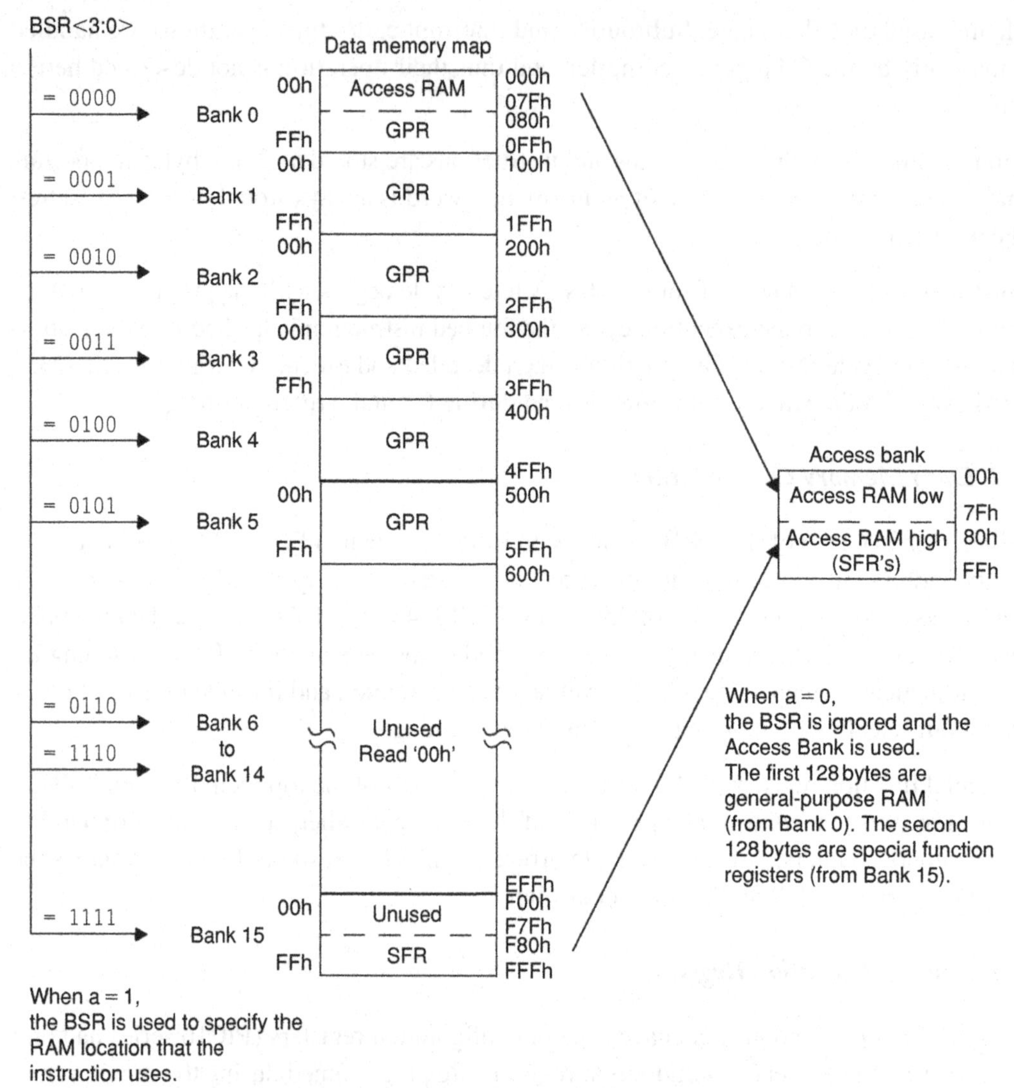

Figure 2.5: The PIC18F452 Data Memory Map

CONFIG2H

This configuration register is at address 300003H and is used to select the watchdog operations. The bit patterns are shown in Figure 2.9.

2.1.4 The Power Supply

The power supply requirements of the PIC18F452 microcontroller are shown in Figure 2.10. As shown in Figure 2.11, PIC18F452 can operate with supply voltage of 4.2–5.5 V at the full speed of 40 MHz. The lower-power version of PIC18LF452 can operate at 2.0–5.5 V.

Address	Name	Address	Name	Address	Name	Address	Name
FFFh	TOSU	FDFh	INDF2[3]	FBFh	CCPR1H	F9Fh	IPR1
FFEh	TOSH	FDEh	POSTINC2[3]	FBEh	CCPR1L	F9Eh	PIR1
FFDh	TOSL	FDDh	POSTDEC2[3]	FBDh	CCP1CON	F9Dh	PIE1
FFCh	STKPTR	FDCh	PREINC2[3]	FBCh	CCPR2H	F9Ch	—
FFBh	PCLATU	FDBh	PLUSW2[3]	FBBh	CCPR2L	F9Bh	—
FFAh	PCLATH	FDAh	FSR2H	FBAh	CCP2CON	F9Ah	—
FF9h	PCL	FD9h	FSR2L	FB9h	—	F99h	—
FF8h	TBLPTRU	FD8h	STATUS	FB8h	—	F98h	—
FF7h	TBLPTRH	FD7h	TMR0H	FB7h	—	F97h	—
FF6h	TBLPTRL	FD6h	TMR0L	FB6h	—	F96h	TRISE[2]
FF5h	TABLAT	FD5h	T0CON	FB5h	—	F95h	TRISD[2]
FF4h	PRODH	FD4h	—	FB4h	—	F94h	TRISC
FF3h	PRODL	FD3h	OSCCON	FB3h	TMR3H	F93h	TRISB
FF2h	INTCON	FD2h	LVDCON	FB2h	TMR3L	F92h	TRISA
FF1h	INTCON2	FD1h	WDTCON	FB1h	T3CON	F91h	—
FF0h	INTCON3	FD0h	RCON	FB0h	—	F90h	—
FEFh	INDF0[3]	FCFh	TMR1H	FAFh	SPBRG	F8Fh	—
FEEh	POSTINC0[3]	FCEh	TMR1L	FAEh	RCREG	F8Eh	—
FEDh	POSTDEC0[3]	FCDh	T1CON	FADh	TXREG	F8Dh	LATE[2]
FECh	PREINC0[3]	FCCh	TMR2	FACh	TXSTA	F8Ch	LATD[2]
FEBh	PLUSW0[3]	FCBh	PR2	FABh	RCSTA	F8Bh	LATC
FEAh	FSR0H	FCAh	T2CON	FAAh	—	F8Ah	LATB
FE9h	FSR0L	FC9h	SSPBUF	FA9h	EEADR	F89h	LATA
FE8h	WREG	FC8h	SSPADD	FA8h	EEDATA	F88h	—
FE7h	INDF1[3]	FC7h	SSPSTAT	FA7h	EECON2	F87h	—
FE6h	POSTINC1[3]	FC6h	SSPCON1	FA6h	EECON1	F86h	—
FE5h	POSTDEC1[3]	FC5h	SSPCON2	FA5h	—	F85h	—
FE4h	PREINC1[3]	FC4h	ADRESH	FA4h	—	F84h	PORTE[2]
FE3h	PLUSW1[3]	FC3h	ADRESL	FA3h	—	F83h	PORTD[2]
FE2h	FSR1H	FC2h	ADCON0	FA2h	IPR2	F82h	PORTC
FE1h	FSR1L	FC1h	ADCON1	FA1h	PIR2	F81h	PORTB
FE0h	BSR	FC0h	—	FA0h	PIE2	F80h	PORTA

(1): Unimplemented registers are read as '0'
(2): This register is not available on PIC18F2X2 devices
(3): This is not a physical register

Figure 2.6: The PIC18F452 SFR Registers

At lower voltages, the maximum clock frequency is 4 MHz and rises to 40 MHz at 4.2 V. The RAM data retention voltage is specified as 1.5 V, and the RAM data will be lost if the power supply voltage becomes lower than this value. In practice, most microcontroller-based systems are operated with a single +5 V supply, which is derived from a suitable voltage regulator.

Table 2.2: PIC18F452 Configuration Registers

File Name		Bit 7	Bit 6	Bit 5	Bit 4	Bit 3	Bit 2	Bit 1	Bit 0	Default/ Unprogrammed Value
300001h	CONFIG1H	–	–	$\overline{\text{OSCSEN}}$	–	–	FOSC2	FOSC1	FOSC0	--1- -111
300002h	CONFIG2L	–	–	–	–	BORV1	BORV0	BOREN	$\overline{\text{PWRTEN}}$	---- 1111
300003h	CONFIG2H	–	–	–	–	WDTPS2	WDTPS1	WDTPS0	WDTEN	---- 1111
300005h	CONFIG3H	–	–	–	–	–	–	–	CCP2MX	---- ---1
300006h	CONFIG4L	$\overline{\text{DEBUG}}$	–	–	–	–	LVP	–	STVREN	1--- -1-1
300008h	CONFIG5L	–	–	–	–	CP3	CP2	CP1	CP0	---- 1111
300009h	CONFIG5H	CPD	CPB	–	–	–	–	–	–	11-- ----
30000Ah	CONFIG6L	–	–	–	–	WRT3	WRT2	WRT1	WRT0	---- 1111
30000Bh	CONFIG6H	WRTD	WRTB	WRTC	–	–	–	–	–	111- ----
30000Ch	CONFIG7L	–	–	–	–	EBTR3	EBTR2	EBTR1	EBTR0	---- 1111
30000Dh	CONFIG7H	–	EBTRB	–	–	–	–	–	–	-1-- ----
3FFFFEh	DEVID1	DEV2	DEV1	DEV0	REV4	REV3	REV2	REV1	REV0	(1)
3FFFFFh	DEVID2	DEV10	DEV9	DEV8	DEV7	DEV6	DEV5	DEV4	DEV3	0000 0100

Shaded cells are unimplemented, read as "0."

Table 2.3: PIC18F452 Configuration Register Descriptions

Configuration Bits	Description
OSCSEN	Clock source switching enable
FOSC2:FOSC0	Oscillator modes
BORV1:BORV0	Brown-out reset voltage
BOREN	Brown-out reset enable
PWRTEN	Power-up timer enable
WDTPS2:WDTPS0	Watchdog timer postscale select
WDTEN	Watchdog timer enable
CCP2MX	CCP2 multiplex
DEBUG	Debug enable
LVP	Low-voltage program enable
STVREN	Stack full/underflow reset enable
CP3:CP0	Code protection
CPD	EEPROM code protection
CPB	Boot block code protection
WRT3:WRT0	Program memory write protection
WRTD	EPROM write protection
WRTB	Boot block write protection
WRTC	Configuration register write protection
EBTR3:EBTR0	Table read protection
EBTRB	Boot block table read protection
DEV2:DEV0	Device ID bits (001 = 18F452)
REV4:REV0	Revision ID bits
DEV10:DEV3	Device ID bits

2.1.5 The Reset

The reset action puts the microcontroller into a known state. Resetting a PIC18F microcontroller starts the execution of the program from address 0000H of the program memory. The microcontroller can be reset during one of the following operations:

- POR
- MCLR reset

U-0	U-0	R/P-1	U-0	U-0	R/P-1	R/P-1	R/P-1
—	—	$\overline{\text{OSCSEN}}$	—	—	FOSC2	FOSC1	FOSC0
bit 7							bit 0

bit 7–6 **Unimplemented:** Read as '0'

bit 5 **$\overline{\text{OSCSEN}}$:** Oscillator system clock switch enable bit

1 = Oscillator system clock switch option is disabled (main oscillator is source)
0 = Oscillator system clock switch option is enabled (oscillator switching is enabled)

bit 4–3 **Unimplemented:** Read as '0'

bit 2–0 **FOSC2:FOSC0:** Oscillator selection bits

111 = RC oscillator w/OSC2 configured as RA6
110 = HS oscillator with PLL enabled/Clock frequency = (4 × FOSC)
101 = EC oscillator w/OSC2 configured as RA6
100 = EC oscillator w/OSC2 configured as divide-by-4 clock output
011 = RC oscillator
010 = HS oscillator
001 = XT oscillator
000 = LP oscillator

Figure 2.7: CONFIG1H Register Bits

U-0	U-0	U-0	U-0	R/P-1	R/P-1	R/P-1	R/P-1
—	—	—	—	BORV1	BORV0	BOREN	$\overline{\text{PWRTEN}}$
bit 7							bit 0

bit 7–4 **Unimplemented:** Read as '0'

bit 3–2 **BORV1:BORV0:** Brown-out reset voltage bits

11 = VBOR set to 2.5 V
10 = VBOR set to 2.7 V
01 = VBOR set to 4.2 V
00 = VBOR set to 4.5 V

bit 1 **BOREN:** Brown-out reset enable bit

1 = Brown-out reset enabled
0 = Brown-out reset disabled

bit 0 **$\overline{\text{PWRTEN}}$:** Power-up timer enable bit

1 = PWRT disabled
0 = PWRT enabled

Figure 2.8: CONFIG2L Register Bits

- WDT reset
- BOR
- Reset instruction
- Stack full reset
- Stack underflow reset

U-0	U-0	U-0	U-0	R/P-1	R/P-1	R/P-1	R/P-1
—	—	—	—	WDTPS2	WDTPS1	WDTPS0	WDTEN
bit 7							bit 0

bit 7–4 **Unimplemented:** Read as '0'

bit 3–1 **WDTPS2: WDTPS0:** Watchdog timer postscale select bits

111 = 1:128
110 = 1:64
101 = 1:32
100 = 1:16
011 = 1:8
010 = 1:4
001 = 1:2
000 = 1:1

bit 0 **WDTEN:** Watchdog timer enable bit

1 = WDT enabled
0 = WDT disabled (control is placed on the SWDTEN bit)

Figure 2.9: CONFIG2H Register Bits

PIC18LFXX2 (Industrial)			**Standard Operating Conditions (unless otherwise stated)** Operating temperature -40°C ≤ TA ≤ +85°C for industrial				
PIC18FXX2 (Industrial, Extended)			**Standard Operating Conditions (unless otherwise stated)** Operating temperature -40°C ≤ TA ≤ +85°C for industrial; -40°C ≤ TA ≤ +125°C for extended				
Param No.	**Symbol**	**Characteristic**	**Min**	**Typ**	**Max**	**Units**	**Conditions**
	VDD	**Supply Voltage**					
D001		PIC18LFXX2	2.0	—	5.5	V	HS, XT, RC and LP Osc mode
D001		PIC18FXX2	4.2	—	5.5	V	
D002	VDR	**RAM Data Retention Voltage**	1.5	—	—	V	
D003	VPOR	**VDD Start Voltage** to ensure internal Power-on Reset signal	—	—	0.7	V	
D004	SVDD	**VDD Rise Rate** to ensure internal Power-on Reset signal	0.05	—	—	V/ms	
	VBOR	**Brown-out Reset Voltage**					
D005		PIC18LFXX2					
		BORV1:BORV0 = 11	1.98	—	2.14	V	85°C ≥ T ≥ 25°C
		BORV1:BORV0 = 10	2.67	—	2.89	V	
		BORV1:BORV0 = 01	4.16	—	4.5	V	
		BORV1:BORV0 = 00	4.45	—	4.83	V	
D005		PIC18FXX2					
		BORV1:BORV0 = 1x	N.A.	—	N.A.	V	Not in operating voltage range of device
		BORV1:BORV0 = 01	4.16	—	4.5	V	
		BORV1:BORV0 = 00	4.45	—	4.83	V	

Legend: Shading of rows is to assist in readability of the table.

Figure 2.10: The PIC8F452 Power Supply Parameters

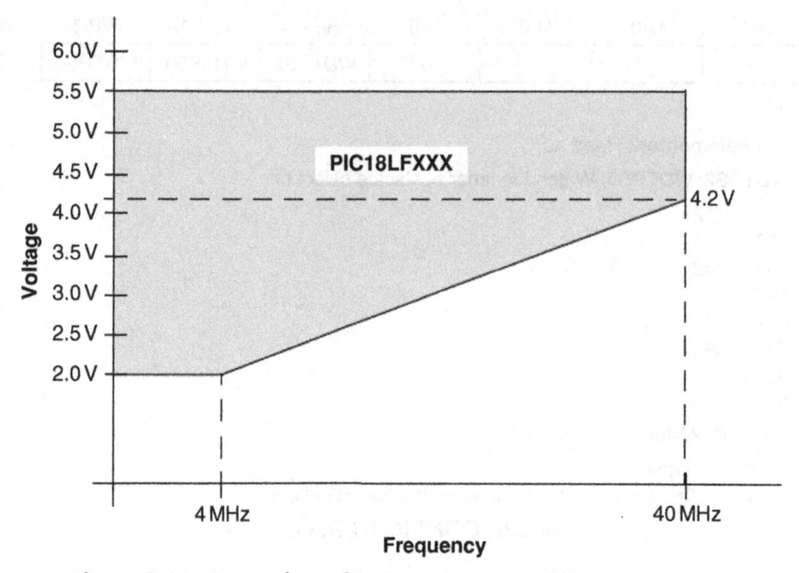

Figure 2.11: Operation of PIC18LF452 at Different Voltages

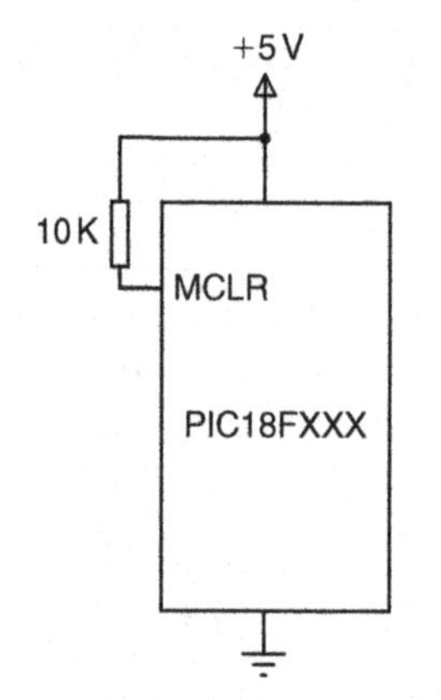

Figure 2.12: Typical Reset Circuit

Generally, two types of resets are commonly used: POR and external reset using the MCLR pin.

Power-On Reset

The POR is generated automatically when power supply voltage is applied to the chip. The MCLR pin should be tied to the supply voltage directly or preferably through a 10-K resistor. Figure 2.12 shows a typical reset circuit.

For applications where the rise time of the voltage is slow, it is recommended to use a diode, a capacitor, and a series resistor, as shown in Figure 2.13.

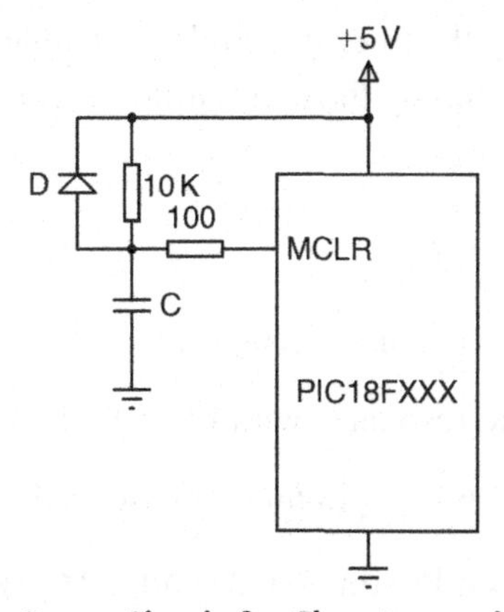

Figure 2.13: Reset Circuit for Slow Increasing Voltages

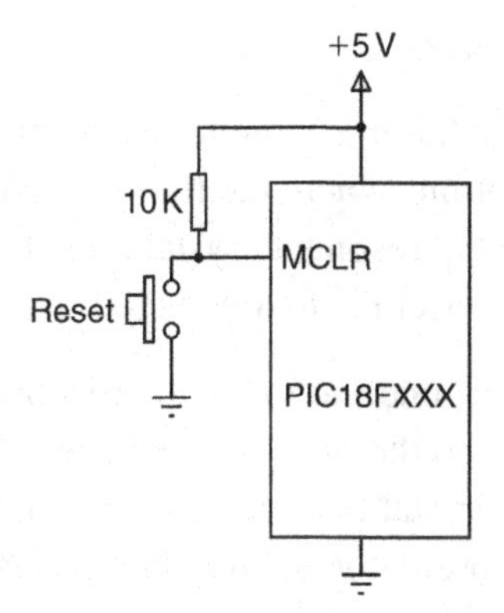

Figure 2.14: External Reset Circuit

In some applications, it may be required to reset the microcontroller externally by pressing a button. Figure 2.14 shows the circuit that can be used to reset the microcontroller externally. Normally, the MCLR input is at logic 1. When the **Reset** button is pressed, this pin goes to logic 0 and resets the microcontroller.

2.1.6 The Clock Sources

The PIC18F452 microcontroller can be operated from an external crystal or a ceramic resonator connected to the OSC1 and OSC2 pins of the microcontroller. In addition, an external resistor and capacitor, external clock source, and in some models, internal oscillators can be used to provide clock pulses to the microcontroller. There are eight clock sources on the PIC18F452 microcontroller, selected by the configuration register CONFIG1H. These are

- Low-power crystal (LP)
- Crystal or ceramic resonator (XT)
- High-speed crystal or ceramic resonator (HS)
- High-speed crystal or ceramic resonator with PLL (HSPLL)
- External resistor/capacitor with $F_{OSC/4}$ output on OSC2 (RC)
- External resistor/capacitor with I/O on OSC2 (port RA6) (RCIO)
- External clock with $F_{OSC/4}$ on OSC2 (EC)
- External clock with I/O on OSC2 (port RA6) (ECIO)

Crystal or Ceramic Resonator Operation

The first mode uses an external crystal or a ceramic resonator, which is connected to the OSC1 and OSC2 pins. For applications where the timing accuracy is important, crystal should be used. If a crystal is used, a parallel resonant crystal must be chosen because series resonant crystals do not oscillate when the system is first powered.

Figure 2.15 shows how a crystal is connected to the microcontroller. The capacitor values depend on the mode of the crystal and the selected frequency. Table 2.4 gives the recommended values. For example, for a 4-MHz crystal frequency, 15-pF capacitors can be used. Higher capacitance not only increases the oscillator stability but also increases the start-up time.

Resonators should be used in low-cost applications, where high-accuracy timing is not required. Figure 2.16 shows how a resonator is connected to the microcontroller.

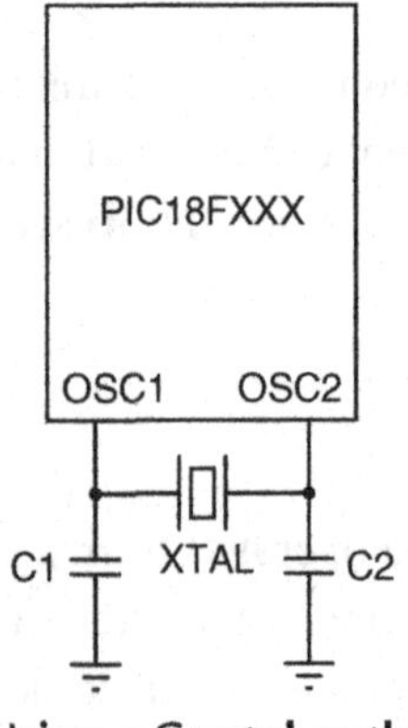

Figure 2.15: Using a Crystal as the Clock Input

Table 2.4 Capacitor Values

Mode	Frequency	C1,C2 (pF)
LP	32 kHz	33
	200 kHz	15
XT	200 kHz	22-68
	1.0 MHz	15
	4.0 MHz	15
HS	4.0 MHz	15
	8.0 MHz	15-33
	20.0 MHz	15-33
	25.0 MHz	15-33

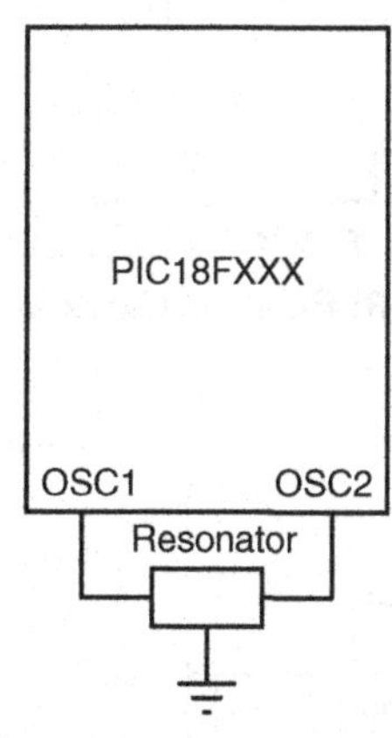

Figure 2.16: Using a Resonator as the Clock Input

LP oscillator mode should be selected in applications in which the clock frequency is up to 200 kHz. XT mode should be selected for up to 4 MHz, and the high-speed HS mode should be selected in applications where the clock frequency is between 4 and 25 MHz.

An external clock source may also be connected to the OSC1 pin in the LP, XT, and HS modes, as shown in Figure 2.17.

External Clock Operation

An external clock source can be connected to the OSC1 input of the microcontroller in EC and ECIO modes. No oscillator start-up time is required after a POR. Figure 2.18 shows the operation with external clock in EC mode. Timing pulses with frequency $F_{OSC/4}$ are available on the OSC2 pin. These pulses can be used for test purposes or to provide clock to external sources.

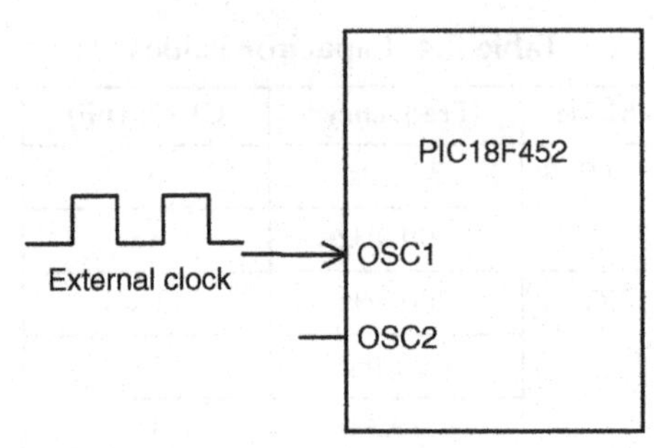

Figure 2.17: Connecting an External Clock in LP, XT, or HS Modes

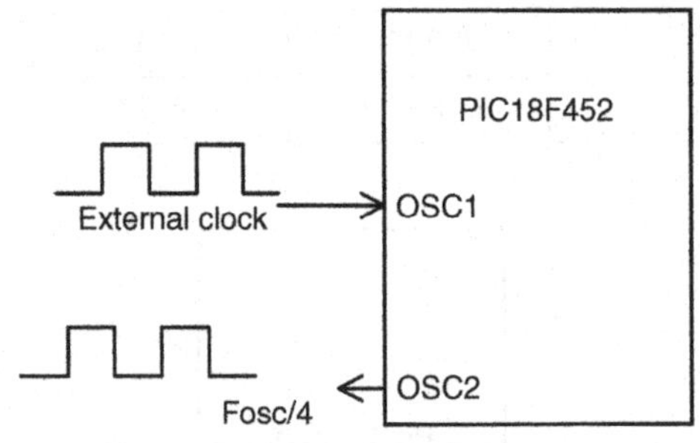

Figure 2.18: External Clock in EC Mode

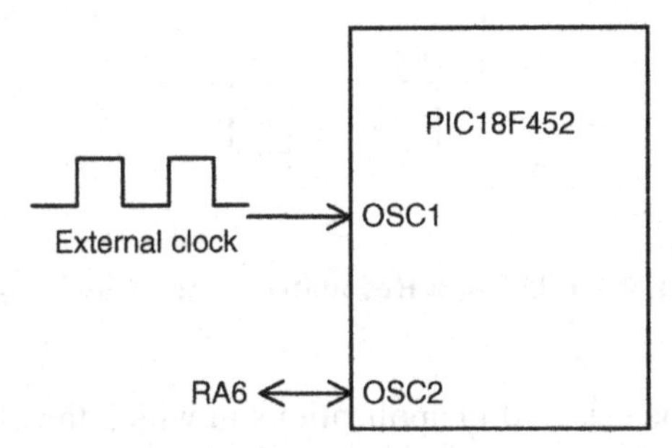

Figure 2.19: External Clock in ECIO Mode

The ECIO mode is similar to the EC mode, except that the OSC2 pin can be used as a general-purpose digital I/O pin. As shown in Figure 2.19, this pin becomes bit 6 of PORTA, i.e., pin RA6.

Resistor/Capacitor Operation

There are many applications where accurate timing is not required. In such applications, we can use an external resistor and a capacitor to provide clock pulses. The clock frequency is a function of the resistor, capacitor, power supply voltage, and temperature. The clock frequency is not accurate and can vary from unit to unit due to manufacturing and component tolerances. Table 2.5 gives the approximate clock frequency with various resistor and capacitor combinations. A close approximation of the clock frequency is 1/(4.2RC), where R should be between 3 and 10 K and C should be greater than 20 pF.

In RC mode, the oscillator frequency divided by 4 ($F_{OSC/4}$) is available on pin OSC2 of the microcontroller. Figure 2.20 shows the operation at a clock frequency of approximately 2 MHz, where $R = 3.9$ K and $C = 30$ pF. In this application, the clock frequency at the output of OSC2 is 2 MHz/4 = 500 KHz.

The RCIO mode is similar to the RC mode, except that the OSC2 pin can be used as a general-purpose digital I/O pin. As shown in Figure 2.21, this pin becomes bit 6 of PORTA. i.e., pin RA6.

Crystal or Resonator with PLL

One of the problems when high-frequency crystals or resonators are used is electromagnetic interference. A PLL circuit that can be enabled to multiply the clock frequency by four is provided. Thus, for a crystal clock frequency of 10 MHz, the internal operation frequency will

Table 2.5: Clock Frequency with RC

C (pF)	*R* (K)	Frequency (MHz)
22	3.3	3.3
	4.7	2.3
	10	1.08
30	3.3	2.4
	4.7	1.7
	10	0.793

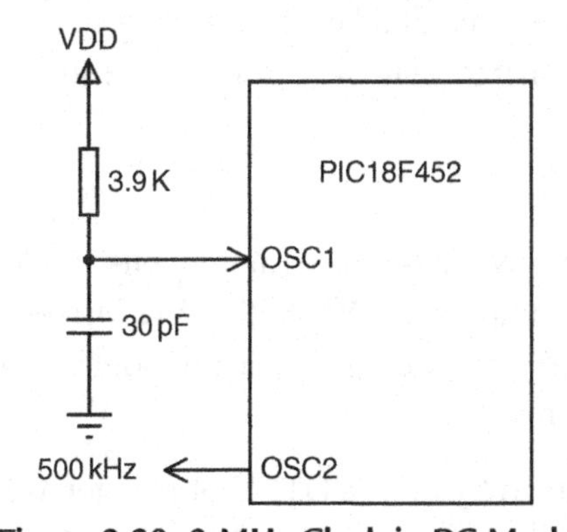

Figure 2.20: 2-MHz Clock in RC Mode

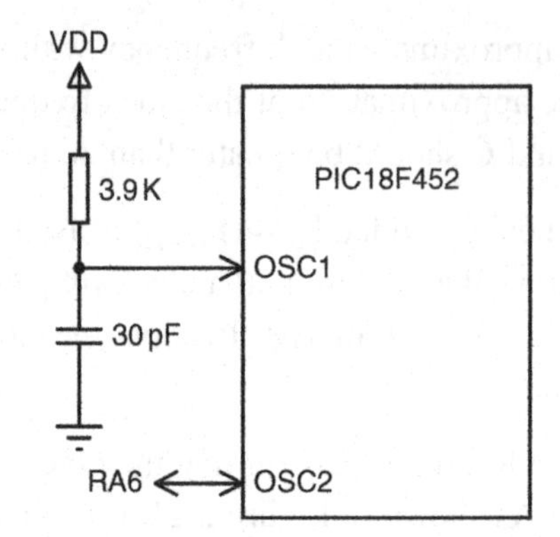

Figure 2.21: 2-MHz Clock in RCIO Mode

be multiplied to 40 MHz. The PLL mode is enabled when the oscillator configuration bits are programmed for HS mode.

Internal Clock

Some PIC18F family members have internal clock modes (PIC18F452 has no internal clock mode). In this mode, OSC1 and OSC2 pins are available for general-purpose I/O (RA6 and RA7) or as $F_{OSC/4}$ and RA7. Internal clock can be from 31 KHz to 8 MHz and is selected by registers OSCCON and OSCTUNE. Figure 2.22 shows the bits of internal clock control registers.

Clock Switching

It is possible to switch the clock from the main oscillator to a low-frequency clock source. For example, the clock can be allowed to run fast for periods of intense activity and slower when there is less activity. In the PIC18F452 microcontroller, this is controlled by bit SCS of the OSCCON register. In some of the other family members that also support internal clock, the clock switching is controlled by bits SCS0 and SCS1 of OSCCON. It is important to ensure that during clock switching, unwanted glitches do not occur in the clock signal. The 18F family contains circuitry to ensure error-free switching from one frequency to another frequency.

2.1.7 Watchdog Timer

In PIC18F family members, the WDT is a free-running on-chip RC-based oscillator, and it does not require any external components. When the WDT times out, a device reset is generated. If the device is in the SLEEP mode, the WDT time-out will wake up the device and continue with normal operation.

The watchdog is enabled/disabled by bit SWDTEN of register WDTCON. Setting SWDTEN = 1 enables the WDT, and clearing this bit turns off the WDT. On the

OSCCON register

IDLEN	IRCF2	IRCF1	IRCF0	OSTS	I0FS	SCSI	SCS0

IDLEN	0	Run mode enabled
	1	Idle mode enabled
IRCF2:IRCF0	000	31 KHz
	001	125 KHz
	010	250 KHz
	011	500 KHz
	100	1 MHz
	101	2 MHz
	110	4 MHz
	111	8 MHz
OSTS	0	Oscillator start-up timer running
	1	Oscillator start-up timer expired
IOFS	0	Internal oscillator unstable
	1	Internal oscillator stable
SCSI:SCS0	00	Primary oscillator
	01	Timer 1 oscillator
	10	Internal oscillator
	11	Internal oscillator

OSCTUNE register

X	X	T5	T4	T3	T2	T1	T0

XX011111	Maximum frequency
XX000001	
XX000000	Center frequency
XX111111	
XX100000	Minimum frequency

Figure 2.22: Internal Clock Control Registers

PIC18F452 microcontroller, an 8-bit postscaler is used to multiply the basic time-out period from 1 to 128 in powers of 2. This postscaler is controlled from configuration register CONFIG2H. The typical basic WDT time-out period is 18 ms for a postscaler value of 1.

2.1.8 Parallel I/O Ports

The parallel ports of the 18F family are very similar to those of the PIC16 series. The number of I/O ports and port pins varies depending on the PIC18F family member used, but all versions have at least PORTA and PORTB. The pins of a port are labeled as RPn, where *P* is the port letter and *n* is the port bit number. For example, PORTA pins are labeled RA0 to RA7, PORTB pins are labeled RB0 to RB7, and so on.

When working with a port, we may want to

- Set port direction
- Set an output value
- Read an input value
- Set an output value and then read back the output value

The first three operations are the same between the PIC16 and the PIC18F series. In some applications, we may want to send a value to the port and then read back the value just sent. In the PIC16 series, there is a weakness in the port design and the value read from a port may be different from the value just written to it. This is because the reading is the actual port bit pin value, and this value could be changed by external devices connected to the port pin. In the PIC18F series, a latch register (e.g., LATA for PORTA) is introduced to the I/O ports to hold the actual value sent to a port pin. From the port, the latched value is read, which is not affected by any external devices.

In this section, we shall be looking at the general structure of I/O ports.

PORTA

In PIC18F452 microcontroller, PORTA is 7 bits wide and port pins are shared with other functions. Table 2.6 shows the PORTA pin functions.

The architecture of PORTA is shown in Figure 2.23. There are three registers associated with PORTA:

- Port data register – PORTA
- Port direction register – TRISA
- Port latch register – LATA

PORTA is the name of the port data register. The TRISA register defines the direction of PORTA pins, where a logic 1 in a bit position defines the pin as an input pin, and a 0 in a bit position defines it as an output pin. LATA is the output latch register, which shares the same data latch as PORTA. Writing to one is equivalent to writing to the other one as well. But reading from LATA activates the buffer at the top of the diagram, and the value held in PORTA/LATA data latch is transferred to the data bus, independent of the state of the actual output pin of the microcontroller.

Table 2.6: PIC18F452 PORTA Pin Functions

Pin	Description
RA0/AN0	
RA0	Digital I/O
AN0	Analog input 0
RA1/AN1	
RA1	Digital I/O
AN1	Analog input 1
RA2/AN2/VREF–	
RA2	Digital I/O
AN2	Analog input 2
VREF–	A/D reference voltage (low) input
RA3/AN3/VREF+	
RA3	Digital I/O
AN3	Analog input 3
VREF+	A/D reference voltage (high) input
RA4/T0CKI	
RA4	Digital I/O
T0CKI	Timer 0 external clock input
RA5/AN4/SS/LVDIN	
RA5	Digital I/O
AN4	Analog input 4
SS	SPI Slave Select input
RA6	Digital I/O

Bits 0 through 3 and 5 of PORTA are also used as analog inputs. After a device reset, these pins are programmed as analog inputs, and RA4 and RA6 are configured as digital inputs. To program the analog inputs as digital I/O, the ADCON1 register (A/D register) must be programmed accordingly. Writing 7 to ADCON1 configures all PORTA pins as digital I/O.

The RA4 pin is multiplexed with the Timer 0 clock input (T0CKI). This is a Schmitt trigger input and an open drain output.

RA6 can be used as a general-purpose I/O pin, or as the OSC2 clock input, or as a clock output providing $F_{OSC/4}$ clock pulses.

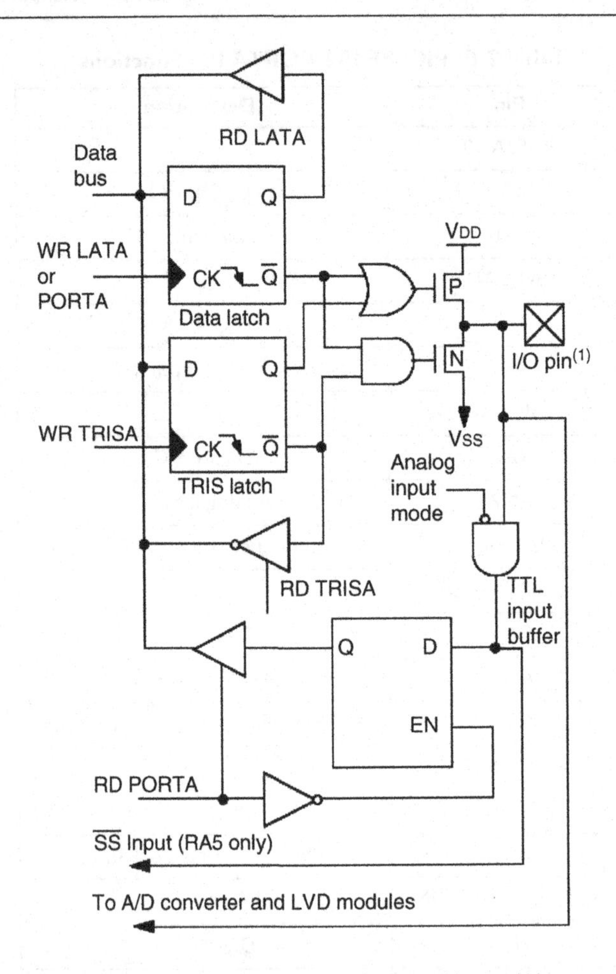

Note (1): I/O pins have protection diodes to VDD and VSS.

Figure 2.23: PIC18F452 PORTA RA0–RA3 and RA5 Pins

PORTB

In the PIC18F452 microcontroller, PORTB is an 8-bit bidirectional port shared with interrupt pins and serial device programming pins. Table 2.7 gives the PORTB bit functions.

PORTB is controlled by registers, and they are as follows:

- Port data register – PORTB
- Port direction register – TRISB
- Port latch register – LATB

The general operation of PORTB is similar to that of PORTA. Figure 2.24 shows the architecture of PORTB. Each port pin has a weak internal pull-up, which can be enabled by clearing bit RBPU

Table 2.7: PIC18F452 PORTB Pin Functions

Pin	Description
RB0/INT0	
RB0	Digital I/O
INT0	External interrupt 0
RB1/INT1	
RB1	Digital I/O
INT1	External interrupt 1
RB2/INT2	
RB2	Digital I/O
INT2	External interrupt 2
RB3/CCP2	
RB3	Digital I/O
CCP2	Capture 2 input, Compare 2 and PWM2 output
RB4	Digital I/O, Interrupt on change pin
RB5/PGM	
RB5	Digital I/O, Interrupt on change pin
PGM	Low-voltage ICSP programming pin
RB6/PGC	
RB6	Digital I/O, Interrupt on change pin
PGC	In-circuit debugger and ICSP programming pin
RB7/PGD	
RB7	Digital I/O, Interrupt on change pin
PGD	In-circuit debugger and ICSP programming pin

of register INTCON2. These pull-ups are disabled on a POR and when the port pin is configured as an output. On POR, PORTB pins are configured as digital inputs. Internal pull-ups allow input devices, such as switches, to be connected to PORTB pins without the use of external pull-up resistors. This saves cost because of the reduced component count and less wiring requirements.

Port pins RB4–RB7 can be used as interrupt on change inputs, whereby a change on any of pins 4–7 causes an interrupt flag to be set. The interrupt enable and flag bits RBIE and RBIF are in register INTCON.

PORTC, D, E, and Beyond

In addition to PORTA and PORTB, PIC18F452 has 8-bit bidirectional ports PORTC and PORTD, and 3-bit PORTE. Each port has its own data register (e.g., PORTC), data direction register (e.g., TRISC), and data latch register (e.g., LATC). The general operation of these ports is similar to PORTA.

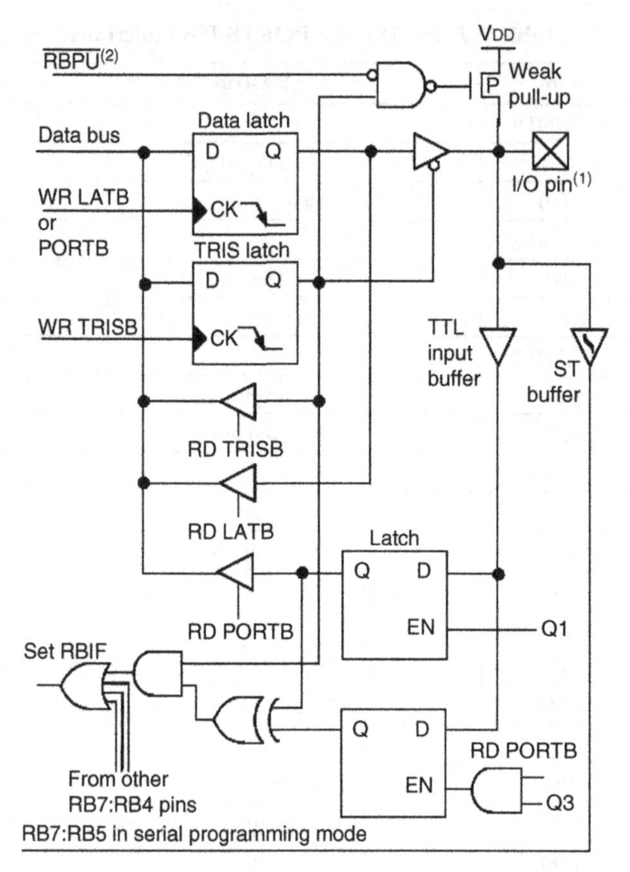

Note (1): I/O pins have diode protection to V_{DD} and V_{SS}.
(2): To enable weak pull-ups, set the appropriate TRIS bit(s) and clear the $\overline{\text{RBPU}}$ bit (INTCON2<7>).

Figure 2.24: PIC18F452 PORTB RB4–RB7 Pins

In the PIC18F452 microcontroller, PORTC is multiplexed with several peripheral functions, as shown in Table 2.8. On a POR, PORTC pins are configured as digital inputs.

In the PIC18F452 microcontroller, PORTD has Schmitt Trigger input buffers. On a POR, PORTD is configured as digital inputs. PORTD can be configured as an 8-bit PSP (i.e., microprocessor port) by setting bit 4 of the TRISE register. Table 2.9 shows functions of PORTD pins.

In the PIC18F452 microcontroller, PORTE is only 3 bits wide. As shown in Table 2.10, port pins are shared with analog inputs and parallel slave port read/write control bits. On a POR, PORTE pins are configured as analog inputs, and register ADCON1 must be programmed to change these pins to digital I/O.

Table 2.8: PIC18F452 PORTC Pin Functions

Pin	Description
RC0/T1OSO/T1CKI	
RC0	Digital I/O
T1OSO	Timer 1 oscillator output
T1CKI	Timer 1/Timer 3 external clock input
RC1/T1OSI/CCP2	
RC1	Digital I/O
T1OSI	Timer 1 oscillator input
CCP2	Capture 2 input, Compare 2 and PWM2 output
RC2/CCP1	
RC2	Digital I/O
CCP1	Capture 1 input, Compare 1 and PWM1 output
RC3/SCK/SCL	
RC3	Digital I/O
SCK	Synchronous serial clock input/output for SPI
SCL	Synchronous serial clock input/output for I2C
RC4/SDI/SDA	
RC4	Digital I/O
SDI	SPI data input
SDA	I^2C data I/O
RC5/SDO	
RC5	Digital I/O
SDO	SPI data output
RC6/TX/CK	
RC6	Digital I/O
TX	USART transmit pin
CK	USART synchronous clock pin
RC7/RX/DT	
RC7	Digital I/O
RX	USART receive pin
DT	USART synchronous data pin

2.1.9 Timers

The PIC18F452 microcontroller has four programmable timers, which can be used in many tasks, such as generating timing signals, causing interrupts to be generated at specific time intervals, measuring frequency and time intervals, and so on.

Table 2.9: PIC18F452 PORTD Pin Functions

Pin	Description
RD0/PSP0	
RD0	Digital I/O
PSP0	Parallel Slave Port bit 0
RD1/PSP1	
RD1	Digital I/O
PSP1	Parallel Slave Port bit 1
RD2/PSP2	
RD2	Digital I/O
PSP2	Parallel Slave Port bit 2
RD3/PSP3	
RD3	Digital I/O
PSP3	Parallel Slave Port bit 3
RD4/PSP4	
RD4	Digital I/O
PSP4	Parallel Slave Port bit 4
RD5/PSP5	
RD5	Digital I/O
PSP5	Parallel Slave Port bit 5
RD6/PSP6	
RD6	Digital I/O
PSP6	Parallel Slave Port bit 6
RD7/PSP7	
RD7	Digital I/O
PSP7	Parallel Slave Port bit 7

Table 2.10: PIC18F452 PORTE Pin Functions

Pin	Description
RE0/RD/AN5	
RE0	Digital I/O
RD	Parallel Slave Port read control pin
AN5	Analog input 5
RE1/WR/AN6	
RE1	Digital I/O
WR	Parallel Slave Port write control pin
AN6	Analog input 6
RE2/CS/AN7	
RE2	Digital I/O
CS	Parallel Slave Port CS
AN7	Analog input 7

This section introduces the timers available in the PIC18F452 microcontroller.

Timer 0

Timer 0 is similar to Timer 0 of the PIC16 series, except that it can operate either in 8-bit or in 16-bit mode. Timer 0 has the following basic features:

- 8-bit or 16-bit operation
- 8-bit programmable prescaler
- External or internal clock source
- Interrupt generation on overflow

Timer 0 control register is T0CON and is shown in Figure 2.25. The lower 6 bits of this register have similar functions to the PIC16 series OPTION register. The top 2 bits are used to select the 8/16-bit mode of operation and to enable/disable the timer.

R/W-1	R/W-1	R/W-1	R/W-1	R/W-1	R/W-1	R/W-1	R/W-1
TMR0ON	T08BIT	T0CS	T0SE	PSA	T0PS2	T0PS1	T0PS0
bit 7							bit 0

bit 7 **TMR0ON:** Timer 0 on/off control
1 = Enables Timer 0
0 = Stops Timer 0

bit 6 **T08BIT:** Timer 0 8-bit/16-bit control bit
1 = Timer 0 is configured as an 8-bit timer/counter
0 = Timer 0 is configured as a 16-bit timer/counter

bit 5 **T0CS:** Timer 0 clock source select bit
1 = Transition on T0CKI pin
0 = Internal instruction cycle clock (CLKO)

bit 4 **T0SE:** Timer 0 source edge select bit
1 = Increment on high-to-low transition on T0CKI pin
0 = Increment on low-to-high transition on T0CKI pin

bit 3 **PSA:** Timer 0 prescaler assignment bit
1 = Timer 0 prescaler is NOT assigned. Timer 0 clock input bypasses prescaler.
0 = Timer 0 prescaler is assigned. Timer 0 clock input comes from prescaler output.

bit 2–0 **T0PS2:T0PS0:** Timer 0 prescaler select bits
111 = 1:256 prescale value
110 = 1:128 prescale value
101 = 1:64 prescale value
100 = 1:32 prescale value
011 = 1:16 prescale value
010 = 1:8 prescale value
001 = 1:4 prescale value
000 = 1:2 prescale value

Figure 2.25: Timer 0 Control Register, T0CON

Timer 0 can be operated either as a timer or as a counter. Timer mode is selected by clearing the T0CS bit, and in this mode, the clock to the timer is derived from $F_{OSC/4}$. Counter mode is selected by setting the T0CS bit, and in this mode, Timer 0 is incremented on the rising or falling edge of input RA4/T0CKI. Bit T0SE of T0CON selects the edge triggering mode.

An 8-bit prescaler can be used to change the rate of the timer clock by a factor of up to 256. The prescaler is selected by bits PSA and T0PS2:T0PS0 of register T0CON.

8-Bit Mode

Figure 2.26 shows Timer 0 in 8-bit mode. The following operations are normally carried out in a timer application:

- Clear T0CS to select clock $F_{OSC/4}$.
- Use bits T0PS2:T0PS0 to select a suitable prescaler value.
- Clear PSA to select the prescaler.
- Load timer register TMR0L.
- Optionally enable Timer 0 interrupts.
- The timer will count up, and an interrupt will be generated when the timer value overflows from FFH to 00H in an 8-bit mode (or from FFFFH to 0000H in 16-bit mode).

By loading a value into the TMR0 register, we can control the count until an overflow occurs. The formula given below can be used to calculate the time it will take for the timer to overflow (or to generate an interrupt) given the oscillator period, the value loaded into the timer, and the prescaler value:

$$\text{Overflow time} = 4 \times T_{OSC} \times \text{Prescaler} \times (256 - \text{TMR0}) \tag{2.1}$$

where Overflow time is in μs, T_{OSC} is the oscillator period in μs, Prescaler is the prescaler value, and TMR0 is the value loaded into TMR0 register.

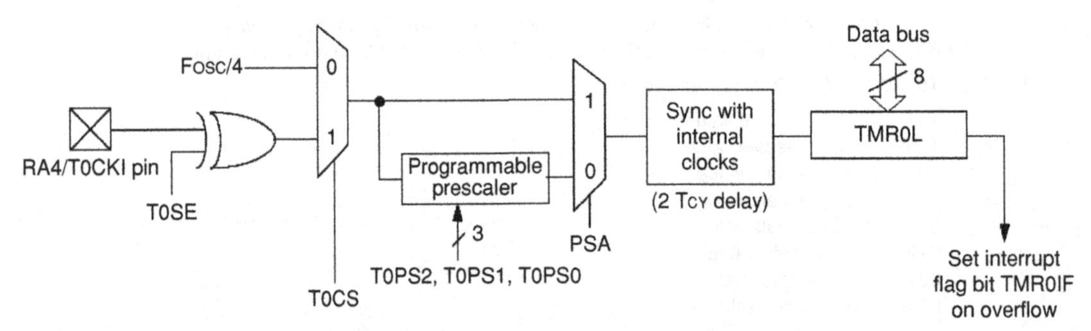

Figure 2.26: Timer 0 in an 8-bit Mode

For example, assume that we are using a 4-MHz crystal and the prescaler is chosen as 1:8 by setting bits PS2:PS0 to "010." In addition, assume that the value loaded into the timer register TMR0 is decimal 100. The overflow time is then given by

$$4\text{-MHz clock has a period, } T = 1/f = 0.25\,\mu s.$$

Using the above formula,

$$\text{Overflow time} = 4 \times 0.25 \times 8 \times (256 - 100) = 1248\ \mu s.$$

Thus, the timer will overflow after 1.248 ms, and a timer interrupt will be generated if the timer interrupt and global interrupts are enabled.

What we normally want is to know what value to load into the TMR0 register for a required overflow time. This can be calculated by modifying equation (2.1) as follows:

$$TMR0 = 256 - (\text{Overflow time})/(4 \times T_{OSC} \times \text{Prescaler}). \quad (2.2)$$

For example, assume that we want an interrupt to be generated after 500 μs and the clock and the prescaler values are same as mentioned above. The value to be loaded into the TMR0 register can be calculated using (2.2) as

$$TMR0 = 256 - 500/(4 \times 0.25 \times 8) = 193.5.$$

The nearest number we can load into TMR0 register is 193.

16-Bit Mode

Timer 0 in 16-bit mode is shown in Figure 2.27. Here, two timer registers named TMR0L and TMR0 are used to store the 16-bit timer value. The low-byte TMR0L is directly loadable from the data bus. The high-byte TMR0 can be loaded through a buffer called TMR0H. During a read of TMR0L, the high byte of the timer (TMR0) is also loaded into TMR0H, and thus, all 16 bits of the timer value can be read. Thus, to read the 16-bit timer value, we have to first read TMR0L and then read TMR0H in a later instruction. Similarly, during a write to TMR0L, the high byte of the timer is also updated with the contents of TMR0H, allowing all 16 bits to be written to the timer. Thus, to write to the timer, the program should first write the required higher byte to TMR0H. When the lower byte is written to TMR0L, the value stored in TMR0H is automatically transferred to TMR0, thus causing all 16 bits to be written to the timer.

Timer 1

Timer 1of the PIC18F452 is a 16-bit timer controlled by register T1CON, as shown in Figure 2.28. Figure 2.29 shows the internal structure of Timer 1.

Timer 1 can be operated either as a timer or as a counter. When bit TMR1CS of register T1CON is low, clock $F_{OSC/4}$ is selected for the timer. When TMR1CS is high, module

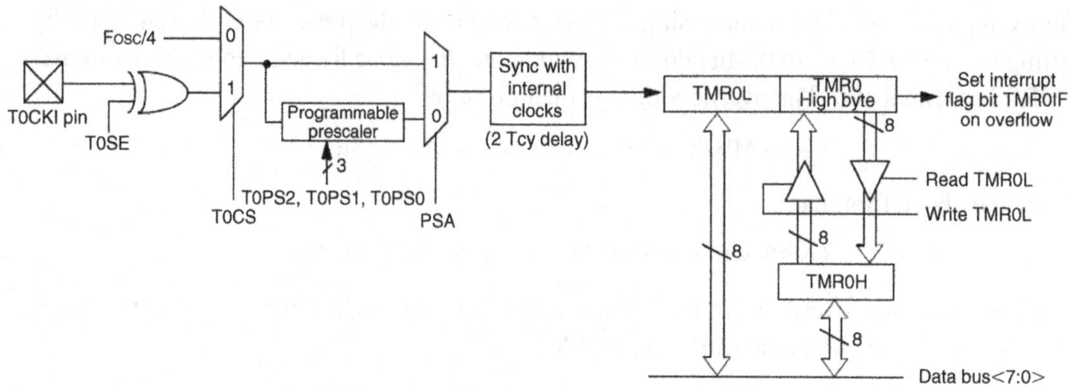

Figure 2.27: Timer 0 in a 16-bit Mode

R/W-0	U-0	R/W-0	R/W-0	R/W-0	R/W-0	R/W-0	R/W-0
RD16	—	T1CKPS1	T1CKPS0	T1OSCEN	$\overline{\text{T1SYNC}}$	TMR1CS	TMR1ON
bit 7							bit 0

bit 7 **RD16:** 16-bit read/write mode enable bit

1 = Enables register read/write of Timer 1 in one 16-bit operation
0 = Enables register read/write of Timer 1 in two 8-bit operations

bit 6 **Unimplemented:** Read as '0'

bit 5–4 **T1CKPS1:T1CKPS0:** Timer 1 input clock prescale select bits

11 = 1:8 prescale value
10 = 1:4 prescale value
01 = 1:2 prescale value
00 = 1:8 prescale value

bit 3 **T1OSCEN:** Timer 1 oscillator enable bit

1 = Timer 1 oscillator is enabled
0 = Timer 1 oscillator is enabled
The oscillator inverter and feedback resistor are turned off to eliminate power drain.

bit 2 $\overline{\text{T1SYNC}}$: Timer 1 external clock input synchronization select bit

when TMR1CS = 1:
1 = Do not synchronize external clock input
0 = Synchronize external clock input

when TMR1CS = 0:
This bit is ignored. Timer 1 uses the internal clock when TMR1CS = 0.

bit 1 TMR1CS: Timer 1 clock source select bit

1 = External clock from pin RC0/T1OSO/T13CKI (on the rising edge)
0 = Internal clock (Fosc/4)

bit 0 TMR1ON: Timer 1 on bit

1 = Enables Timer 1
0 = Stops Timer 1

Figure 2.28: Timer 1 Control Register, T1CON

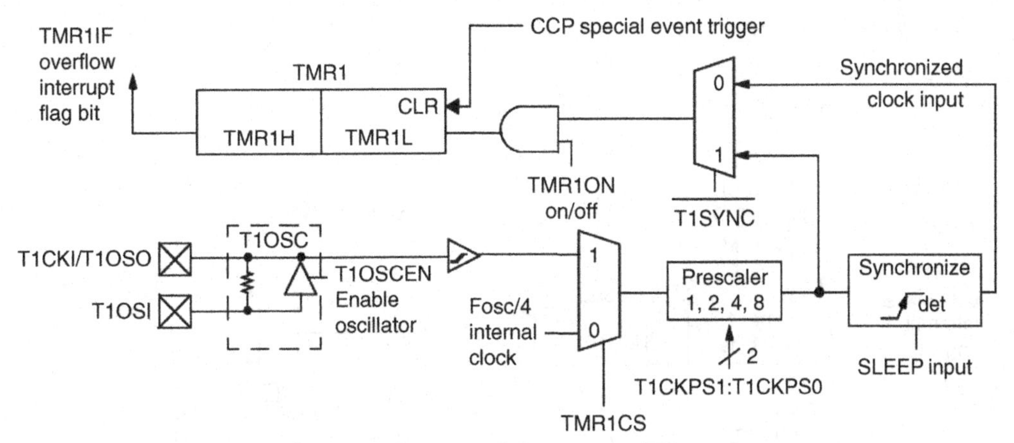

Figure 2.29: Internal Structure of Timer 1

operates as a counter clocked from input T1OSI. A crystal oscillator circuit, enabled from bit T1OSCEN of T1CON, is built between pins T1OSI and T1OSO, where a crystal up to 200 KHz can be connected between these pins. This oscillator is primarily intended for a 32-KHz crystal operation in real-time clock applications. A prescaler is used in Timer 1, which can change the timing rate as a factor of 1, 2, 4, or 8.

Timer 1 can be configured so that the read/write can be performed either in a 16-bit mode or in two 8-bit modes. Bit RD16 of register T1CON controls the mode. When RD16 is low, timer's read and write are performed as two 8-bit operations. When RD16 is high, the timer's read and write operations are as in Timer 0 16-bit mode. That is, a buffer is used between the timer register and the data bus (see Figure 2.30).

If the Timer 1 interrupts are enabled, an interrupt will be generated when the timer value rolls over from FFFFH to 0000H.

Timer 2

Timer 2 is an 8-bit timer with the following features:

- 8-bit timer (TMR2)
- 8-bit period register (PR2)
- Programmable prescaler
- Programmable postscaler
- Interrupt when TM2 matches PR2

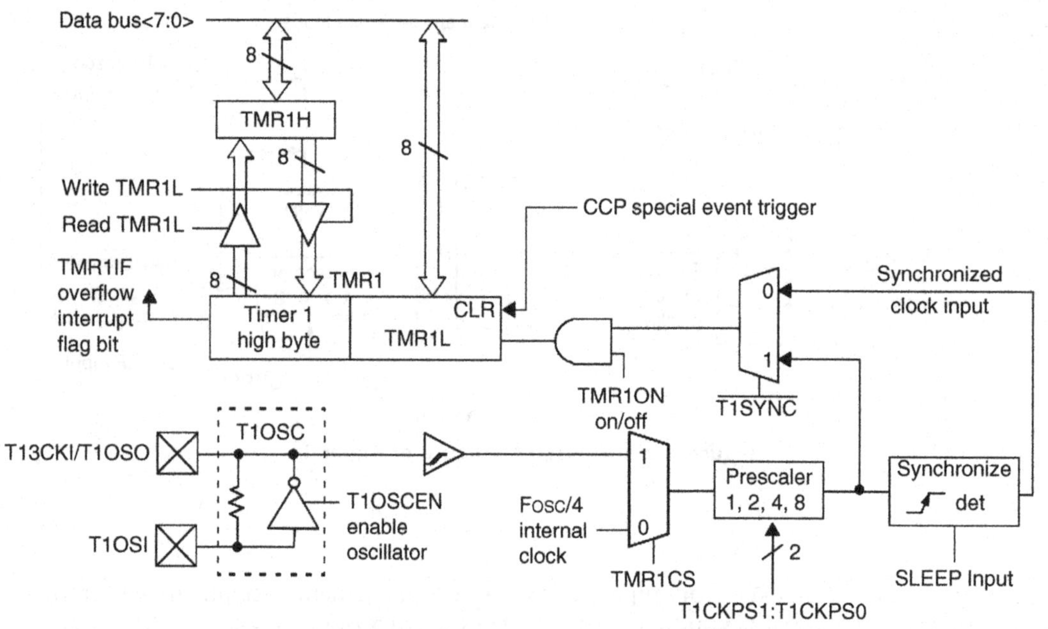

Figure 2.30: Timer 1 in a 16-bit Mode

U-0	R/W-0	R/W-0	R/W-0	R/W-0	R/W-0	R/W-0	R/W-0
—	TOUTPS3	TOUTPS2	TOUTPS1	TOUTPS0	TMR2ON	T2CKPS1	T2CKPS0
bit 7							bit 0

bit 7 **Unimplemented:** Read as '0'

bit 6–3 **TOUTPS3:TOUTPS0:** Timer 2 output postscale select bits

0000 = 1:1 postscale
0001 = 1:2 postscale
•
•
•
1111 = 1:16 postscale

bit 2 **TMR2ON:** Timer 2 on bit

1 = Timer 2 is on
0 = Timer 2 is off

bit 1–0 **T2CKPS1:T2CKPS0:** Timer 2 clock prescale select bits

00 = Prescaler is 1
01 = Prescaler is 4
1x = Prescaler is 16

Figure 2.31: Timer 2 Control Register, T2CON

Timer 2 is controlled from register T2CON, as shown in Figure 2.31. Bits T2CKPS1: T2CKPS0 set the prescaler for a scaling of 1, 4, and 16. Bits TOUTPS3:TOUTPS0 set the postscaler for a scaling of 1:1 to 1:16. The timer can be turned on or off by setting or clearing bit TMR2ON.

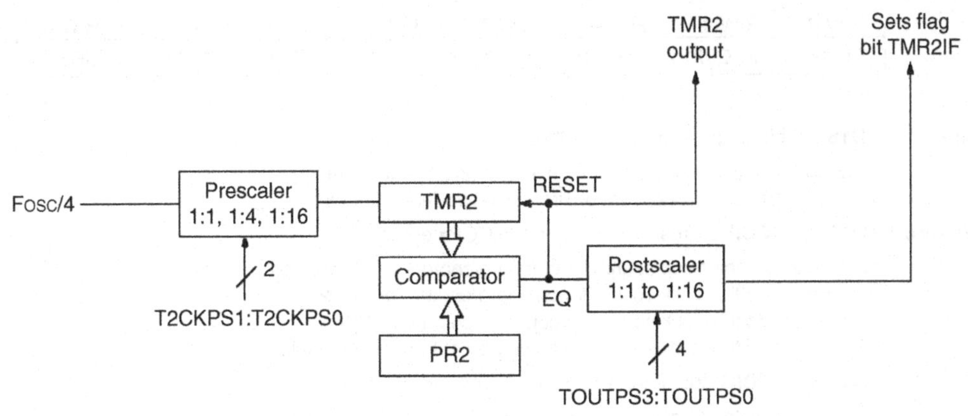

Figure 2.32: Timer 2 Block Diagram

The block diagram of Timer 2 is shown in Figure 2.32. Timer 2 can be used for the PWM mode of the CCP module. The output of Timer 2 can be selected by the SSP module as a baud clock using software. Timer 2 increases from 00H until it matches PR2 and sets the interrupt flag. It then resets to 00H on the next cycle.

Timer 3

The structure and operation of Timer 3 are the same as for Timer 1, having registers TMR3H and TMR3L. This timer is controlled from register T3CON, as shown in Figure 2.33.

The block diagram of Timer 3 is shown in Figure 2.34.

2.1.10 Capture/Compare/PWM Modules

In the PIC18F452 microcontroller, there are two Capture/Compare/PWM (CCP) modules, and they work with Timers 1, 2, and 3 to capture and compare and for PWM operations. Each module has two 8-bit registers. Module 1 registers are CCPR1L and CCPR1H, and module 2 registers are CCPR2L and CCPR2H. Together, each register pair form a 16-bit register and can be used to capture, compare, or generate waveforms with a specified duty cycle. Module 1 is controlled by register CCP1CON, and module 2 is controlled by CCP2CON. Figure 2.35 shows the bit allocations of the CCP control registers.

Capture Mode

In capture mode, the registers operate like a stopwatch, and when an event occurs, the time of the event is recorded, and the clock continues running (in a stopwatch, the watch stops when the event time is recorded).

R/W-0	R/W-0	R/W-0	R/W-0	R/W-0	R/W-0	R/W-0	R/W-0
RD 16	T3CCP2	T3CKPS1	T3CKPS0	T3CCP1	$\overline{\text{T3SYNC}}$	TMR3CS	TMR3ON
bit 7							bit 0

bit 7 **RD16:** 16-bit read/write mode enable bit

1 = Enables register read/write of Timer 3 in one 16-bit operation
1 = Enables register read/write of Timer 3 in two 8-bit operations

bit 6 and 3 **T3CCP2:T3CCP1:** Timer 3 and Timer 1 to CCPx enable bits

1x = Timer 3 is the clock source for compare/capture CCP modules
01 = Timer 3 is the clock source for compare/capture of CCP2,
Timer 1 is the clock source for compare/capture of CCP1
00 = Timer 1 is the clock source for compare/capture CCP modules

bit 5–4 **T3CKPS1:T3CKPS0:** Timer 3 input clock prescale select bits

11 = 1:8 prescale value
10 = 1:4 prescale value
01 = 1:2 prescale value
00 = 1:1 prescale value

bit 2 **$\overline{\text{T3SYNC}}$:** Timer 3 external clock input synchronization control bit
(Not usable if the system clock comes from Timer 1/Timer 3)
When TMR3CS = 1:
1 = Do not synchronize external clock input
0 = Synchronize external clock input
When TMR3CS = 0:
This bit is ignored. Timer 3 uses the internal clock when TMR3CS = 0.

bit 1 **TMR3CS:** Timer 3 clock source select bit
1 = External clock input from Timer 1 oscillator or T1CK1
(on the rising edge after the first falling edge)
0 = Internal clock ($Fosc_{/4}$)

bit 0 **TMR3ON:** Timer 3 on bit
1 = Enables Timer 3
0 = Stops Timer 3

Figure 2.33: Timer 3 Control Register, T3CON

Figure 2.36 shows the capture mode of operation. Here, CCP1 will be considered, but the operation of CCP2 is identical, with the register and port names change accordingly. In this mode, CCPR1H:CCPR1L captures the 16-bit value of the TMR1 or TMR3 registers when an event occurs on pin RC2/CCP1 (pin RC2/CCP1 must be configured as an input pin using TRISC). An external signal can be prescaled by 4 or 16. The event is selected by control bits CCP1M3:CCP1M0, and an event can be selected to be one of the following:

- Every falling edge
- Every rising edge
- Every fourth rising edge
- Every sixteenth rising edge

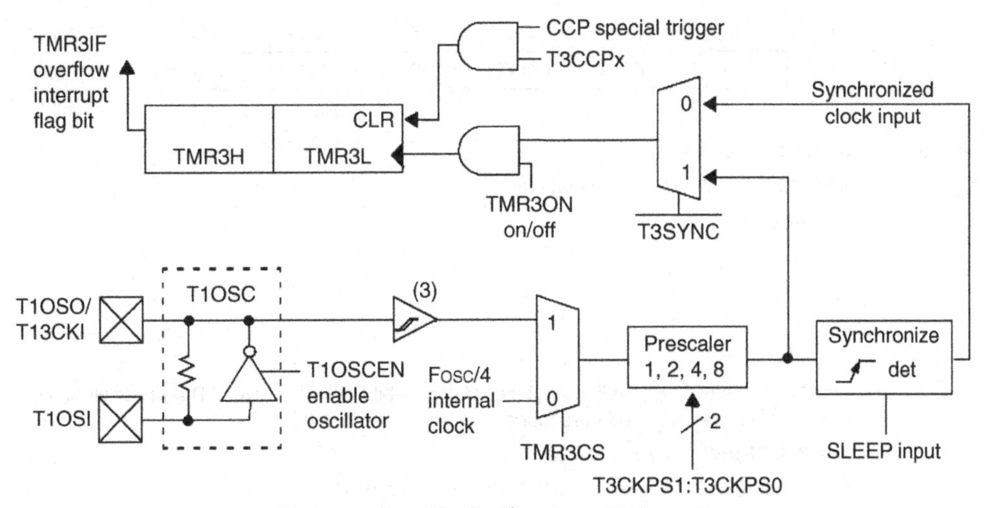

Figure 2.34: Block Diagram of Timer 3

If the capture interrupt is enabled, the occurrence of an event causes an interrupt to be generated in software. If another capture occurs before the value in register CCPR1 is read, the old captured value is overwritten by the new captured value.

Either of Timers 1 or 3 can be used in capture mode, and these timers must be running in timer mode or in Synchronized Counter mode selected by register T3CON.

Compare Mode

In compare mode, a digital comparator is used to compare the value of Timer 1 or Timer 3 with the value in a 16-bit register pair, and when a match occurs, the output state of a pin is changed. Figure 2.37 shows the block diagram of compare mode of operation.

Here, only module CCP1 is considered, but the operation of module CCP2 is identical.

The value of 16-bit register pair CCPR1H:CCPR1L is continuously compared against the Timer 1 or Timer 3 value. When a match occurs, the state of the RC2/CCP1 pin is changed depending on the programming of bits CCP1M2:CCP1M0 of register CCP1CON. The following changes can be programmed:

- Force RC2/CCP1 high
- Force RC2/CCP1 low
- Toggle RC2/CCP1 pin (low-to-high or high-to-low)
- Generate interrupt when a match occurs
- No change

U-0	U-0	R/W-0	R/W-0	R/W-0	R/W-0	R/W-0	R/W-0
—	—	DC×B1	DC×B0	CCP×M3	CCP×M2	CCP×M1	CCP×M0
bit 7							bit 0

bit 7–6 **Unimplemented:** Read as '0'

bit 5–4 **DC×B1:DC×B0:** PWM duty cycle bit 1 and bit 0

Capture mode:

Unused

Compare mode:

Unused

PWM mode:

These bits are the two LSbs (bit 1 and bit 0) of the 10-bit PWM duty cycle. The upper eight bits (DC×9:DC×2) of the duty cycle are found in CCPRxL.

bit 3–0 **CCP×M3: CCP×M0:** CCPx mode select bits

0000 = Capture/compare/PWM disabled (resets CCPx module)
0001 = Reserved
0010 = Compare mode, toggle output on match (CCPxIF bit is set)
0011 = Reserved
0100 = Capture mode, every falling edge
0101 = Capture mode, every rising edge
0110 = Capture mode, every 4th rising edge
0111 = Capture mode, every 16th rising edge
1000 = Compare mode,
Initialize CCP pin Low, on compare match force CCP pin High (CCPIF bit is set)
1001 = Compare mode,
Initialize CCP pin High, on compare match force CCP pin Low (CCPIF bit is set)
1010 = Compare mode,
Generate software interrupt on compare match (CCPIF bit is set, CCP pin is unaffected)
1011 = Compare mode,
Trigger special event (CCPIF bit is set)
11xx = PWM mode

Figure 2.35: CCPxCON Register Bit Allocations

Timer 1 or Timer 3 must be running in Timer mode or in Synchronized Counter mode selected by register T3CON.

2.1.11 Pulse Width Modulation Module

The Pulse Width Modulation (PWM) mode produces a PWM output at 10-bit resolution. A PWM output is basically a square waveform with a specified period and duty cycle. Figure 2.38 shows a typical PWM waveform.

Figure 2.39 shows the PWM module block diagram. The module is controlled by Timer 2. The PWM period is given by

$$\text{PWM period} = (\text{PR2} + 1) \times \text{TMR2PS} \times 4 \times T_{OSC} \quad (2.3)$$

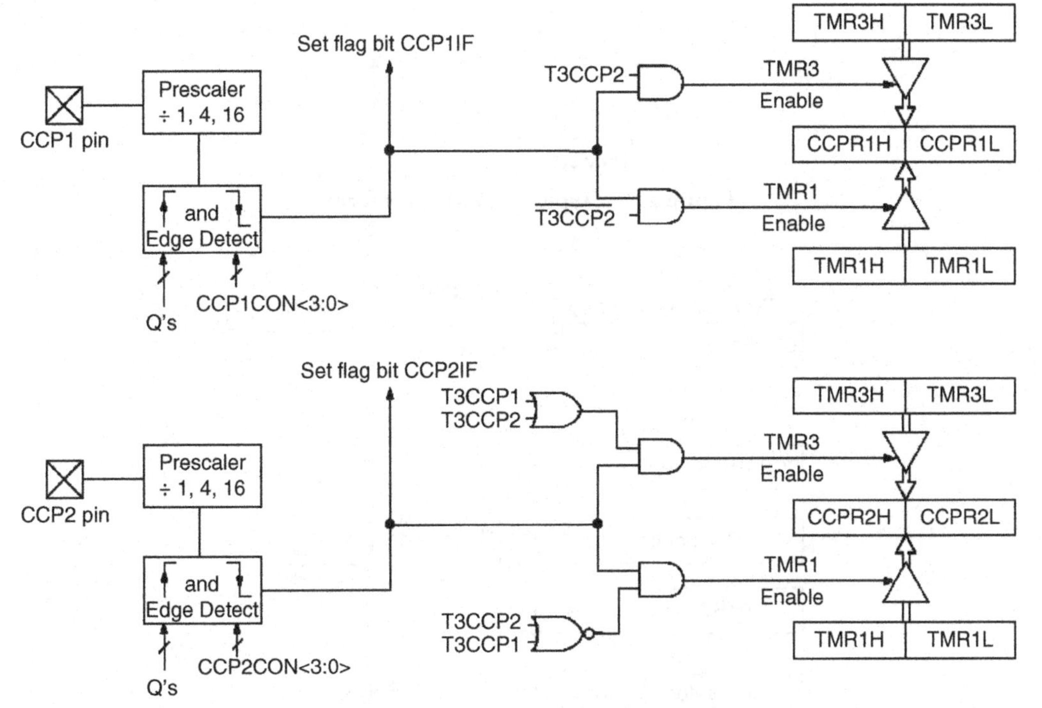

Figure 2.36: Capture Mode of Operation

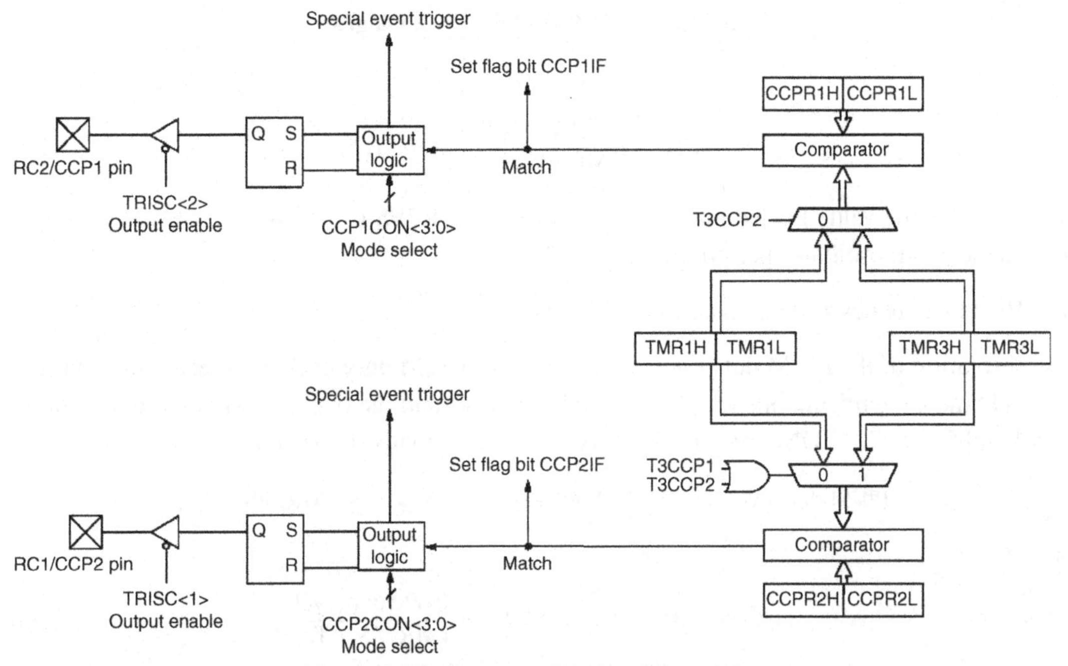

Figure 2.37: Compare Mode of Operation

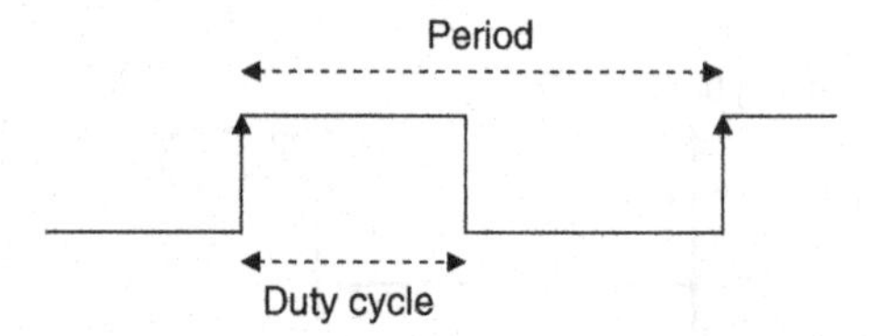

Figure 2.38: Typical PWM Waveform

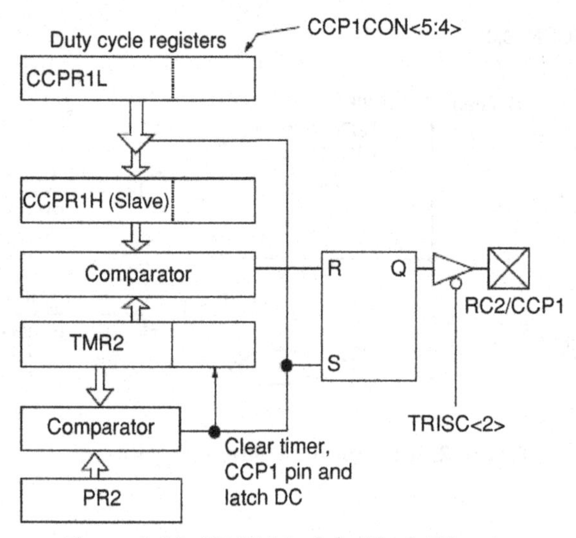

Figure 2.39: PWM Module Block Diagram

or

$$PR2 = \frac{\text{PWM period}}{\text{TMR2PS} \times 4 \times T_{OSC}} - 1, \tag{2.4}$$

where PR2 is the value loaded into Timer 2 register, TMR2PS is the Timer 2 prescaler value, T_{OSC} is the clock oscillator period (seconds).

The PWM frequency is defined as 1/(PWM period).

The resolution of the PWM duty cycle is 10 bits. The PWM duty cycle is selected by writing the eight most significant bits into the CCPR1L register and the two least significant bits into bits 4 and 5 of CCP1CON register. The duty cycle (in seconds) is given by

$$\text{PWM duty cycle} = (\text{CCPR1L:CCP1CON} < 5{:}\ 4 >) \times \text{TMR2PS} \times T_{OSC} \tag{2.5}$$

or

$$\text{CCPR1L:CCP1CON} < 5{:}4 > = \frac{\text{PWM duty cycle}}{\text{TMR2PS} \times T_{OSC}} \tag{2.6}$$

The steps to configure the PWM are as follows:

- Specify the required period and duty cycle.
- Choose a value for Timer 2 prescaler (TMR2PS).
- Calculate the value to be written into PR2 register using equation (2.2).
- Calculate the value to be loaded into CCPR1L and CCP1CON registers using equation (2.6).
- Clear the bit 2 of TRISC to make CCP1 pin an output pin.
- Configure the CCP1 module for PWM operation using register CCP1CON.

An example is given below to show how the PWM can be set up.

Example 2.1

It is required to generate PWM pulses from pin CCP1 of a PIC18F452 microcontroller. The required pulse period is 44 μs, and the required duty cycle is 50%. Assuming that the microcontroller operates with a 4-MHz crystal, calculate the values to be loaded into various registers.

Solution

Using a 4-MHz crystal, calculate $T_{OSC} = 1/4 = 0.25 \times 10^{-6}$.

The required PWM duty cycle is 44/2 = 22 μs.

From equation (2.4), assuming a timer prescaler factor of 4, we have

$$PR2 = \frac{\text{PWM period}}{TMR2PS \times 4 \times T_{OSC}} - 1$$

or

$$PR2 = \frac{44 \times 10^{-6}}{4 \times 4 \times 0.25 \times 10^{-6}} - 1 = 10, \text{ i.e., } 0AH$$

and from equation (2.6)

$$CCPR1L{:}CCP1CON<5{:}4> = \frac{\text{PWM duty cycle}}{TMR2PS \times T_{OSC}}$$

or

$$CCPR1L{:}CCP1CON<5{:}4> = \frac{22 \times 10^{-6}}{4 \times 0.25 \times 10^{-6}} = 22.$$

But the equivalent of number 22 in 10-bit binary is 00 00010110.

Therefore, the value to be loaded into bits 4 and 5 of CCP1CON is 00. Bits 2 and 3 of CCP1CON must be set to high for PWM operation. Therefore, CCP1CON must be set to bit pattern ("X" is don't care): XX001100

Considering the don't care entries as 0, we can set CCP1CON to hexadecimal 0CH.

The value to be loaded into CCPR1L is 00010110, i.e., hexadecimal number 16H.

The required steps are summarized below.

- Load Timer 2 with prescaler of 4, i.e., load T2CON with 00000101, which is 05H.
- Load 0AH into PR2.
- Load 16H into CCPR1L.
- Load 0 into TRISC (make CCP1 pin output).
- Load 0CH into CCP1CON.

One period of the generated PWM waveform is shown in Figure 2.40. ■

2.1.12 Analog-to-Digital Converter Module

An A/D converter is another important peripheral component of a microcontroller. The A/D converts an analog input voltage into a digital number so that it can be processed by a microcontroller or any other digital system. There are many A/D chips available in the market, and an embedded designer should understand the characteristics of such chips so that the chips can be used efficiently.

A/D converters can be classified into two types as far as the input and output voltage are concerned. These are unipolar and bipolar. Unipolar A/D converters accept unipolar input voltages in the range 0 to +V, and bipolar A/D converters accept bipolar input voltages in the range ±V. Bipolar converters are frequently used in signal processing applications, where the signals by nature are bipolar. Unipolar converters are usually cheaper, and they are used in many control and instrumentation applications.

Figure 2.41 shows the typical steps involved in reading and converting an analog signal into digital form. This is also known as the process of *signal conditioning*. Signals received from sensors usually need to be processed before being fed to an A/D converter. The processing

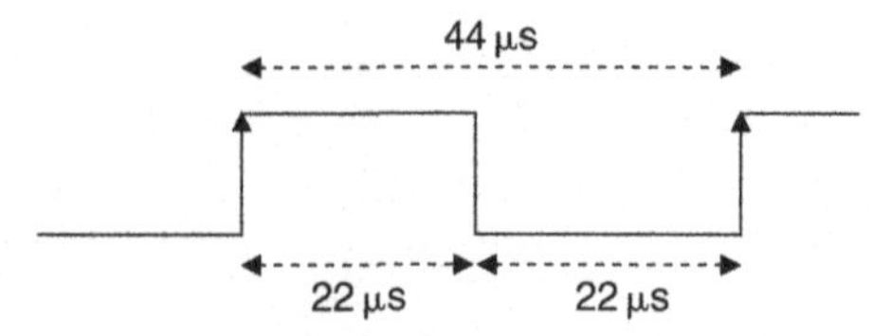

Figure 2.40: Generated PWM Waveform

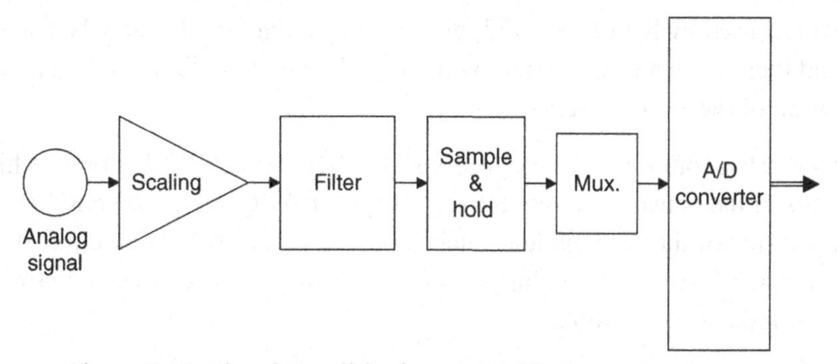

Figure 2.41: Signal Conditioning and A/D Conversion Process

usually consists of scaling the signal to the correct value. Unwanted signal components are then removed by filtering the signal using classical filters (e.g., low-pass filter). The final processing stage before feeding the signal to an A/D converter is to pass the signal through a sample-and-hold device. This is particularly important for fast real-time signals whose value may be changing between the sampling instants. A sample-and-hold device ensures that the signal stays at a constant value during the actual conversion process. In many applications, it is usually required to have more than one A/D, and this is normally done by using an analog multiplexer at the input of the A/D. The multiplexer selects only one signal at any time and presents this signal to the actual A/D converter. An A/D converter usually has a single analog input and a digital parallel output. The conversion process is as follows:

- Apply the processed signal to the A/D input.
- Start the conversion.
- Wait until conversion is complete.
- Read the converted digital data.

The A/D conversion starts by triggering the converter. Depending on the speed of the converter, the actual conversion process can take several microseconds. At the end of the conversion, the converter either raises a flag or generates an interrupt to indicate that the conversion is complete. The converted parallel output data can then be read by the digital device connected to the A/D converter.

Most PIC18F family members contain a 10-bit A/D converter. If the voltage reference is chosen to be +5 V, the voltage step value is

$$\left(\frac{5\,\text{V}}{1023}\right) = 0.00489\,\text{V} \quad \text{or} \quad 4.89\,\text{mV}.$$

Therefore, for example, if the input voltage is 1.0 V, the converter will generate a digital output of 1.0/0.00489 = 205 decimal. Similarly, if the input voltage is 3.0 V, the converter will generate 3.0/0.00489 = 613.

The A/D converter used by the PIC18F452 microcontroller has eight channels, called AN0–AN7, and these channels are shared by the PORTA and PORTE pins. Figure 2.42 shows the block diagram of the A/D converter.

The A/D converter has four registers. Registers ADRESH and ADRESL store the higher and lower results of the conversion, respectively. Register ADCON0, shown in Figure 2.43, controls the operation of the A/D module, such as selecting the conversion clock (together with register ADCON1), selecting an input channel, starting a conversion, and powering-up and shutting-down the A/D converter.

The ADCON1 register (see Figure 2.44) is used for selecting the conversion format, configuring the A/D channels for analog input, selecting the reference voltage, and selecting the conversion clock (together with ADCON0).

The A/D conversion starts by setting the GO/DONE bit of ADCON0. When the conversion is complete, the 2 bits of the converted data are written into register ADRESH, and the remaining 8 bits are written into register ADRESL. At the same time, the GO/DONE bit is cleared to indicate the end of conversion. If required, interrupts can be enabled so that a software interrupt is generated when the conversion is complete.

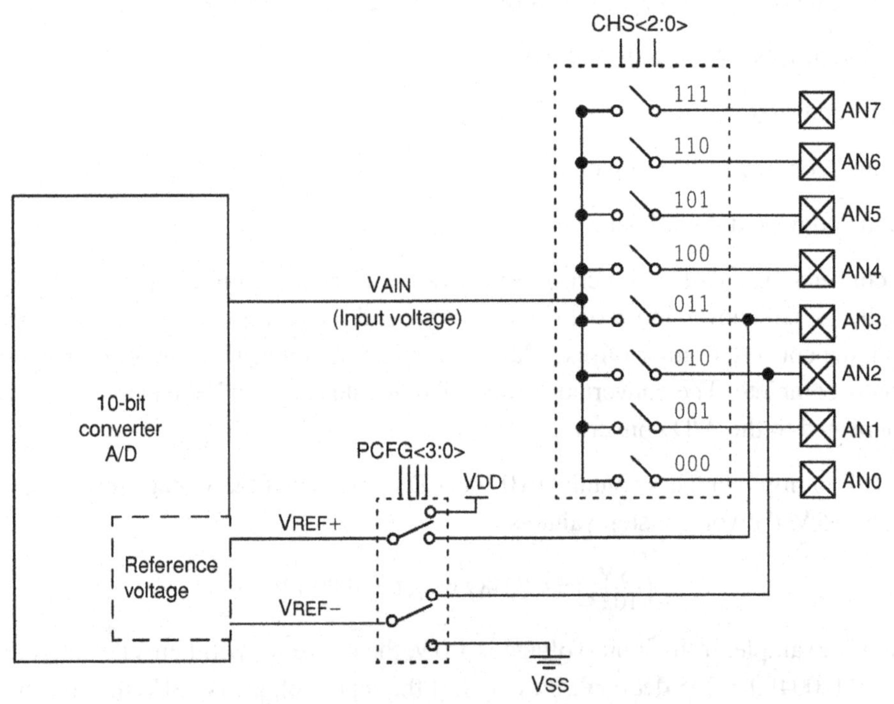

Figure 2.42 Block Diagram of the PIC18F452 A/D Converter

The steps in carrying out an A/D conversion are listed below.

- Use ADCON1 to configure required channels as analog and configure the reference voltage.
- Set the TRISA or TRISE bits so that the required channel is an input port.
- Use ADCON0 to select the required analog input channel.

R/W-0	R/W-0	R/W-0	R/W-0	R/W-0	R/W-0	U-0	R/W-0
ADCS1	ADCS0	CHS2	CHS1	CHS0	GO/$\overline{\text{DONE}}$	—	ADON
bit 7							bit 0

bit 7–6 **ADCS1:ADCS0:** A/D conversion clock select bits

ADCON1 <ADCS2>	ADCON0 <ADCS1:ADCS0>	Clock Conversion
0	00	Fosc/2
0	01	Fosc/8
0	10	Fosc/32
0	11	FRC (clock derived from the internal A/D RC oscillator)
1	00	Fosc/4
1	01	Fosc/16
1	10	Fosc/64
1	11	FRC (clock derived from the internal A/D RC oscillator)

bit 5–3 **CHS2:CHS0:** Analog channel select bits
000 = channel 0, (AN0)
001 = channel 1, (AN1)
010 = channel 2, (AN2)
011 = channel 3, (AN3)
100 = channel 4, (AN4)
101 = channel 5, (AN5)
110 = channel 6, (AN6)
111 = channel 7, (AN7)

Note: The PIC18F2x2 devices do not implement the full eight A/D channels; the unimplemented selections are reserved. Do not select any unimplemented channel.

bit 2 **GO/$\overline{\text{DONE}}$:** A/D conversion status bit

bit 1 **Unimplemented:** Read as '0'

When ADON = 1:
1 = A/D conversion in progress (setting this bit starts the A/D conversion, which is automatically cleared by hardware when the A/D conversion is complete)
0 = A/D conversion not in progress

bit 0 **ADON:** A/D on bit
1 = A/D converter module is powered up
0 = A/D converter module is shut off and consumes no operating current

Figure 2.43: ADCON0 Register

R/W-0	R/W-0	U-0	U-0	R/W-0	R/W-0	R/W-0	R/W-0
ADFM	ADCS2	—	—	PCFG3	PCFG2	PCFG1	PCFG0
bit 7							bit 0

bit 7 **ADFM:** A/D result format select bit
1 = Right justified. Six (6) most significant bits of ADRESH are read as '0'.
0 = Left justified. Six (6) least significant bits of ADRESL are read as '0'.

bit 6 **ADCS2:** A/D conversion clock select bit

ADCON1 <ADCS2>	ADCON0 <ADCS1:ADCS0>	Clock Conversion
0	00	Fosc/2
0	01	Fosc/8
0	10	Fosc/32
0	11	FRC (clock derived from the internal A/D RC oscillator)
1	00	Fosc/4
1	01	Fosc/16
1	10	Fosc/64
1	11	FRC (clock derived from the internal A/D RC oscillator)

bit 5–4 **Unimplemented:** Read as '0'

bit 3–0 **PCFG3:PCFG0:** A/D port configuration control bits

PCFG <3:0>	AN7	AN6	AN5	AN4	AN3	AN2	AN1	AN0	VREF+	VREF-	C/R
0000	A	A	A	A	A	A	A	A	VDD	Vss	8/0
0001	A	A	A	A	VREF+	A	A	A	AN3	Vss	7/1
0010	D	D	D	A	A	A	A	A	VDD	Vss	5/0
0011	D	D	D	A	VREF+	A	A	A	AN3	Vss	4/1
0100	D	D	D	D	A	D	A	A	VDD	Vss	3/0
0101	D	D	D	D	VREF+	D	A	A	AN3	Vss	2/1
011x	D	D	D	D	D	D	D	D	—	—	0/0
1000	A	A	A	A	VREF+	VREF–	A	A	AN3	AN2	6/2
1001	D	D	A	A	A	A	A	A	VDD	Vss	6/0
1010	D	D	A	A	VREF+	A	A	A	AN3	Vss	5/1
1011	D	D	A	A	VREF+	VREF–	A	A	AN3	AN2	4/2
1100	D	D	D	A	VREF+	VREF–	A	A	AN3	AN2	3/2
1101	D	D	D	D	VREF+	VREF–	A	A	AN3	AN2	2/2
1110	D	D	D	D	D	D	D	A	VDD	Vss	1/0
1111	D	D	D	D	VREF+	VREF–	D	A	AN3	AN2	1/2

A = Analog input D = Digital I/O

Figure 2.44: ADCON1 Register

- Use ADCON0 and ADCON1 to select the conversion clock.
- Use ADCON0 to turn on the A/D module.
- Configure the A/D interrupt (if desired).
- Set GO/DONE bit to start conversion.
- Wait until GO/DONE bit is cleared or until a conversion complete interrupt is generated.
- Read the converted data from ADRESH and ADRESL.
- Repeat above steps as required.

For correct A/D conversion, the A/D conversion clock must be selected to ensure a minimum bit conversion time of 1.6 μs. Table 2.11 gives the recommended A/D clock sources for the chosen microcontroller operating frequency. For example, if the microcontroller is operated at a 10-MHz clock, the A/D clock source should be selected as $F_{OSC/16}$ or higher (e.g., $F_{OSC/32}$).

Bit ADFM of register ADCON1 controls format of a conversion. When ADFM is cleared, the 10-bit result is left justified (see Figure 2.45) and lower 6 bits of ADRESL are cleared to zero. When ADFM is set to 1, the result is right justified and the upper 6 bits of ADRESH are cleared to zero. This is the commonly used mode, and, here, ADRESL contains the low-order 8 bits, and bits 0 and 1 of ADRESH contain the upper 2 bits of the 10-bit result.

Analog Input Model and the Acquisition Time

An understanding of the A/D analog input model is necessary to interface the A/D to external devices. Figure 2.46 shows the analog input model of the A/D. On the left-hand side of the diagram, we see the analog input voltage V_{AIN} and the source resistance R_S. It is recommended that the source resistance not be greater than 2.5 K. The analog signal is applied to the pin labeled ANx. There is a small capacitance (5 pF) and a leakage current to the ground of

Table 2.11: A/D Conversion Clock Selection

A/D Clock Source		
Operation	ADCS2:ADCS0	Maximum Microcontroller Frequency
2 T_{OSC}	000	1.25 MHz
4 T_{OSC}	100	2.50 MHz
8 T_{OSC}	001	5.0 MHz
16 T_{OSC}	101	10.0 MHz
32 T_{OSC}	010	20.0 MHz
64 T_{OSC}	110	40.0 MHz
RC	011	-

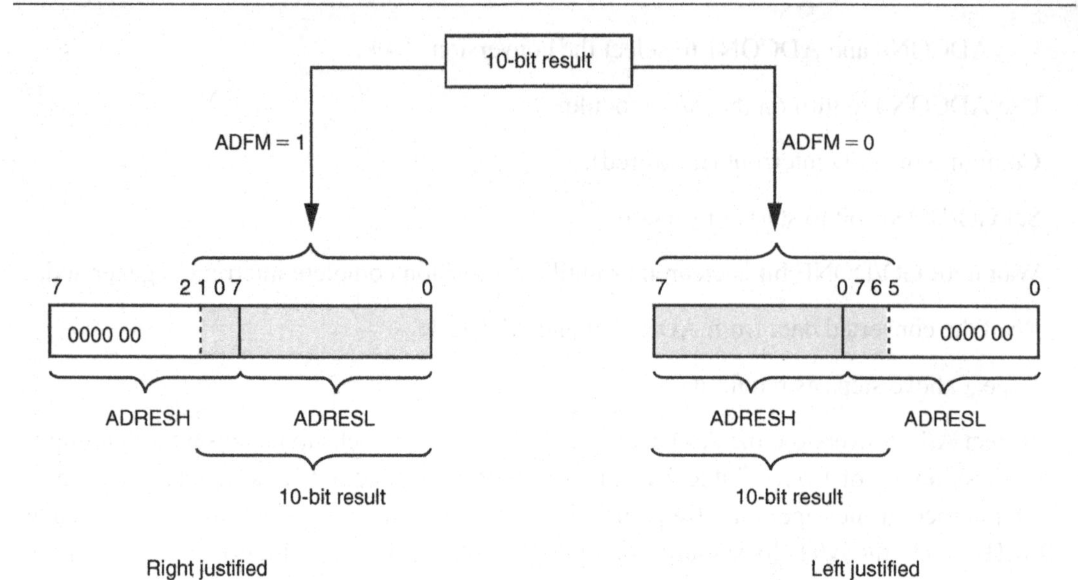

Figure 2.45: Formatting the A/D Conversion Result

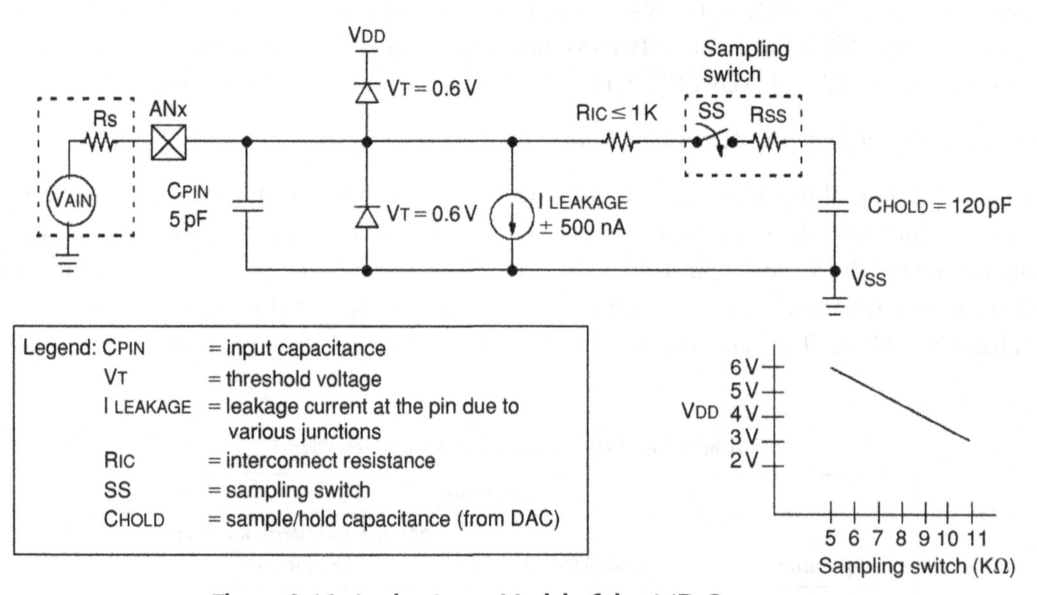

Figure 2.46: Analog Input Model of the A/D Converter

approximately 500 nA. R_{IC} is the interconnect resistance, which has a value of less than 1 K. The sampling process is shown with switch SS, having a resistance R_{SS} whose value depends on the voltage as shown in the small graph in the bottom of Figure 2.46. The value of R_{SS} is approximately 7 K at 5 V of supply voltage.

The A/D converter is based on a switched capacitor principle, and capacitor C_{HOLD} shown in Figure 2.46 must be charged fully before the start of a conversion. This is a 120-pF capacitor, which is disconnected from the input pin once the conversion is started.

The acquisition time can be calculated by using equation (2.7) from Microchip Inc.:

$$T_{ACQ} = \text{Amplifier settling time} + \text{Holding capacitor charging time} + \text{temperature coefficient.} \quad (2.7)$$

The amplifier settling time is specified as a fixed 2 μs. The temperature coefficient is only applicable if the temperature is above 25°C, and it is specified as

$$\text{Temperature coefficient} = (\text{Temperature} - 25°\text{C})(0.05\,\mu\text{s}/°\text{C}). \quad (2.8)$$

Equation (2.8) shows that the effect of the temperature is very small, creating an approximately 0.5-μs delay for every 10°C above 25°C. Thus, assuming a working environment between 25 and 35°C, the maximum delay due to temperature will be 0.5 μs, which can be neglected for most practical applications.

The holding capacitor charging time is specified by Microchip Inc. as

$$\text{Holding capacitor charging time} = -(120\,\text{pF})(1\,\text{K} + R_{SS} + R_S)\text{Ln}(1/2048). \quad (2.9)$$

Assuming that $R_{SS} = 7\,\text{K}$ and $R_S = 2.5\,\text{K}$, equation (2.9) gives the holding capacitor charging time as 9.6 μs.

The acquisition time is then calculated as

$$T_{ACQ} = 2 + 9.6 + 0.5 = 12.1\,\mu\text{s}.$$

A full 10-bit conversion takes 12 A/D cycles, and each A/D cycle is specified to be a minimum of 1.6 μs. Thus, the fastest conversion time is 19.2 μs. Addition of this to the best possible acquisition time gives the total time to complete a conversion to be 19.2 + 12.1 = 31.3 μs.

When a conversion is complete, it is specified that the converter should wait for two conversion periods before starting a new conversion. This corresponds to 2 × 1.6 = 3.2 μs. Addition of this to the best possible conversion time of 31.3 μs gives the complete conversion time to be 34.5 μs. Assuming that the A/D converter is to be used successively and ignoring the software overheads, this implies a maximum sampling frequency of approximately 29 KHz.

2.1.13 Interrupts

An interrupt is an external or internal event that requires the CPU to stop normal program execution and then execute a program code related to the event causing the interrupt. Interrupts can be generated internally (by some event inside the chip) or externally (by some external event). An example of an internal interrupt is a timer overflowing or the A/D completing a conversion. An example of an external interrupt is an I/O pin changing state.

Interrupts can be useful in many applications and some are as follows:

- *Time critical applications*: Applications that require immediate attention of the CPU can use interrupts. For example, emergency events, such as power failure or fire in a plant, may require the CPU to shut down the system immediately in an orderly manner. Using an external interrupt in such applications will force the CPU to stop whatever it is doing and take immediate action.
- *Performing routine tasks*: There are many applications that require the CPU to perform routine work. For example, it may be required to check the state of a peripheral device exactly at every millisecond. By creating a timer interrupt with the required scheduling time, the CPU can be diverted from normal program execution, and the state of the peripheral can be checked exactly at required times.
- *Task switching in multitasking applications*: In multitasking applications, each task may be given a finite time to execute its code. Interrupt mechanisms can be used to stop a task if more time is consumed than allocated for it.
- *To service peripheral devices quickly*: In some applications, we may have to wait for the completion of a task, such as the completion of the A/D conversion. This can be done by continuously checking the completion flag of the A/D converter. A more elegant solution would be to enable the A/D completion interrupt so that the CPU is forced to read the converted data as soon as it becomes available.

The PIC18F452 microcontroller has core and peripheral interrupt sources. The core interrupt sources are as follows:

- External edge-triggered interrupt on INT0, INT1, and INT2 pins
- PORTB pins change interrupts (any one of RB4–RB7 pins changing state)
- Timer 0 overflow interrupt

The PIC18F452 peripheral interrupt sources are as follows:

- PSP read/write interrupt
- A/D conversion complete interrupt
- USART receive interrupt
- USART transmit interrupt
- Synchronous serial port interrupt
- CCP1 interrupt
- TMR1 overflow interrupt

- TMR2 overflow interrupt
- Comparator interrupt
- EEPROM/FLASH write interrupt
- Bus collision interrupt
- Low-voltage detect interrupt
- Timer 3 overflow interrupt
- CCP2 interrupt

Interrupts in the PIC18F family can be divided into two groups: a high-priority group and a low-priority group. Applications that require more attention can be placed in the high-priority group. A high-priority interrupt can stop a low-priority interrupt in progress and gain access to the CPU. High-priority interrupts cannot be stopped by low-priority interrupts. If the application does not need to set priorities for interrupts, the user can choose to disable the priority scheme so that all interrupts are at the same priority level. High-priority interrupts are vectored to address 00008H and low-priority ones to address 000018H of the program memory. Normally, a user program code (interrupt service routine [ISR]) should be at the interrupt vector address to service the interrupting device.

In the PIC18F452 microcontroller, there are 10 registers that control interrupt operations. These registers are as follows:

- RCON
- INTCON
- INTCON2
- INTCON3
- PIR1, PIR2
- PIE1, PIE2
- IPR1, IPR2

Each interrupt source (except INT0) has 3 bits to control its operation. These bits are

- A flag bit to indicate whether an interrupt has occurred. This bit has a name ending in IF.
- An interrupt enable bit to enable or disable the interrupt source. This bit has a name ending in IE.
- A priority bit to select high or low priority. This bit has a name ending in IP.

R/W-0	U-0	U-0	R/W-1	R-1	R-1	R/W-0	R/W-0
IPEN	—	—	$\overline{RI}$	$\overline{TO}$	$\overline{PD}$	$\overline{POR}$	$\overline{BOR}$
bit 7							bit 0

bit 7 **IPEN:** Interrupt priority enable bit
1 = Enable priority levels on interrupts
0 = Disable priority levels on interrupts (16CXXX compatibility mode)

bit 6–5 **Unimplemented:** Read as '0'

bit 4 **$\overline{RI}$:** Reset instruction flag bit

bit 3 **$\overline{TO}$:** Watchdog time-out flag bit

bit 2 **$\overline{PD}$:** Power-down detection flag bit

bit 1 **$\overline{POR}$:** Power-on reset status bit

bit 0 **$\overline{BOR}$:** Brown-out reset operation

Figure 2.47: RCON Register Bits

RCON Register

The top bit of the RCON register, called IPEN, is used to enable the interrupt priority scheme. When IPEN = 0, interrupt priority levels are disabled, and the microcontroller interrupt structure is similar to the PIC16 series. When IPEN = 1, interrupt priority levels are enabled. Figure 2.47 shows the bits of register RCON.

Enabling/Disabling Interrupts – No Priority Structure

When the IPEN bit is cleared, the priority feature is disabled. All interrupts branch to address 00008H of the program memory. In this mode, bit PEIE of register INTCON enables/disables all peripheral interrupt sources. Similarly, bit GIE of INTCON enables/disables all interrupt sources. Figure 2.48 shows the bits of register INTCON.

For an interrupt to be accepted by the CPU, the following conditions must be satisfied:

- Interrupt enable bit of the interrupt source must be enabled. For example, if the interrupt source is an external interrupt pin INT0, then bit INT0IE of register INTCON must be set to 1.
- The interrupt flag of the interrupt source must be cleared. For example, if the interrupt source is an external interrupt pin INT0, then bit INT0IF of register INTCON must be cleared to 0.
- Peripheral interrupt enable/disable bit PEIE of INTCON must be set to 1 if the interrupt source is a peripheral.
- Global interrupt enable/disable bit GIE of INTCON must be set to 1.

R/W-0	R/W-0	R/W-0	R/W-0	R/W-0	R/W-0	R/W-0	R/W-x
GIE/GIEH	PEIE/GIEL	TMR0IE	INT0IE	RBIE	TMR0IF	INT0IF	RBIF
bit 7							bit 0

bit 7 **GIE/GIEH:** Global interrupt enable bit

When IPEN = 0:
1 = Enables all unmasked interrupts
0 = Disables all interrupts

When IPEN = 1:
1 = Enables all high-priority interrupts
0 = Disables all interrupts

bit 6 **PEIE/GIEL:** Peripheral interrupt enable bit

When IPEN = 0:
1 = Enables all unmasked peripheral interrupts
0 = Disables all peripheral interrupts

When IPEN = 1:
1 = Enables all low-priority peripheral interrupts
0 = Disables all low-priority peripheral interrupts

bit 5 **TMR0IE:** TMR0 overflow interrupt enable bit
1 = Enables the TMR0 overflow interrupt
0 = Disables the TMR0 overflow interrupt

bit 4 **INT0IE:** INT0 external interrupt enable bit
1 = Enables the INT0 external interrupt
1 = Disables the INT0 external interrupt

bit 3 **RBIE:** RB port change interrupt enable bit
1 = Enables the RB port change interrupt
0 = Disables the RB port change interrupt

bit 2 **TMR0IF:** TMR0 overflow interrupt flag bit
1 = TMR0 register has overflowed (must be cleared in software)
0 = TMR0 register did not overflow

bit 1 **INT0IF:** INT0 external interrupt flag bit
1 = The INT0 external interrupt occurred (must be cleared in software)
0 = The INT0 external interrupt did not occur

bit 0 **RBIF:** RB port change interrupt flag bit
1 = At least one of the RB7:RB4 pins changed state (must be cleared in software)
0 = None of the RB7:RB4 pins has changed state

Note: A mismatch condition will continue to set this bit. Reading PORTB will end the mismatch condition and allow the bit to be cleared.

Figure 2.48: INTCON Register Bits

For an external interrupt source, we normally have to define whether the interrupt should occur on the low-to-high or high-to-low transition of the interrupt source. For example, for INT0 interrupts, this is done by setting/clearing bit INTEDG0 of register INTCON2.

When an interrupt occurs, the CPU stops its normal flow of execution, pushes the return address onto the stack, and jumps to address 00008H in the program memory, where the user interrupt service routine program resides. Once in the interrupt service routine, the global interrupt enable bit (GIE) is cleared to disable further interrupts. When multiple interrupt sources are enabled, the source of the interrupt can be determined by polling the interrupt flag bits. The interrupt flag bits must be cleared in software before re-enabling interrupts to avoid recursive interrupts. On return from the interrupt service routine, the global interrupt bit GIE is set automatically by the software.

Enabling/Disabling Interrupts – Priority Structure

When the IPEN bit is set to 1, the priority feature is enabled and interrupts are grouped into low-priority interrupts and high-priority interrupts. Low-priority interrupts branch to address 00008H, and high-priority interrupts branch to address 000018H of the program memory. Setting the priority bit makes the interrupt source a high-priority interrupt, and clearing this bit makes the interrupt source a low-priority interrupt. Setting the GIEH bit of INTCON enables all high-priority interrupts that have the priority bit set. Similarly, setting GIEL of INTCON enables all low-priority interrupts (priority bit cleared).

For a high-priority interrupt to be accepted by the CPU, the following conditions must be satisfied:

- Interrupt enable bit of the interrupt source must be enabled. For example, if the interrupt source is an external interrupt pin INT1, then bit INT1IE of register INTCON3 must be set to 1.
- The interrupt flag of the interrupt source must be cleared. For example, if the interrupt source is an external interrupt pin INT1, then bit INT1IF of register INTCON3 must be cleared to 0.
- The priority bit must be set to 1. For example, if the interrupt source is an external interrupt INT1, then bit INT1P of register INTCON3 must be set to 1.
- Global interrupt enable/disable bit GIEH of INTCON must be set to 1.

For a low-priority interrupt to be accepted by the CPU, the following conditions must be satisfied:

- Interrupt enable bit of the interrupt source must be enabled. For example, if the interrupt source is an external interrupt pin INT1, then bit INT1IE of register INTCON3 must be set to 1.

Table 2.12: PIC18F452 Interrupt Bits and Registers

Interrupt Source	Flag Bit	Enable Bit	Priority Bit
INT0 external	INT0IF	INT0IE	-
INT1 external	INT1IF	INT1IE	INTI1P
INT2 external	INT2IF	INT2IE	INT2IP
RB port change	RBIF	RBIE	RBIP
TMR0 overflow	TMR0IF	TMR0IE	TMR0IP
TMR1 overflow	TMR1IF	TMR1IE	TMR1IP
TMR2 match PR2	TMR2IF	TMR2IE	TMR2IP
TMR3 overflow	TMR3IF	TMR3IE	TMR3IP
A/D complete	ADIF	ADIE	ADIP
CCP1	CCP1IF	CCP1IE	CCP1IP
CCP2	CCP2IF	CCP2IE	CCP2IP
USART RCV	RCIF	RCIE	RCIP
USART TX	TXIF	TXIE	TXIP
Parallel slave port	PSPIF	PSPIE	PSPIP
Sync serial port	SSPIF	SSPIE	SSPIP
Low-voltage detect	LVDIF	LVDIE	LVDIP
Bus collision	BCLIF	BCLIE	BCLIP
EEPROM/FLASH write	EEIF	EEIE	EEIP

- The interrupt flag of the interrupt source must be cleared. For example, if the interrupt source is an external interrupt pin INT1, then bit INT1IF of register INTCON3 must be cleared to 0.
- The priority bit must be cleared to 0. For example, if the interrupt source is an external interrupt INT1, then bit INT1P of register INTCON3 must be cleared to 0.
- Low-priority interrupts must be enabled by setting bit GIEL of INTCON to 1.
- Global interrupt enable/disable bit GIEH of INTCON must be set to 1.

Table 2.12 gives a listing of PIC18F452 microcontroller interrupt bit names and register names for every interrupt source.

Figures 2.49 to 2.56 show the bit definitions of interrupt registers INTCON2, INTCON3, PIR1, PIR2, PIE1, PIE2, IPR1, and IPR2.

Some examples are given in this section to illustrate how the CPU can be programmed for an interrupt.

R/W-1	R/W-1	R/W-1	R/W-1	U-0	R/W-1	U-0	R/W-1
RBPU	INTEDG0	INTEDG1	INTEDG2	—	TMR0IP	—	RBIP
bit 7							bit 0

bit 7 **$\overline{\text{RBPU}}$:** PORTB Pull-up enable bit
1 = All PORTB pull-ups are disabled
0 = PORTB pull-ups are enabled by individual port latch values

bit 6 **INTEDG0:** External interrupt0 edge select bit
1 = Interrupt on the rising edge
0 = Interrupt on the falling edge

bit 5 **INTEDG1:** External interrupt1 edge select bit
1 = Interrupt on the rising edge
0 = Interrupt on the falling edge

bit 4 **INTEDG2:** External interrupt2 edge select bit
1 = Interrupt on the rising edge
0 = Interrupt on the falling edge

bit 3 **Unimplemented:** Read as '0'

bit 2 **TMR0IP:** TMR0 overflow interrupt priority bit
1 = High priority
0 = Low priority

bit 1 **Unimplemented:** Read as '0'

bit 0 **RBIP:** RB port change interrupt priority bit
1 = High priority
0 = Low priority

Figure 2.49: INTCON2 Bit Definitions

R/W-1	R/W-1	U-0	R/W-0	R/W-0	U-0	R/W-0	R/W-0
INT2IP	INT1IP	—	INT2IE	INT1IE	—	INT2IF	INT1IF
bit 7							bit 0

bit 7 **INT2IP:** INT2 external interrupt priority bit
1 = High priority
0 = Low priority

bit 6 **INT1IP:** INT1 external interrupt priority bit
1 = High priority
0 = Low priority

bit 5 **Unimplemented:** Read as '0'

bit 4 **INT2IE:** INT2 external interrupt enable bit
1 = Enables the INT2 external interrupt
0 = Disables the INT2 external interrupt

bit 3 **INT1IE:** INT1 external interrupt enable bit
1 = Enables the INT1 external interrupt
0 = Disables the INT1 external interrupt

bit 2 **Unimplemented:** Read as '0'

bit 1 **INT2IF:** INT2 external interrupt flag bit
1 = The INT2 external interrupt occurred (must be cleared in software)
0 = The INT2 external interrupt did not occur

bit 0 **INT1IF:** INT1 external interrupt flag bit
1 = The INT1 external interrupt occurred (must be cleared in software)
0 = The INT1 external interrupt did not occur

Figure 2.50: INTCON3 Bit Definitions

R/W-0	R/W-0	R-0	R-0	R/W-0	R/W-0	R/W-0	R/W-0
PSPIF(1)	ADIF	RCIF	TXIF	SSPIF	CCP1IF	TMR2IF	TMR1IF
bit 7							bit 0

bit 7 **PSPIF(1):** Parallel slave port read/write interrupt flag bit
1 = A read or a write operation has taken place (must be cleared in software)
0 = No read or write has occurred

bit 6 **ADIF:** A/D converter interrupt flag bit
1 = An A/D conversion completed (must be cleared in software)
0 = The A/D conversion is not complete

bit 5 **RCIF:** USART receive interrupt flag bit
1 = The USART receive buffer, RCREG, is full (cleared when RCREG is read)
0 = The USART receive buffer is empty

bit 4 **TXIF:** USART transmit interrupt flag bit
1 = The USART transmit buffer, TXREG is empty (cleared when TXREG is written)
0 = The USART transmit buffer is full

bit 3 **SSPIF:** Master synchronous serial port interrupt flag bit
1 = The transmission/reception is complete (must be cleared in software)
0 = Waiting to transmit/receive

bit 2 **CCP1IF:** CCP1 interrupt flag bit
Capture mode:
1 = A TMR1 register capture occurred (must be cleared in software)
0 = No TMR1 register capture occurred
Compare mode:
1 = A TMR1 register compare match occurred (must be cleared in software)
0 = No TMR1 register compare match occurred
PWM mode:
Unused in this mode

bit 1 **TMR2IF:** TMR2 to PR2 match interrupt flag bit
1 = TMR2 to PR2 match occurred (must be cleared in software)
0 = No TMR2 to PR2 match occurred

bit 0 **TMR1IF:** TMR1 overflow interrupt flag bit
1 = TMR1 register overflowed (must be cleared in software)
0 = TMR1 register did not overflow

Note (1): Bit reserved on the PIC18F2x2 devices.

Figure 2.51: PIR1 Bit Definitions

Example 2.2

Set up INT1 as a falling-edge triggered interrupt input having low priority.

Solution

The following bits should be set up before the INT1 falling-edge triggered interrupts can be accepted by the CPU in low-priority mode:

- Enable priority structure. Set IPEN = 1

- Make INT1 an input pin. Set TRISB = 1
- Set INT1 interrupts for falling edge. SET INTEDG1 = 0
- Enable INT1 interrupts. Set INT1IE = 1
- Enable low priority. Set INT1IP = 0
- Clear INT1 flag. Set INT1IF = 0
- Enable low-priority interrupts. Set GIEL = 1
- Enable all interrupts. Set GIEH = 1

When an interrupt occurs, the CPU jumps to address 00008H in the program memory to execute the user program at the interrupt service routine. ■

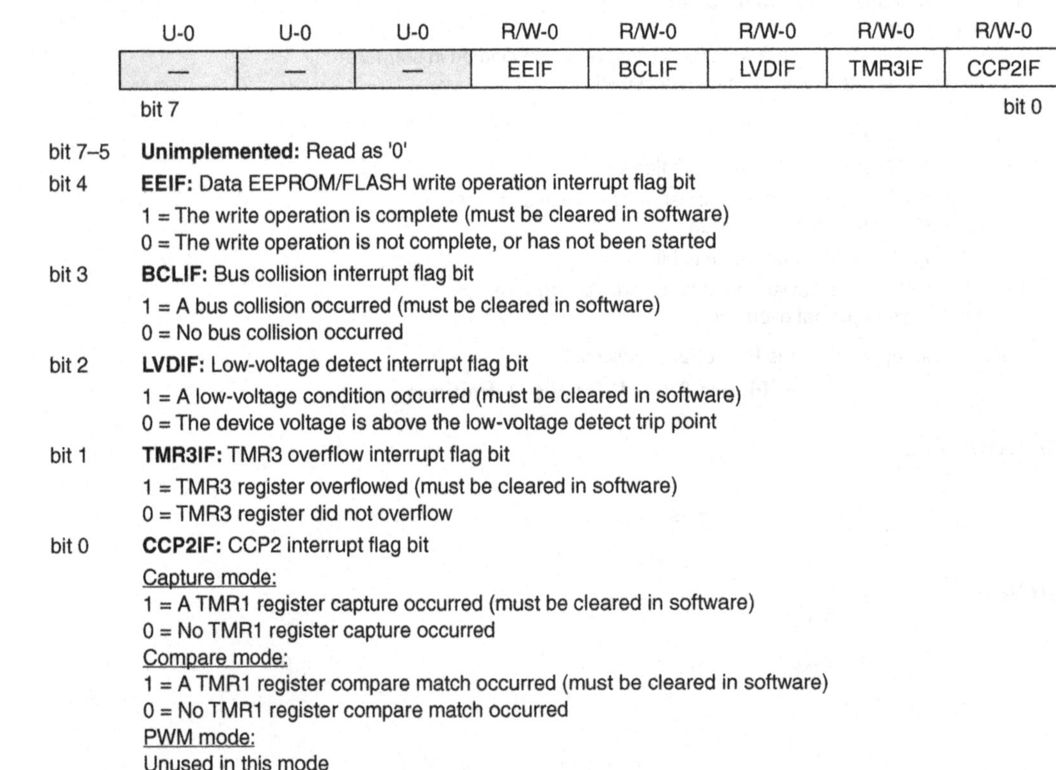

U-0	U-0	U-0	R/W-0	R/W-0	R/W-0	R/W-0	R/W-0
—	—	—	EEIF	BCLIF	LVDIF	TMR3IF	CCP2IF
bit 7							bit 0

bit 7–5 **Unimplemented:** Read as '0'

bit 4 **EEIF:** Data EEPROM/FLASH write operation interrupt flag bit
1 = The write operation is complete (must be cleared in software)
0 = The write operation is not complete, or has not been started

bit 3 **BCLIF:** Bus collision interrupt flag bit
1 = A bus collision occurred (must be cleared in software)
0 = No bus collision occurred

bit 2 **LVDIF:** Low-voltage detect interrupt flag bit
1 = A low-voltage condition occurred (must be cleared in software)
0 = The device voltage is above the low-voltage detect trip point

bit 1 **TMR3IF:** TMR3 overflow interrupt flag bit
1 = TMR3 register overflowed (must be cleared in software)
0 = TMR3 register did not overflow

bit 0 **CCP2IF:** CCP2 interrupt flag bit
Capture mode:
1 = A TMR1 register capture occurred (must be cleared in software)
0 = No TMR1 register capture occurred
Compare mode:
1 = A TMR1 register compare match occurred (must be cleared in software)
0 = No TMR1 register compare match occurred
PWM mode:
Unused in this mode

Figure 2.52: PIR2 Bit Definitions

R/W-0	R/W-0	R/W-0	R/W-0	R/W-0	R/W-0	R/W-0	R/W-0
PSPIE(1)	ADIE	RCIE	TXIE	SSPIE	CCP1IE	TMR2IE	TMR1IE
bit 7							bit 0

bit 7 **PSPIE(1):** Parallel slave port read/write interrupt enable bit
1 = Enables the PSP read/write interrupt
0 = Disables the PSP read/write interrupt

bit 6 **ADIE:** A/D converter interrupt enable bit
1 = Enables the A/D interrupt
0 = Disables the A/D interrupt

bit 5 **RCIE:** USART receive interrupt enable bit
1 = Enables the USART receive interrupt
0 = Disables the USART receive interrupt

bit 4 **TXIE:** USART transmit interrupt enable bit
1 = Enables the USART transmit interrupt
0 = Disables the USART transmit interrupt

bit 3 **SSPIE:** Master synchronous serial port interrupt enable bit
1 = Enables the MSSP interrupt
0 = Disables the MSSP interrupt

bit 2 **CCP1IE:** CCP1 interrupt enable bit
1 = Enables the CCP1 interrupt
0 = Disables the CCP1 interrupt

bit 1 **TMR2IE:** TMR2 to PR2 match interrupt enable bit
1 = Enables the TMR2 to PR2 match interrupt
0 = Disables the TMR2 to PR2 match interrupt

bit 0 **TMR1IE:** TMR1 overflow interrupt enable bit
1 = Enables the TMR1 overflow interrupt
0 = Disables the TMR1 overflow interrupt

Note (1): Bit reserved on the PIC18F2x2 devices.

Figure 2.53: PIE1 Bit Definitions

U-0	U-0	U-0	R/W-0	R/W-0	R/W-0	R/W-0	R/W-0
—	—	—	EEIE	BCLIE	LVDIE	TMR3IE	CCP2IE
bit 7							bit 0

bit 7–5 **Unimplemented:** Read as '0'

bit 4 **EEIE:** Data EEPROM/FLASH write operation interrupt enable bit
1 = Enabled
0 = Disabled

bit 3 **BCLIE:** Bus collision interrupt enable bit
1 = Enabled
0 = Disabled

bit 2 **LVDIE:** Low-voltage detect interrupt enable bit
1 = Enabled
0 = Disabled

bit 1 **TMR3IE:** TMR3 overflow interrupt enable bit
1 = Enables the TMR3 overflow interrupt
0 = Disables the TMR3 overflow interrupt

bit 0 **CCP2IE:** CCP2 interrupt enable bit
1 = Enables the CCP2 interrupt
0 = Disables the CCP2 interrupt

Figure 2.54: PIE2 Bit Definitions

R/W-1	R/W-1	R/W-1	R/W-1	R/W-1	R/W-1	R/W-1	R/W-1
PSPIP(1)	ADIP	RCIP	TXIP	SSPIP	CCP1IP	TMR2IP	TMR1IP
bit 7							bit 0

bit 7 **PSPIP(1):** Parallel slave port read/write interrupt priority bit
1 = High priority
0 = Low priority

bit 6 **ADIP:** A/D converter interrupt priority bit
1 = High priority
0 = Low priority

bit 5 **RCIP:** USART receive interrupt priority bit
1 = High priority
0 = Low priority

bit 4 **TXIP:** USART transmit interrupt priority bit
1 = High priority
0 = Low priority

bit 3 **SSPIP:** Master synchronous serial port interrupt priority bit
1 = High priority
0 = Low priority

bit 2 **CCP1IP:** CCP1 interrupt priority bit
1 = High priority
0 = Low priority

bit 1 **TMR2IP:** TMR2 to PR2 match interrupt priority bit
1 = High priority
0 = Low priority

bit 0 **TMR1IP:** TMR1 overflow interrupt priority bit
1 = High priority
0 = Low priority

Note (1): Bit reserved on the PIC18F2x2 devices.

Figure 2.55: IPR1 Bit Definitions

U-0	U-0	U-0	R/W-1	R/W-1	R/W-1	R/W-1	R/W-1
—	—	—	EEIP	BCLIP	LVDIP	TMR3IP	CCP2IP
bit 7							bit 0

bit 7–5 **Unimplemented:** Read as '0'

bit 4 **EEIP:** Data EEPROM/FLASH write operation interrupt priority bit
1 = High priority
0 = Low priority

bit 3 **BCLIP:** Bus collision interrupt priority bit
1 = High priority
0 = Low priority

bit 2 **LVDIP:** Low-voltage detect interrupt priority bit
1 = High priority
0 = Low priority

bit 1 **TMR3IP:** TMR3 overflow interrupt priority bit
1 = High priority
0 = Low priority

bit 0 **CCP2IP:** CCP2 interrupt priority bit
1 = High priority
0 = Low priority

Figure 2.56: IPR2 Bit Definitions

Example 2.3

Set up INT1 as a rising-edge triggered interrupt input having high priority.

Solution

The following bits should be set up before the INT1 rising-edge triggered interrupts can be accepted by the CPU in the high-priority mode:

- Enable priority structure. Set IPEN = 1
- Make INT1 an input pin. Set TRISB = 1
- Set INT1 interrupts for rising edge. SET INTEDG1 = 1
- Enable INT1 interrupts. Set INT1IE = 1
- Enable high priority. Set INT1IP = 1
- Clear INT1 flag. Set INT1IF = 0
- Enable all interrupts. Set GIEH = 1

When an interrupt occurs, the CPU jumps to address 000018H of the program memory to execute the user program at the interrupt service routine. ■

2.2 Summary

This chapter has described the architecture of the PIC18F family of microcontrollers. PIC18F452 was considered as a typical example microcontroller in the family. Some other members of the family, such as the PIC18F242, have smaller pin counts and less functionality. In addition, some other members of the family, such as the PIC18F6680, have larger pin counts and more functionality.

Various important parts and peripheral circuits of the PIC18F series have been described, including the data memory, program memory, clock circuits, reset circuits, WDT, general-purpose timers, capture and compare module, PWM module, A/D converter, and the interrupt structure.

2.3 Exercises

1. Describe the data memory structure of the PIC18F452 microcontroller. What is a bank? How many banks are there?
2. Explain the differences between a general-purpose register (GPR) and a special function register (SFR).

3. Explain the various ways that the PIC18F microcontroller can be reset. Draw a circuit diagram to show how an external push-button switch can be used to reset the microcontroller.
4. Describe the various clock sources that can be used to provide clock to a PIC18F452 microcontroller. Draw a circuit diagram to show how a 10-MHz crystal can be connected to the microcontroller.
5. Draw a circuit diagram to show how a resonator can be connected to a PIC18F microcontroller.
6. In a nontime-critical application, it is required to provide clock to a PIC18F452 microcontroller using an external resistor and a capacitor. Draw a circuit diagram to show how this can be done and find the component values if the required clock frequency is 5 MHz.
7. Explain how an external clock can be used to provide clock pulses to a PIC18F microcontroller.
8. What are the registers of PORTA? Explain the operation of the port by drawing the port block diagram.
9. It is required to set the WDT to provide automatic reset every 0.5 s. Describe how this can be done and list the appropriate register bits.
10. It is required to generate PWM pulses from pin CCP1 of a PIC18F452 microcontroller. The required pulse period is 100 μs, and the required duty cycle is 50%. Assuming that the microcontroller operates with a 4-MHz crystal, calculate the values to be loaded into various registers.
11. It is required to generate PWM pulses from pin CCP1 of a PIC18F452 microcontroller. The required pulse frequency is 40 KHz, and the required duty cycle is 50%. Assuming that the microcontroller operates with a 4-MHz crystal, calculate the values to be loaded into various registers.
12. An LM35DZ type analog temperature sensor is connected to analog port AN0 of a PIC18F452 microcontroller. The sensor provides an analog output voltage proportional to the temperature, i.e., Vo = 10 mV/°C. Show the steps required to read the temperature.
13. Explain the differences between a priority interrupt and a nonpriority interrupt.
14. Show the steps required to set up INT2 as a falling-edge triggered interrupt input having low priority. What is the interrupt vector address?
15. Show the steps required to set up both INT1 and INT2 as falling-edge triggered interrupt inputs having low priorities.

16. Show the steps required to set up INT1 as falling-edge triggered and INT2 as rising-edge triggered interrupt inputs having high priorities. Explain how you can find the source of the interrupt when an interrupt occurs.

17. Show the steps required to set up Timer 0 to generate high-priority interrupts every millisecond. What is the interrupt vector address?

18. In an application, the CPU registers have been configured to accept interrupts from external sources INT0, INT1, and INT2. An interrupt has been detected. Explain how you can find the source of the interrupt.

CHAPTER 3

Memory Cards

3.1 Memory Card Types

A memory card (also called a flash memory card) is a solid-state electronic data storage device. First invented by Toshiba in the 1980s, memory cards save the stored data even after the memory device is disconnected from its power source. This ability to retain data is the key for flash memory card applications, for example, in digital cameras, where the saved pictures are not lost after the memory card is removed from the camera.

Nowadays, memory cards are used in consumer electronics and industrial applications. In consumer devices, we see the use of memory cards in applications like

- Personal computers
- Digital cameras
- Mobile phones
- Video cameras
- Notebook computers
- Global positioning systems
- MP3 players
- Personal digital assistants

In industrial applications, we see the use of memory cards in

- Embedded computers
- Networking products
- Military systems
- Communication devices
- Medical products

D.O.I.: 10.1016/B978-1-85617-719-1.00007-5

- Security systems
- Handheld scanners

Memory cards are based on two technologies: NOR technology and NAND technology. NOR technology provides high-speed random access capabilities, where data as small as a single byte can be retrieved. NOR technology-based memory cards are often found in mobile phones, personal digital assistants, and computers. NAND technology was invented after the NOR technology, and it allows sequential access to the data in single pages but cannot retrieve single bytes of data like NOR flash. NAND technology-based memory cards are commonly found in digital cameras, mobile phones, audio and video devices, and other devices where the data is written and read sequentially.

There are many different types of memory cards available in the market. Some of the most commonly known memory cards are

- Smart media (SM) card
- Multimedia card (MMC)
- Compact flash (CF) card
- Memory stick (MS) card
- Microdrive
- xD card
- Secure digital (SD) card

The specifications and details of each card are summarized in the following sections.

3.2 Smart Media Card

The SM card was first developed in 1995 by Toshiba and was also called the Solid State Floppy Disc Card (SSFDC). The SM card consists of a single NAND flash chip embedded in a thin plastic card, and it is the thinnest card of all. Figure 3.1 shows a typical SM card. The dimensions of the card are $45.0 \times 37.0 \times 0.76$ mm, and it weighs only 1.8 g. The card consists of a flat electrode terminal with 22 pins.

The SM card was mainly used in Fuji and Olympus cameras, where it had approximately 50% of the memory card share in 2001. The capacities of these cards ranged from 0.5 to 128 MB, and the data transfer rate was approximately 2 MB/s. SM cards started having problems as camera resolution increased considerably and cards greater than 128 MB were

Figure 3.1: Smart Media Card

Figure 3.2: Multimedia Card

not available. SM cards were designed to operate at either 3.3 or 5 V, and a small notched corner was used to protect 3.3-V cards from being inserted into 5-V readers.

SM cards incorporated a copy protection mechanism known as the "ID," which gave every card a unique identification number for use with copy protection systems.

SM cards are no longer manufactured, and Fujifilm and Olympus both switched to xD cards. Now, no devices are designed to use SM cards, but 128-MB cards for old devices can still be obtained from memory card suppliers.

3.3 Multimedia Card

MMCs were first developed in the late 1970s by Intgenix and SanDisk. These cards were initially used in mobile phone and pager devices, but today they are commonly used in many other electronic devices. MMCs are backward compatible with SD cards, and they can be plugged into SD card slots. The reverse is not possible because SD cards are thicker (2.1 mm) and will not fit into MMC slots. Figure 3.2 is a picture of a typical MMC. The card dimensions are 24.0 × 32.0 × 1.4 mm and it has 8 pins.

The MMC operating voltage is 3.3 V, and the data transfer rate is approximately 2.5 MB/s. MMCs are available with capacities up to 4 GB. The older MMCs have been replaced by new multimedia *mobile* cards. These cards offer higher performance than older MMCs, and they offer lower working voltages (1.8–3.3 V) to reduce power consumption in portable devices.

3.4 Compact Flash Card

CF cards were first developed in 1994 by SanDisk. These are the cards offering the highest capacities, from 2 MB to 100 GB. Today, CF cards are used in expensive professional digital cameras and other professional mass storage devices. Low-capacity cards (up to 2 GB) use the FAT16 filing system, and cards with capacities higher than 2 GB use the FAT32 filing system.

There are two versions of CF cards: Type I and Type II. The only physical difference between the two types is that Type II cards are thicker than Type I cards. Type I card dimensions are 43.0 × 36.0 × 3.3 mm and Type II cards are 43.0 × 36.0 × 5.0 mm. Both cards have 50 pins. The Type I interface can supply up to 70 mA to the card, and the Type II interface can supply up to 500 mA. The card operating voltage is 3.3 or 5 V. Figure 3.3 shows a typical CF card.

There are four speeds of CF cards: a Standard CF, a CF High Speed (CF 2.0 with a data transfer rate of 16 MB/s), a faster CF 3.0 with a data transfer rate of 66 MB/s, and the fastest CF 4.0 standard with a data transfer rate of 133 MB/s. A future version of the CF cards, known as CFast, will be based on the Serial ATA bus with an expected speed of 300 MB/s.

The memory card speed is usually specified using the "x" rating. This is the speed of the first audio CD-ROM, which was 150 KB/s. For example, a card with a speed of 10x corresponds to a data transfer rate of 10 × 150 KB/s = 1.5 MB/s. Table 3.1 lists some of the commonly used speed ratings in memory cards.

The advantages of CF cards are as follows:

- CF cards are rugged and more durable than other types of cards, and they can withstand more physical damage than other cards.
- CF cards are available at very high storage capacities.
- CF cards operate at high speeds.
- CF cards are compatible with the IDE/ATA hard-disk standards, and thus they can be used in many embedded systems to replace hard disks.

Besides the advantages, the CF cards have some disadvantages, such as lack of a mechanical write protection switch or notch, and its large dimensions in comparison with other cards limit its use in slim devices.

Figure 3.3: Compact Flash Card

Table 3.1: Memory Card Speed Ratings

Speed Rating	Speed (MB/s)
6x	0.9
32x	4.8
40x	6.0
66x	10.0
100x	15.0
133x	20.0
150x	22.5
200x	30.0
266x	40.0
300x	45.0

Figure 3.4: Memory Stick Card

3.5 Memory Stick Card

The MS cards were first developed by Sony in 1998. Although the original MS was only 128 MB, the largest capacity currently available is 16 GB. Figure 3.4 shows a typical MS card.

The original MS, although it is not manufactured any more, is approximately the size and thickness of a stick of chewing gum. MS has now been replaced with Memory Stick PRO, Memory Stick Duo, and Memory Stick PRO-HG.

Memory Stick PRO was introduced in 2003 as a joint effort between Sony and SanDisk, and it allows a greater storage capacity and faster file transfer rate than the original MS.

Memory Stick Duo was developed as a result of the need for smaller and faster memory cards. It is smaller than the standard MS card, but an adapter allows it to be used in original MS applications.

Memory Stick PRO-HG was developed by Sony and SanDisk together in 2006. The data transfer speed of this card is 60 MB/s, which exceeds the speeds of all previous memory cards and is approximately three times faster than the Memory Stick PRO cards.

3.6 Microdrive

Microdrive is basically a hard disk designed to fit into a Type II CF card size enclosure. Although the size of a microdrive is same as that of a CF card, its power consumption is much higher than flash memories and therefore cannot be used in low-power applications. The capacity of microdrives is 8 GB or more. Figure 3.5 shows a typical microdrive.

The physical dimensions of microdrives are 42.8 × 36.4 × 5.0 mm, and they weigh approximately 16 g. The first microdrive was developed by IBM in 1999 with a capacity of 170 and 340 MB. Soon after, the capacity was increased to greater than 2 GB in the year 2003 by Hitachi. Microdrives with a capacity of 8 GB were introduced in 2008 by Hitachi and Seagate.

The advantage of microdrives is that they allow more write cycles than the memory cards. In addition, microdrives are better at handling power loss in the middle of writing. One of the disadvantages of microdrives is that they do not survive if dropped from a height of 1.2 m. In addition, their transfer speeds are around 5 MB/s, which is lower than most of the present day high-end memory cards. In addition, they are not designed to operate at high altitudes, and their power consumption is high compared with memory cards.

3.7 xD Card

xD stands for e**x**treme **D**igital, and these cards are mainly used in digital cameras, digital voice recorders, and MP3 players. xD cards were developed by Olympus and Fujifilm in 2002 and then manufactured by Toshiba Corporation and Samsung Electronics. Figure 3.6 is a picture of a typical xD card. xD cards are available in three types: Type M, Type H, and Type M+.

Figure 3.5: Microdrive

Figure 3.6: xD Card

Type M cards were developed in 2005 and are available in capacities up to 2 GB. The read and write speeds of these cards are 4 and 2.5 MB/s, respectively.

Type H xD cards were first released in 2005 have the advantage of a higher data transfer speed. These cards are also available in capacities up to 2 GB with read and write speeds of 5 and 4 MB/s, respectively. Unfortunately, the production of Type H cards has now been discontinued due to their high production costs.

Type M+ xD cards were first released in 2008, and their capacities are up to 2 GB. These cards are the fastest xD cards, with read and write speeds of 6 and 3/75 MB/s, respectively.

The advantage of xD cards is that they are faster than SM cards, MMCs, and MS cards. In addition, their small size makes them attractive in portable low-power applications. Some of the disadvantages of xD cards are their higher cost, bigger size than some other memory cards (such as microSD), and the fact that they are proprietary to Fujifilm and Olympus. This means that there is no publicly available documentation on their design and implementation.

3.8 Secure Digital Card

SD cards are probably the most widely used memory cards today. The SD card was originally developed by Matsushita, SanDisk, and Toshiba in 2000. SD cards nowadays a used in many portable devices, such as digital cameras, mobile phones, PDAs, handheld computers, video recorders, GPS receivers, video game consoles, and so on.

Standard SD cards are available with capacities from 4 MB to 4 GB. Recently, a new type of SD card called the high-capacity SD card (SDHC) has been developed with capacities ranging from 4 to 32 GB. It has been announced that a new specification called eXtended Capacity (SDXC) will allow capacities to reach 2 TB.

SD cards are based on MMC, but they have a number of differences: SD cards are physically thicker than MMCs and would not fit into MMC slots. The MMC, on the other hand, can be easily inserted into SD card slots. In addition, SD cards are shaped asymmetrically to prevent them being inserted upside down, whereas an MMC would go in either direction, although it will not make contact if inserted upside down. In addition, the internal register structures of the two types of cards are not the same.

3.8.1 Standard SD Cards

SD cards are available in three different sizes: normal SD, miniSD, and microSD. Figure 3.7 shows the three types of SD cards.

Normal SD cards have the dimensions 24.0 × 32.0 × 2.1 mm and a weight of 2 g. A write-protect switch is provided on the card to stop accidental deletion of the contents of

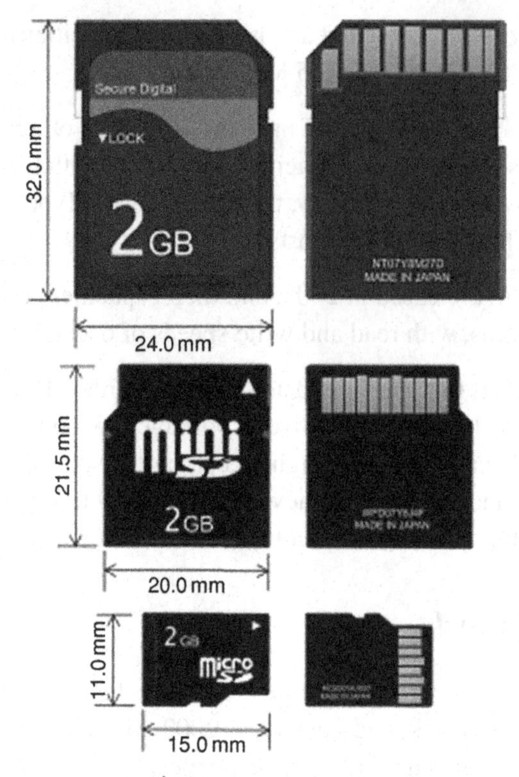

Figure 3.7: SD Cards

the card. The data transfer speed is approximately 15–20 MB/s. Normal SD cards operate at 2.7–3.6 V and have 9 pins.

miniSD cards were first released in 2003. They have the dimensions 20.0 × 21.5 × 1.4 mm and a weight of 1 g. A write-protect switch is not provided on the card. The data transfer speed is approximately 15 MB/s. miniSD cards operate at 2.7–3.6 V and have 11 pins. miniSD cards are available in capacities ranging from 16 MB to 8 GB.

microSD cards were released in 2008, and they have the dimensions 11.0 × 15.0 × 1.0 mm and a weight of 0.5 g. As in the miniSD cards, no write-protect switch is provided on the card. The data transfer rate and the card operating voltages are same as in miniSD cards, and they have 8 pins. microSD cards are available in capacities ranging from 64 MB to 4 GB.

Standard SD cards are available up to a capacity of 2 GB. Table 3.2 shows a comparison of all three standard SD cards. miniSD and microSD cards can be used with adapters in normal SD card applications. Figure 3.8 shows a typical miniSD card adapter.

Table 3.2: Comparison of Standard SD Cards

Property	SD	miniSD	microSD
Width	24 mm	20 mm	11 mm
Length	32 mm	21.5 mm	15 mm
Thickness	2.1 mm	1.4 mm	1 mm
Weight	2 g	1 g	0.5 g
Operating voltage	2.7–3.6 V	2.7–3.6 V	2.7–3.6 V
No. of pins	9	11	8

Figure 3.8: miniSD Card Adapter

Standard SD cards (up to 2 GB) are usually shipped with the FAT16 file system preloaded on the card.

3.8.2 High-Capacity SD Cards

Secure Digital SDHC was released in 2006 and is an extension of the standard SD card format. SDHC cards provide capacities f 2 GB up to 32 GB. It is important to realize that although the SDHC cards have the same physical dimensions as the standard SD cards, they use different protocols and as such will only work in SDHC-compatible devices and not in standard SD card applications. SDHC cards should not be used in standard SD compatible devices. Standard SD cards are, however, forward compatible with SDHC host devices, making standard SD cards compatible with both SD and SDHC host devices.

SDHC cards offer

- Larger data capacities
- Larger number of files
- FAT32 filing system (instead of the FAT16)
- Higher data transfer rates
- Content protection for recordable media (CPRM) copyright protection
- Standard SD card physical size compatibility

Figure 3.9: Class 6 SDHC Card

Figure 3.10: A Typical Memory Card Reader

SDHC cards have Speed Class Ratings defined by the SD Association. The defined classes are

- Class 2: data transfer rate 2 MB/s
- Class 4: data transfer rate 4 MB/s
- Class 6: data transfer rate 6 MB/s

The Speed Class Rating of a card is labeled on the card. Figure 3.9 shows a typical Class 6 SDHC card. SDHC cards are identified with the letters "HC" labeled on the card as a logo.

SDHC cards are normally shipped with the FAT32 filing system preloaded on the card. These cards are used in applications requiring high capacities, such as video recorders, MP3 players, and general large-volume data storage, and by users in general who want higher performance from their high-end digital devices.

Like standard SD cards, SDHC cards come in three types: normal SDHC, miniSDHC, and microSDHC.

3.9 Memory Card Readers

Memory card readers are usually in the form of small devices with many different types of sockets compatible with various memory cards. Figure 3.10 shows a typical memory card reader. The card reader is normally connected to the USB port of the PC, and most card readers accept most of the popular cards available on the market. Old card readers are based on the USB 1.1 specification with 12 Mb/s, whereas new card readers are based on the USB 2.0 specification, with a maximum data transfer rate of 480 Mb/s. When a memory card is inserted into it, the device automatically detects the card and assigns a drive letter to the slot where the card is inserted. Data on the card can be read as files or files can be copied to the card using the standard Windows Explorer functions.

3.10 Memory Card Physical Properties

Table 3.3 gives a comparison of the physical properties of commonly used memory cards. As can be seen from the table, the microSD card has the smallest form factor.

3.11 Memory Card Technical Properties

Table 3.4 gives a comparison of the technical properties of commonly used memory cards.

Table 3.3: Physical Properties of Memory Cards

Card	Width (mm)	Length (mm)	Thickness (mm)	Weight (g)
CF - Type I	43.0	36.0	3.3	3.3
CF - Type II	43.0	36.0	5.0	5.0
SM	37.0	45.0	0.76	2.0
MMC	24.0	32.0	1.4	1.3
RS-MMC	24.0	16.0	1.4	1.3
MMC-micro	14.0	12.0	1.1	1.0
MS	21.5	50.0	2.8	4.0
MS PRO Duo	20.0	31.0	1.6	2.0
SD	24.0	32.0	2.1	2.0
miniSD	20.0	21.5	1.4	0.5
microSD	15.0	11.0	1.0	0.27
xD	25.0	20.0	1.78	2.8

Table 3.4: Technical Properties of Memory Cards

Card	Max Capacity (2009)	Max Write Speed (MB/s)	Max Read Speed (MB/s)	Operating Voltage (V)	Pin Count
CF - Type I	32 GB	133	133	3.5 and 5.0	50
CF - Type II	32 GB	133	133	3.3 and 5.0	50
SM	128 MB	20	20	3.3 and 5.0	22
MMC	4 GB	52	52	3.3	7
RS-MMC	2 GB	52	52	3.3	7
MMC micro	2 GB	40	40	3.3	13
MS	128 MB	160	160	3.3	10
MS PRO Duo	16 GB	160	160	3.3	10
SD	4 GB	150	150	3.3	9
miniSD	4 GB	100	100	3.3	11
microSD	4 GB	100	100	3.3	8
SDHC	64 GB	48	48	3.3	9

3.12 Detailed SD Card Structure

As the topic of this book is SD cards, the internal structure and the use of these cards in PIC microcontroller-based systems will be described in this section.

3.12.1 SD Card Pin Configuration

Figure 3.11 shows the pin configuration of a standard SD card. The card has nine pins, as shown in the figure, and a write-protect switch to enable/disable writing onto the card.

A standard SD card can be operated in two modes: the *SD Bus mode* and the *SPI Bus mode*. SD Bus mode is the native operating mode of the card, and all the pins are used in this mode. Data is transferred using four pins (D0–D3), a clock (CLK) pin, and a command line (CMD). Data can be transferred from the card to the host or vice versa over the four data lines. Figure 3.12 shows the SD card connection in SD Bus mode.

SPI Bus mode is the more commonly used mode, and it allows data to be transferred on two lines (DO and DI) in serial format using a chip select (CS) and a CLK line. The SPI mode is easier to use, but it has the disadvantage of reduced performance compared with the SD mode of operation. Figure 3.13 shows the SD card connections in SPI mode.

SD card pins have different meanings depending upon the mode of operation. Table 3.5 shows the pin assignments when the card is operated in SD Bus and SPI Bus modes.

The projects in this book are based on the operations in the SPI Bus mode. The following pins are used in SPI Bus mode:

- Chip select – Pin 1
- Data in – Pin 2
- Clock – Pin 5
- Data out – Pin 7

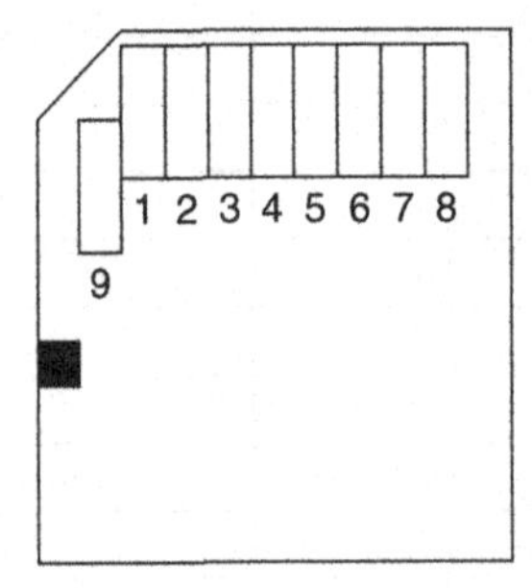

Figure 3.11: Standard SD Card Pin Configuration

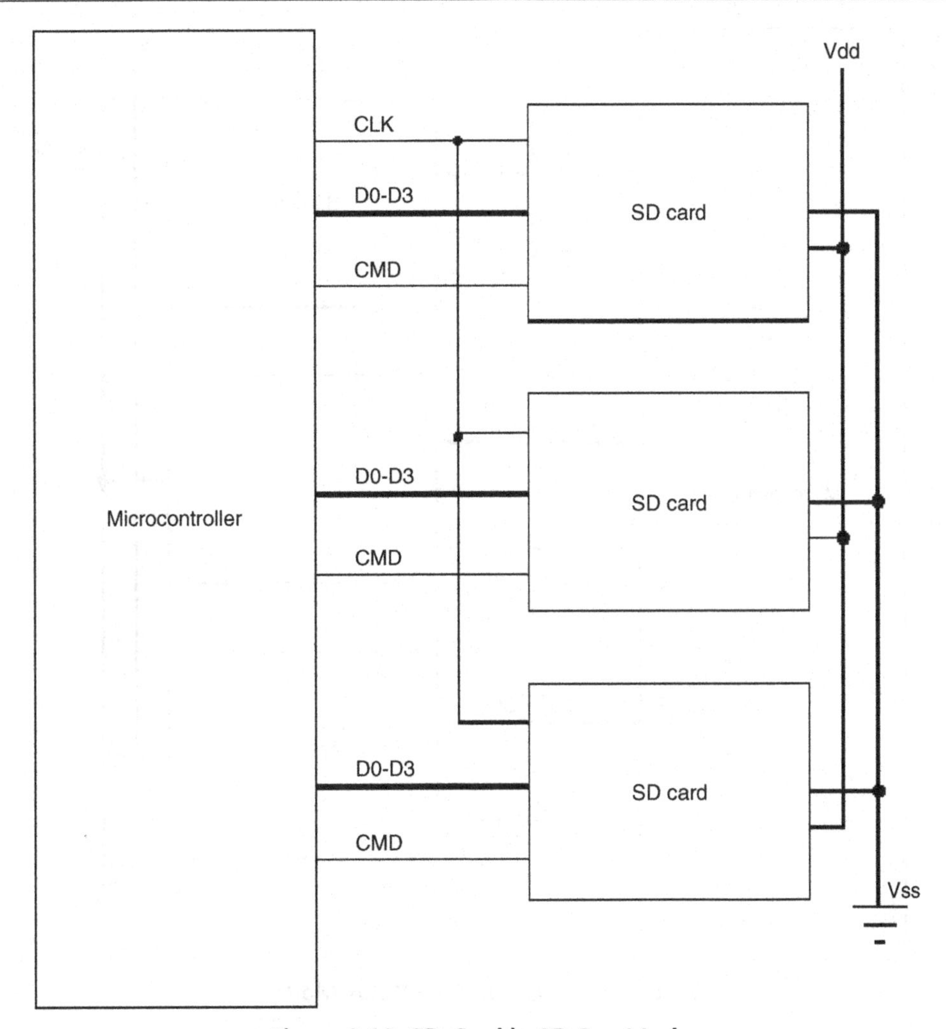

Figure 3.12: SD Card in SD Bus Mode

In addition, pin 4 must be connected to the supply voltage, and pins 3 and 6 must be connected to the supply ground.

3.12.2 SD Card Interface

Before we can use an SD card in an electronic circuit, we have to know the interface signal levels. Table 3.6 shows the input–output voltage levels of the standard SD cards.

According to Table 3.6,

Minimum logic 1 output voltage, VOH = 2.475 V

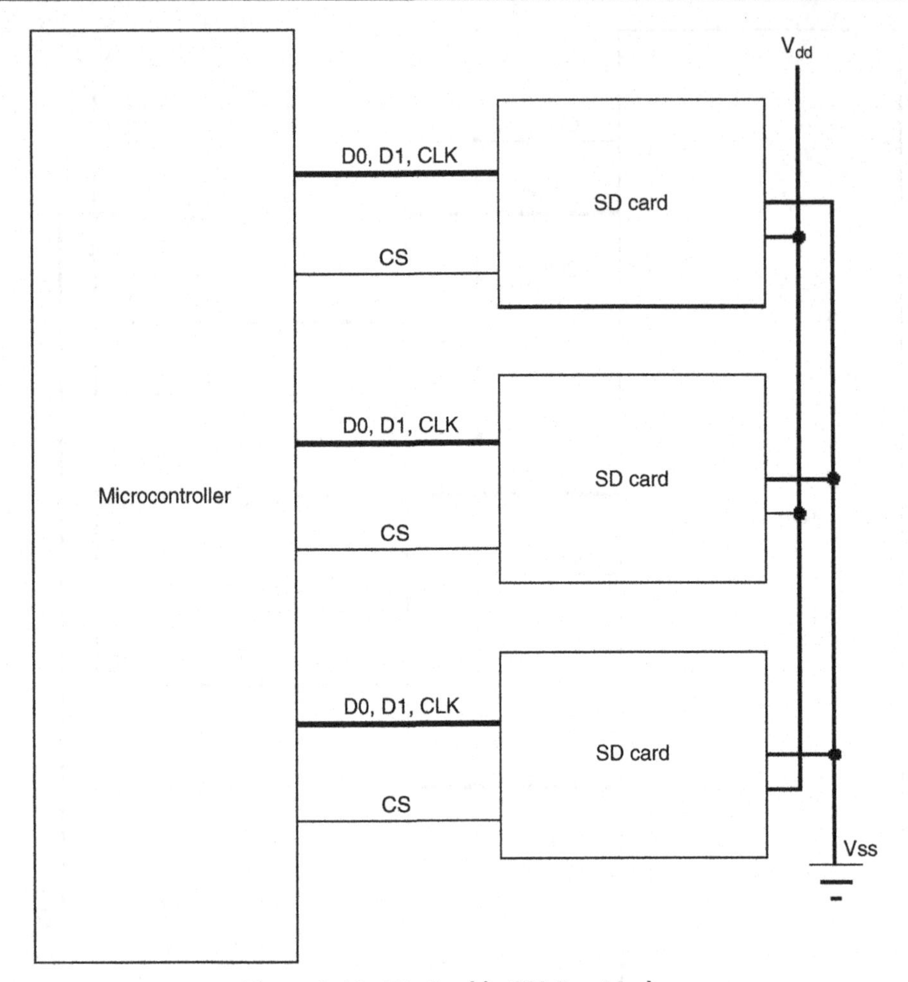

Figure 3.13: SD Card in SPI Bus Mode

Table 3.5: SD Card Pin Assignments

Pin No	Name	SD Mode	SPI Mode
1	CD/DAT3	Card detect/Data line	Chip select
2	CMD	Command response	Data in
3	Vss	Ground	Ground
4	Vdd	Supply voltage	Supply voltage
5	CLK	Clock	Clock
6	Vss	Ground	Ground
7	DAT0	Data line	Data out
8	DAT1	Data line	Reserved
9	DAT2	Data line	Reserved

Table 3.6: SD Card Input–Output Voltage Levels

	Symbol	Minimum	Maximum
Logic 1 output voltage	VOH	0.75 × Vdd	
Logic 0 output voltage	VOL		0.125 × Vdd
Logic 1 input voltage	VIH	0.625 × Vdd	Vdd + 0.3
Logic 0 input voltage	VIL	Vss − 0.3	0.25 × Vdd

Maximum logic 0 output voltage, VOL = 0.4125 V

Minimum required logic 1 input voltage, VIH = 2.0625 V

Maximum logic 1 input voltage = 3.6 V

Maximum required logic 0 input voltage, VIL = 0.825 V

When connected to a PIC microcontroller, the output voltage (2.475 V) of the SD card is enough to drive the input circuit of the microcontroller. The typical logic 1 output voltage of a PIC microcontroller pin is 4.3 V, and this is too high when applied as an input to a microcontroller pin, where the maximum voltage should not exceed 3.6 V. As a result of this, it is required to use resistors at the inputs of the SD card to lower the input voltage. Figure 3.14 shows a typical SD card interface to a PIC microcontroller. In this figure, 2.2- and 3.3-K resistors are used as a potential divider circuit to lower the SD card input voltage to approximately 2.48 V, as shown below.

$$\text{SD card input voltage} = 4.3\,\text{V} \times 3.3\,\text{K}/(2.2\,\text{K} + 3.3\,\text{K}) = 2.48\,\text{V}.$$

In Figure 3.14, the SD card is connected to PORTC pins of the microcontroller as follows:

SD Card Pin	Microcontroller Pin
CS	RC2
CLK	RC3
DO	RC4
DI	RC5

This is the recommended connection because it uses the SPI Bus port pins of the microcontroller (RC3, RC4, and RC5).

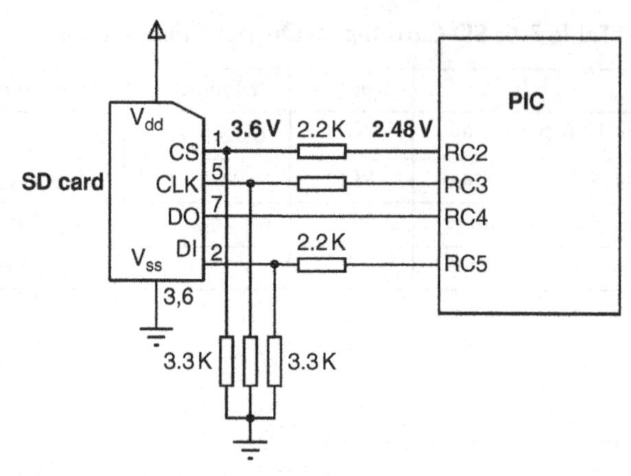

Figure 3.14: PIC Microcontroller SD Card Interface

SD cards support "hot" insertion of the card, i.e., the card can be inserted into the circuit without powering down the host. This is usually achieved through the card connector. Connector manufacturers usually provide sockets that have power pins long enough to power the card before any contact is made with the other pins.

A feature of most SD cards is the automatic entry and exit from sleep mode. After an operation, cards usually enter a sleep mode to conserve power-in, if no more commands are received within 5 ms. Although the host does not need to do anything for this to happen, it is recommended that the host shut the clock generation. Any command sent to the card will force it to exit from the sleep mode.

SD cards can consume up to 100–200 mA while reading or writing onto the card. This is usually a high current, and an appropriate voltage regulator capable of supplying the required current must be used in the design. The card consumes approximately 150 μA in sleep mode.

3.13 SD Card Internal Registers

The operations of SD cards are controlled by a number of internal registers. Some registers are 16 bits wide, some are 32 bits wide, and some are 128 bits wide. Table 3.7 gives a list of all the registers.

Detailed information on the functions and bit definitions of all the registers can be found in the product manuals of card manufacturers (e.g., *SanDisk Secure Digital Card, Product Manual, Document no: 80-13-00169, 2003*). The details of the important registers and their bit definitions are given in this section.

Table 3.7: SD Card Registers

Register	Width (Bits)	Description
OCR	32	Operation condition
CID	128	Card information
CSD	128	Card specific information
RCA	16	Relative card address
DSR	16	Driver stage register
SCR	64	Special features
Status	512	Status bits

3.13.1 OCR Register

The OCR register is 32 bits wide, and it describes the operating voltage range and status bits in the power supply. Table 3.8 shows the bit definitions of the OCR register. In summary,

- Bits 0–3 are reserved
- Bits 4–23 describe the SD card voltage
- Bits 24–30 are reserved
- Bit 31 is the power-up busy status bit. This bit is set to "1" after the power-up initialization of the card has been completed.

The initial value of the OCR register is usually set to binary value:

"*000 0000 1111 1111 1000 0000 0000 0000," which corresponds to 2.7–3.6 V operation (OCR bits 15–23 are all logic 1). Bit "*" indicates the busy status of the card at power-up.

3.13.2 CID Register

This is a 128-bit register that contains the card identification information specific to card manufacturers. Table 3.9 shows the bit definitions of the CID register. In summary,

- Bit 0 is reserved and is always "1."
- **CRC:** Bits 1–7 are CRC bits.
- **MDT:** Bits 8–19 are the manufacturing date.
  - Bits 8–11 are the Month field (01h = January).
  - Bits 12–19 are the Year field (00h = 2000).
  - Bits 20–23 are reserved (all "0"s).
- **PSN:** Bits 24–55 are the serial number (unsigned integer).

Table 3.8: OCR Register Bit Definitions

OCR Bit	Card Voltage	Initial Value
31	Card busy bit	"0" = busy, "1" = ready
30–24	Reserved	All "0"s
23	3.6–3.5	1
22	3.5–3.4	1
21	3.4–3.3	1
20	3.3–3.2	1
19	3.2–3.1	1
18	3.1–3.0	1
17	3.0–2.9	1
16	2.9–2.8	1
15	2.8–2.7	1
14	2.7–2.6	0
13	2.6–2.5	0
12	2.5–2.4	0
11	2.4–2.3	0
10	2.3–2.2	0
9	2.2–2.1	0
8	2.1–2.2	0
7	2.0–1.9	0
6	1.9–1.8	0
5	1.8–1.7	0
4	1.7–1.6	0
3–0	Reserved	All "0"s

Table 3.9: CID Register Bit Definitions

Field	Width	Bit Position	Description
MID	8	127–120	Manufacturer's ID
OID	16	119–104	Card OEM
PNM	40	103–64	Product code
PRV	8	63–56	Product revision
PSN	32	55–24	Serial number
–	4	23–20	0
MDT	12	19–8	Manufacturing date
CRC	7	7–1	Checksum
–	1	0	1

- **PRV:** Bits 56–63 are the Product Revision of the card.
- **PNM:** Bits 64–103 are the 5-ASCII-character Product Code, for example,
    - SD064 is 64 MB card.
    - SD128 is 128 MB card.
    - SD256 is 256 MB card.
- **OID:** Bits 104–119 are the card OEM, allocated by the SD Card Association, for example,
    - TM indicates Toshiba.
    - SD indicates SanDisk.
- **MID:** Bits 120–127 are the Manufacturer's ID, for example,
    - 02h indicates Toshiba.
    - 03h indicates SanDisk.

As an example, the Toshiba 64-MB card has the following initial values in its 128-bit CID register (in hexadecimal, "*" depends on the card, and "#" depends on values on the card):

MID: 02

OID: 54 4D

PNM: 53 44 30 36 34

PRV: * *

PSN: * * * * * * * *

0

MDT: * * *

CRC: # #

3.13.3 CSD Register

CSD is the 128-bit Card Specific Data register that contains information required to access the data on the card. Some fields of the CSD register are read only, whereas some other fields are writeable. Table 3.10 shows bit definitions of the CSD register. In summary,

- **CSD_STRUCTURE:** Bits 126–127 are the CSD structure version number.
- Bits 120–125 are reserved (all "0").

Table 3.10: CSD Register Bit Definitions

Field	Description	Width	Bits	*Value	Code
CSD_STRUCTURE	CSD structure	2	127–126	1.0	00b
–		6	125–120	–	000000b
TAAC	Data read access time	8	119–112	10 ms	00001111b
NSAC	Data read access time	8	111–104	0	00000000b
TRAN_SPEED	Max data transfer rate	8	103–96	25 MHz	00110010b
CCC	Command classes	12	95–84	All	1F5h
READ_BL_LEN	Max read block length	4	83–80	512 bytes	1001h
READ_BL_PARTIAL	Partial read blocks allowed	1	79–79	Yes	1b
WRITE_BLK_MISALIGN	Write block misalignment	1	78–78	No	0b
READ_BLK_MISALIGN	Read block misalignment	1	77–77	No	0b
DSR_IMP	DSR implemented	1	76–76	No	0b
–	Reserved	2	75–74	–	00b
C_SIZE	Device size	12	73–62	899	383h
VDD_R_CURR_MIN	Max. Read current at Vdd min	3	61–59	100 mA	111b
VDD_R_CURR_MAX	Max. Read current at Vdd max	3	58–56	80 mA	110b
VDD_W_CURR_MIN	Max. Write current at Vdd min	3	55–53	100 mA	111b
VDD_W_CURR_MAX	Max. Write current at Vdd max	3	52–50	80 mA	110b
C_SIZE_MULT	Device size multiplier	3	49–47	32	011b
ERASE_BLK_EN	Erase single block enable	1	46–46	Yes	1b
SECTOR_SIZE	Erase sector size	7	45–39	32 blocks	0011111b
WP_GRP_SIZE	Write protect group size	7	38–32	128 sectors	1111111b
WP_GRP_ENABLE	Write protect group enable	1	31–31	Yes	1b
–	Reserved	2	30–29	–	00b
R2W_FACTOR	Write speed factor	3	28–26	X16	100b
		3	28–26	X4	010b
WRITE_BL_LEN	Max write block length	4	25–22	512 bytes	1001b
WRITE_BL_PARTIAL	Partial write allowed	1	21–21	No	0b
–	Reserved	5	20–16	–	00000b
FILE_FORMAT_GRP	File format group	1	15–15	0	0b
COPY	Copy flag	1	14–14	Not original	0b
PERM_WRITE_PROTECT	Permanent write protection	1	13–13	Not protected	0b
TMP_WRITE_PROTECT	Temporary write protection	1	12–12	Not protected	0b
FILE_FORMAT	File format	2	11–10	HD w/ partition	00b
–	Reserved	2	9–8	–	00b

Table 3.10: CSD Register Bit Definitions —cont'd

Field	Description	Width	Bits	*Value	Code
CRC	CRC	7	7-1	–	
–	Always 1	1	0-0	–	1b

*values are based on a 16 MB SanDisk card.

- **TAAC:** Bits 112–119 define the asynchronous part of the read access time of the card. The bits are decoded as follows:

TAAC bit	Code
2–0	Time unit. 0 = 1 ns, 1 = 10 ns, 2 = 100 ns, 3 = 1 μs, 4 = 10 μs, 5 = 100 μs
6–3	Time value. 0 = Reserved, 1 = 1.0, 2 = 1.2, 3 = 1.3, 4 = 1.5, 5 = 2.0, 6 = 2.5, 7 = 3.0, 8 = 3.5, 9 = 4.0, A = 4.5, B = 5.0, C = 5.5, D = 6.0, E = 7.0, F = 8.0
7	Reserved

- **NSAC:** Bits 104–111 define the worst case for the clock-dependent factor of the data access time. The unit is 100 clock cycles. The total access time is equal to TAAC plus NSAC.

- **TRAN_SPEED:** Bits 96–103 define the maximum data transfer rate. The bits are decoded as follows:

TRAN_SPEED bit	Code
2–0	Transfer Rate Unit. 0 = 100 kb/s, 1 = 1 Mb/s, 2 = 10 Mb/s, 3 = 100 Mb/s, 4–7 = Reserved
6–3	Time Value. 0 = Reserved, 1 = 1.0, 2 = 1.2, 3 = 1.3, 4 = 1.5, 5 = 2.0, 6 = 2.5, 7 = 3.0, 8 = 3.5, 9 = 4.0, A = 4.5, B = 5.0, C = 5.5, D = 6.0, E = 7.0, F = 8.0
7	Reserved

- **CCC:** Bits 84–95 define the command classes that are supported by the card. The bit definitions are as follows:

CCC bit	Supported card command class
0	Class 0
1	Class1

...	...
11	Class 11

- **READ_BL_LEN:** Bits 80–83 define the maximum read data block length, which is equal to $2^{READ\_BL\_LEN}$. The data block length is specified as follows:

READ_BL_LEN	**Block Length**
0–8	Reserved
9	$2^9 = 512$ bytes
...	...
10	$2^{11} = 2048$ bytes
12–15	Reserved

- Bit 79 is always "1."
- **WRITE_BLK_MISALIGN:** Bit 78 defines whether the data block to be written by one command can be spread over more than one physical block:

WRITE_BLK_MISALIGN	**Access Block boundary write**
0	Not allowed
1	Allowed

- **READ_BLK_MISALIGN:** Bit 77 defines whether the data block to be read by one command can be spread over more than one physical block:

READ_BLK_MISALIGN	**Access Block boundary read**
0	Not allowed
1	Allowed

- **DSR_IMP:** Bit 76, if set, a driver stage register (DSR) is implemented.
- Bits 74–75 are reserved.
- **C_SIZE:** Bits 62–73 define the user's data card capacity as follows:

$$\text{Memory capacity} = \text{BLOCKNR} * \text{BLOCK\_LEN},$$

where

$$\text{BLOCKNR} = (\text{C\_SIZE} + 1) * \text{MULT}$$

$$\text{MULT} = 2^{\text{C\_SIZ\_MULT}+2} \qquad \text{if C\_SIZE\_MULT} < 8$$

and

$$\text{BLOCK\_LEN} = 2^{\text{READ\_BL\_LEN}} \qquad \text{if READ\_BL\_LEN} < 12.$$

- Bits 50–61 define the maximum and minimum values for read/write currents

- **C_SIZE_MULT:** Bits 47–49 are used to compute the user's data card capacity multiply factor:

C_SIZE_MULT	MULT
0	$2^2 = 4$
1	$2^3 = 8$
2	$2^4 = 16$
3	$2^5 = 32$
4	$2^6 = 64$
5	$2^7 = 128$
6	$2^8 = 256$
7	$2^9 = 512$

- **ERASE_BLK_EN:** Bit 46 defines if host can erase by WRITE_BL_LEN:

ERASE_BLK_EN	Description
0	Host cannot erase by WRITE_BL_LEN
1	Host can erase by WRITE_BL_LEN

- **SECTOR_SIZE:** Bits 39–45 define the minimum erasable size as the number of write blocks.
- **WP_GRP_SIZE:** Bits 32–38 define the minimum number of sectors that can be set for the write protect group.
- **WP_GRP_ENABLE:** Bit 31 defines the write protect group functions:

WP_GRP_ENABLE	Description
0	Not implemented
1	Implemented

- Bits 29–30 are reserved.
- **R2W_FACTOR:** Bits 26–28 define a multiple number for a typical write time as a multiple of the read access time.
- **WRITE_BL_LEN:** Bits 22–25 define the maximum write block length, which is calculated as $2^{\text{WRITE\_BL\_LEN}}$. The data block length is specified as follows:

WRITE_BL_LEN	Block Length
0–9	Reserved
11	$2^9 = 512$ bytes
...	...
12	$2^{11} = 2048$ bytes
12–16	Reserved

- **WRITE_BL_PARTIAL:** Bit 21 defines whether partial block write is available:

WRITE_BL_PARTIAL	Write data size
0	Only WRITE_BL_LEN size of 512 bytes is available
1	Partial size write available

- Bits 16–20 are reserved.
- **FILE_FORMAT_GRP:** Bit 15 indicates the selected group of file format group and file format:

FILE_FORMAT_GRP	FILE_FORMAT	Kinds
0	0	Hard disk-like file system with partition table
0	1	DOS FAT with boot sector only (no partition table)
0	2	Universal File Format
0	3	Others
1	0, 1, 2, 3	Reserved

- **COPY:** Bit 14 defines the contents of the card as original or duplicated. The bit definition is:

COPY	Description
0	Original
1	Copy

- **PERM_WRITE_PROTECT:** Bit 13, if set, permanently write protects the card.
- **TMP_WRITE_PROTECT:** Bit 12, if set, temporarily write protects the card.
- **FILE_FORMAT:** Bits 10–11 define the file format on the card. This field is used together with field FILE_FORMAT_GRP as in the above table.
- Bits 8–9 are reserved.
- **CRC:** Bits 1–7 are the CRC error checking bits.
- Bit 0 is not used and is always "1."

3.13.4 RCA Register

This 16-bit register carries the card addresses in SD card mode.

3.13.5 DSR Register

This register is not implemented in many cards.

3.13.6 SCR Register

This 64-bit register provides information on the SD card's special features, such as the structure version number, the physical layer specification, the security algorithm used, and the bus width.

3.13.7 SD Status Register

This 512-bit register defines the card status bits and card features.

3.14 Calculating the SD Card Capacity

An example is given in this section to show how the capacity of an SD card can be calculated.

Example 3.1

The following CSD register fields are given by a card manufacturer:

C_SIZE = E27h (or decimal 3623)
C_SIZE_MULT = 3
READ_BL_LEN = 9

Calculate the capacity of this card.

Solution

The card capacity is defined by two fields within the CSD register: C_SIZE and C_SIZE_MULT. C_SIZE is a 12-bit value with an offset of 1 (1–4096), and C_SIZE_MULT is a 3-bit value with an offset of 2 (2–9).

The number of blocks on the card is given by

$$\text{BLOCKNR} = (\text{C\_SIZE} + 1) \times 2^{(\text{C\_SIZE\_MULT} + 2)},$$

where

$$\text{C\_SIZE\_MULT} < 8.$$

The default block length is 512 bytes (but it can also be specified as 1024 or 2048 bytes). The block length is calculated from

$$\text{BLOCK\_LEN} = 2^{\text{READ\_BL\_LEN}},$$

where

$$\text{READ\_BL\_LEN} = 9,\ 10,\ \text{or}\ 11.$$

Combining the two equations, we get the card capacity as

$$\text{Card Capacity (in bytes)} = \text{BLOCKNR} \times \text{BLOCK\_LEN}$$

or

$$\text{Card Capacity (in bytes)} = (\text{C\_SIZE} + 1) \times 2^{(\text{C\_SIZE\_MULT} + 2)} \times 2^{\text{READ\_BL\_LEN}}.$$

The capacity is usually shown in MB and

$$\text{Card Capacity (Megabyte)} = (\text{C\_SIZE} + 1) \times 2^{(\text{C\_SIZE\_MULT} + 2)} \times 2^{\text{READ\_BL\_LEN}}/(1024 \times 1024)$$

Using the CSD parameters given in this example, we get

$$\text{Card Capacity (Megabyte)} = 3624 \times 32 \times 512/(1024 \times 1024) = 56.525 \text{ Megabytes}$$

It is interesting to note that when a block length of 1024 bytes is used (READ_BL_LEN = 10), cards up to 2 GB can be specified, and with a block length of 2048 bytes (READ_BL_LEN = 11), cards up to 4 GB can be specified.

■

3.15 SD Card SPI Bus Protocol

All communications between the host and the card are controlled by the host. Messages in the SPI bus protocol consist of commands, responses, and tokens. The card returns a response to every command received and also a data response token for every write command.

The SD card wakes up in SD card mode, and it will enter the SPI mode if its CS line is held low when a reset command is sent to the card. The card can only be returned to the SD mode after a power-down and power-up sequence.

When the SPI mode is entered, the card is in the nonprotected mode, where CRC checking is not used (CRC checking can be turned on and off by sending command CRC_ON_OFF, command name CMD59, to the card).

3.15.1 Data Read

Data can be read in either single or multiple blocks. The basic unit of data size is blocks, defined by field READ_BL_LEN of the CSD register. In this book, we shall be using only single-block reads. Single-block reads are initiated by issuing the command READ_SINGLE_BLOCK (CMD17) to the card. Any valid address can be used as the starting address.

3.15.2 Data Write

Data can be written in either single or multiple blocks. After receiving a valid write command, the card sends a response token and then waits for the data block to be sent from the host. The starting address can be any valid address. After receiving a data block from the host, the card returns a data response token and writes the data on the card if the data contains no errors.

Table 3.11: Some Important SD Card Commands

Command	Abbreviation	Argument	Response	Description
CMD0	GO_IDLE_STATE	None	R1	Reset the SD card
CMD1	SEND_OP_COND	None	R1	Initialize card
CMD9	SEND_CSD	None	R1	Get CSD register data
CMD10	SEND_CID	None	R1	Get CID register data
CMD17	READ_SINGLE_BLOCK	Data address (0:31)	R1	Read a block of data
CMD24	WRITE_BLOCK	Data address (0:31)	R1	Write a block of data

Table 3.12: Command Format

Byte 1			Bytes 2–5	Byte 6	
7	6	5 4 3 2 1 0	31..0	7 6 5 4 3 2 1	0
0	**1**	**Command**	**Command Argument**	**CRC**	**1**

There are a large number of commands available in SPI mode for reading the card registers, reading and writing single and multiple blocks of data, erasing blocks, etc. Table 3.11 gives a list of the important SD card commands. All SPI mode commands are 6 bytes long (48 bits). As shown in Table 3.12, the commands start with the most significant bit (MSB) as logic 0, a transmission bit as logic 1, 6 bits of command index, 32 bits of argument (not all commands need arguments), 7 bits of CRC, and an end-bit (logic 1). The commands are divided into classes. If no argument is required in a command, the value of the argument filed should be all "0"s. The command index contains the actual command number. For example, the command index value for command CMD8 is binary number 8 in 6 bits, i.e., "001000."

3.15.3 Response Tokens

There are several types of response tokens that can be sent by the card. A token is transmitted with the MSB bit sent first. The response tokens are as follows:

R1 Format: This response token is 1 byte long and is sent by the card after every command (except the SEND_STATUS command). The MSB bit is "0," and other bits indicate an error ("1" bit). For example, if bit "0" is set, it indicates that the card is in Idle State and running initialization sequence. Table 3.13 shows bit definitions of the R1 Format.

R1b Format: This format is similar to R1 Format with the addition of the busy signal.

Table 3.13: R1 Format bits

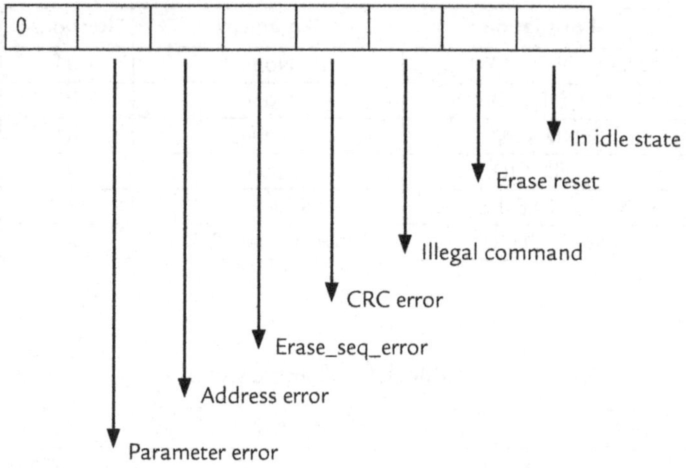

0							

R2 Format: This response token is 2 bytes long and is sent as a response to command SEND_STATUS.

R3 Format: This response token is 5 bytes long and is sent in response to command READ_OCR. The first byte is identical to R1 format, whereas the other bytes contain the OCR register data.

Data Response Token: Whenever a data block is written to the card, the card acknowledges with a data response. This token is 1 byte long and has the following bit definitions:

Bit 0	Always 1
Bits 1–2	Status
Bit 3	Always 0
Bit 4	Reserved

The Status bits are defined as follows:

010	Data accepted
101	CRC error, data rejected
110	Write error, data rejected

3.16 Data Tokens

Data is received or transmitted via data tokens with all data bytes transmitted with the MSB first.

Data tokens are 4–515 bytes long and have the following format (for single-block operations):

- First Byte: START BLOCK

This block is identified by data "11111110," i.e., FEh.

- Bytes 2–513: USER DATA
- Last 2 bytes (byte 514 and byte 515): CRC

3.17 Card Reset State

After power-up, the SD card is in the Idle State. Sending command CMD0 also puts the card in the Idle State. At least 74 clock cycles should be sent to the card with the Data Out and CS lines set to logic "1" before starting to communicate with the card.

The SD card is initially in the SD Bus mode. It will enter the SPI mode if the CS line is held low while sending the CMD0 command. When the card switches to the SPI mode, it will respond with the SPI mode R1 response format. In SPI mode, CRC checking is disabled by default. However, because the card powers-up in the SD Bus mode, CMD0 command must be sent with a valid CRC byte before the card is put into SPI mode. When sending the CMD0 command, the CRC byte is fixed and is equal to 95h. The following hexadecimal 6-byte command sequence can then be used to send the CMD0 command after a power-up (see Table 3.12 with the command field set to "000000" for CMD0):

40 00 00 00 00 95

The steps to switch the SD card into SPI mode should therefore be as follows:

- Power-up.
- Send at least 74 clock pulses to the card with CS and Data Out lines set to logic "1."
- Set CD line low.
- Send 6-byte CMD0 command "40 00 00 00 00 95" to put the card in SPI mode.
- Check R1 response to make sure there are no error bits set.
- Send command CMD1 repeatedly until the "in-idle-state" bit in R1 response is set to "0," and there are no error bits set.
- The card is now ready for read/write operations.

During the reset state, the card clock frequency should be between 10–400 KHz. After the reset state, the maximum clock frequency can be increased to 25 MHz (20 MHz for the MMC).

3.18 Summary

The brief details of commonly used memory cards are given in this chapter. SD cards are currently the most widely used memory cards. The technical details and communication methods of these cards have been described in detail in the chapter.

3.19 Exercises

1. Explain the main differences between the standard SD cards and the new SDHC cards. Which card would you choose in a long video recording application?
2. How many types of standard SD cards are there? Explain their main differences.
3. Which memory card would you choose in very high-speed data transfer applications?
4. Explain how you could read the data stored on a memory card using your PC.
5. What are the names of the internal registers of a standard SD card?
6. Explain functions of the CID register of an SD card.
7. Explain functions of the CSD register of an SD card.
8. The TAAC field of the CSD register of an SD card is binary "00101101." Explain what this means.
9. The TRAN_SPEED field of the CSD register of an SD card is binary "00110010." Explain what this means.
10. Explain the two operating modes of SD cards. Which mode is commonly used?
11. What is the operating voltage range of an SD card? Explain how this voltage can be obtained from a standard +5 V regulated supply.
12. Draw a circuit diagram to show how an SD card can be connected to a PIC microcontroller in SPI mode.
13. Explain how an SD card can be put into the SPI mode after power-up.
14. Explain how an SD card can be put into SD card mode after operating in the SPI mode.
15. How many types of response tokens are there? Explain where each token is used and also give their differences.

CHAPTER 4

Programming with the MPLAB C18 Compiler

4.1 C Programming Languages for PIC18 Microcontrollers

There are several C compilers on the market for the PIC18 series of microcontrollers. Most of the features of these compilers are similar, and they can all be used to develop C-based high-level programs for the PIC18 series of microcontrollers.

Some of the popular C compilers used in the development of commercial, industrial, and educational PIC18 microcontroller applications are as follows:

- mikroC C compiler
- PICC18 C compiler
- CCS C compiler
- MPLAB C18 C compiler

mikroC C compiler has been developed by *MikroElektronika* (Web site: http://www.mikroe.com) and is one of the easy-to-learn compilers with rich resources, such as a large number of library functions and an integrated development environment with built-in simulator and an in-circuit-debugger (e.g., mikroICD). A demo version of the compiler with a 2K-program limit is available from MikroElektronika.

PICC18 C compiler is another popular C compiler developed by *Hi-Tech Software* (Web site: http://www.htsoft.com). This compiler has two versions: the standard compiler and the professional version. A powerful simulator and an integrated development environment (Hi-Tide) are provided by the company. PICC18 is supported by the Proteus simulator (http://www.labcenter.co.uk), which can be used to simulate PIC microcontroller-based systems. A limited-period demo version of this compiler is available from the developer's Web site.

CCS C compiler has been developed by *Custom Computer Systems Inc.* (Web site: http://www.ccsinfo.com). The company provides a limited-period demo version of their compiler. CCS compiler provides a large number of built-in functions and supports an in-circuit-debugger (e.g., ICD-U40), which aids greatly in the development of PIC18 microcontroller-based systems.

D.O.I.: 10.1016/B978-1-85617-719-1.00008-7

MPLAB C18 C compiler is a product of *Microchip Inc*. (Web site: http://www.microchip.com). An evaluation version with limited functionality is available from the Microchip Web site. MPLAB C18 includes a simulator and supports hardware and software development tools such as in-circuit-emulators (e.g., ICE2000) and in-circuit-debuggers (e.g., ICD2 and ICD3). In this book, we will be using the MPLAB C18 compiler for all the projects.

4.2 MPLAB C18 Compiler

MPLAB C18 compiler (or the C18 compiler) is one of the most popular C compilers available for the PIC18 series of microcontrollers. This compiler has been developed by Microchip Inc. An evaluation version (student version) of the compiler is available from the Microchip Web site free of charge. This evaluation version includes all functionality of the full version of the compiler for the first 60 days. However, some optimization routines are disabled after 60 days, and PIC18 extended mode (extended instruction set and indexed with literal offset addressing) is not supported after 60 days.

MPLAB C18 is a cross compiler, where programs are developed on a PC and are then loaded to the memory of the target microcontroller using a suitable programming device.

The installation and use of the MPLAB C18 compiler and details of programming using the compiler are given in this chapter.

4.2.1 Installing the MPLAB C18 Compiler

Before installing the MPLAB C18 compiler, it is necessary to install the MPLAB IDE integrated development environment software. MPLAB IDE supports many development tools, such as assemblers, linkers, C18 and third-party compilers; built-in device programming software; and many more tools.

Installing the MPLAB IDE

The steps for installing the MPLAB IDE are given below:

- Copy MPLAB IDE zip file from the Microchip Web site. At the time of writing, the IDE had the filename **mplab__v820.zip**
- Create a subdirectory called **MPLAB** under your main **C:** drive
- Unzip MPLAB IDE into directory **MPLAB**
- Start the installation program by double-clicking on icon **Install_MPLAB_V820.exe**
- Accept **Complete** installation
- Restart the computer after installation
- When the system comes up, you can choose to view an MPLAB document

The installation program creates a shortcut icon in the desktop with the name **MPLAB IDE v8.20**.

Installing the C18 Compiler

The steps for installing the MPLAB C18 compiler are given below:

- Copy MPLAB C18 compiler installation file from the Microchip Web site. At the time of writing, the compiler had the filename **MPLAB-C18-Evaluation-v3_30.exe**.
- Double-click to start the installation.
- You can see the welcome page that displays the version number of the compiler. Click **Next** to see the license agreement.
- The next display, as shown in Figure 4.1, is the folder name where the compiler will be installed. Accept the default as **C:\MCC18** and click **Next**.
- The products that will be installed are displayed next, as in Figure 4.2. Accept the default and click **Next.**
- Figure 4.3 shows the next display, which is about the environment variables to be added to the system path. Tick all the boxes and click **Next**.
- Tick the MPLAB IDE options shown in Figure 4.4 so that the compiler becomes integrated into the MPLAB IDE. Click **Next**.
- The final form just before the installation is displayed as shown in Figure 4.5. Click **Next** to start the installation.

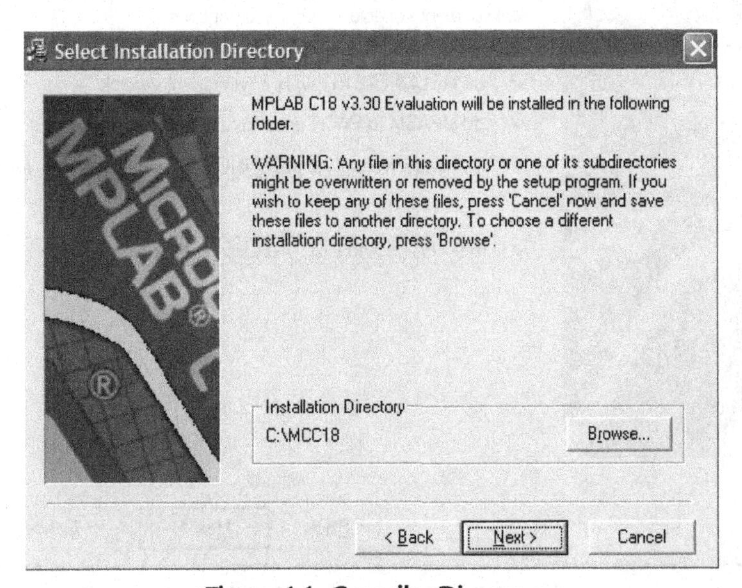

Figure 4.1: Compiler Directory

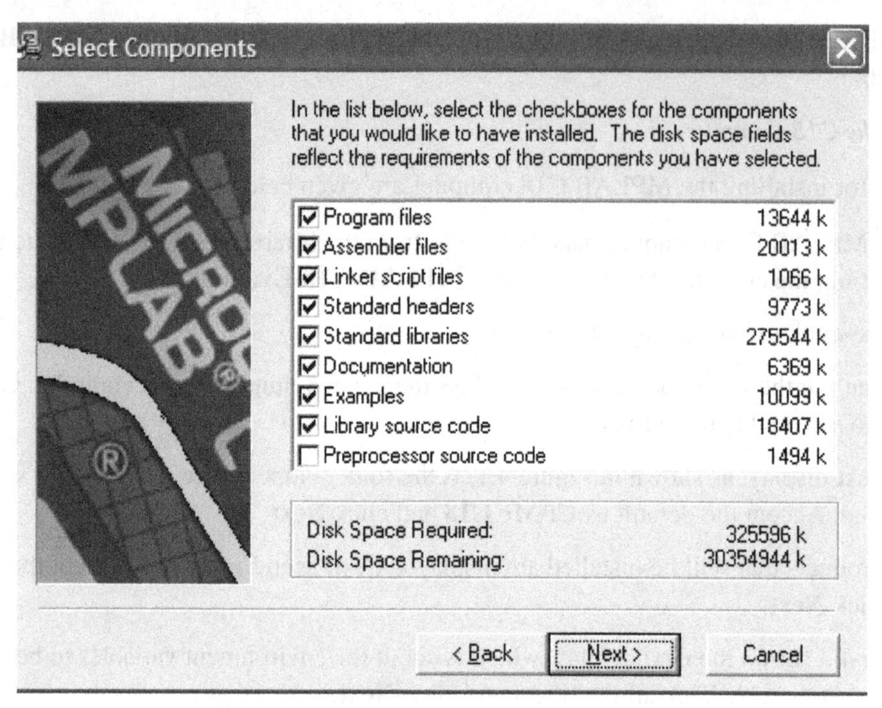

Figure 4.2: Products That Will Be Installed

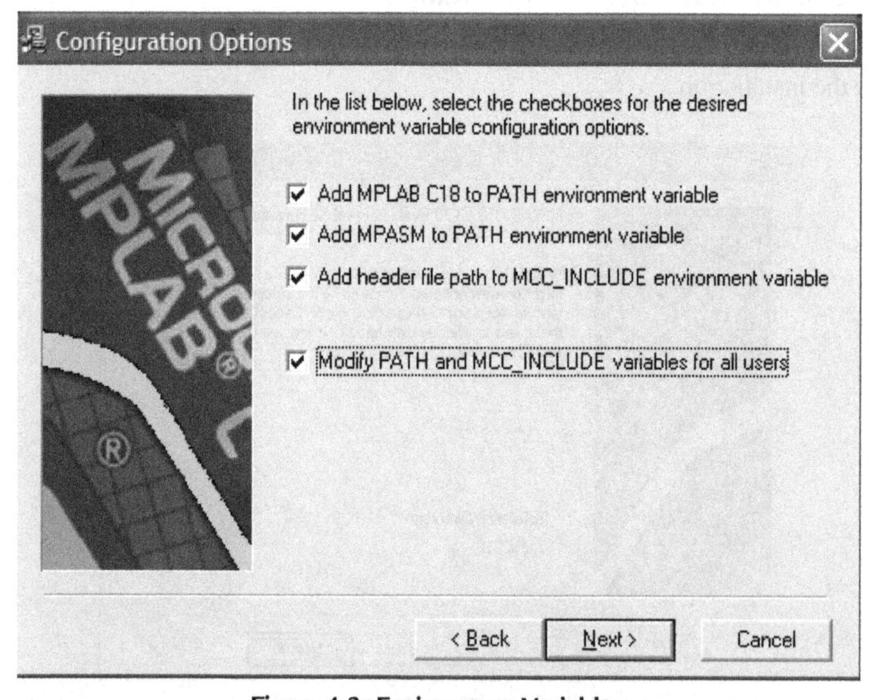

Figure 4.3: Environment Variables

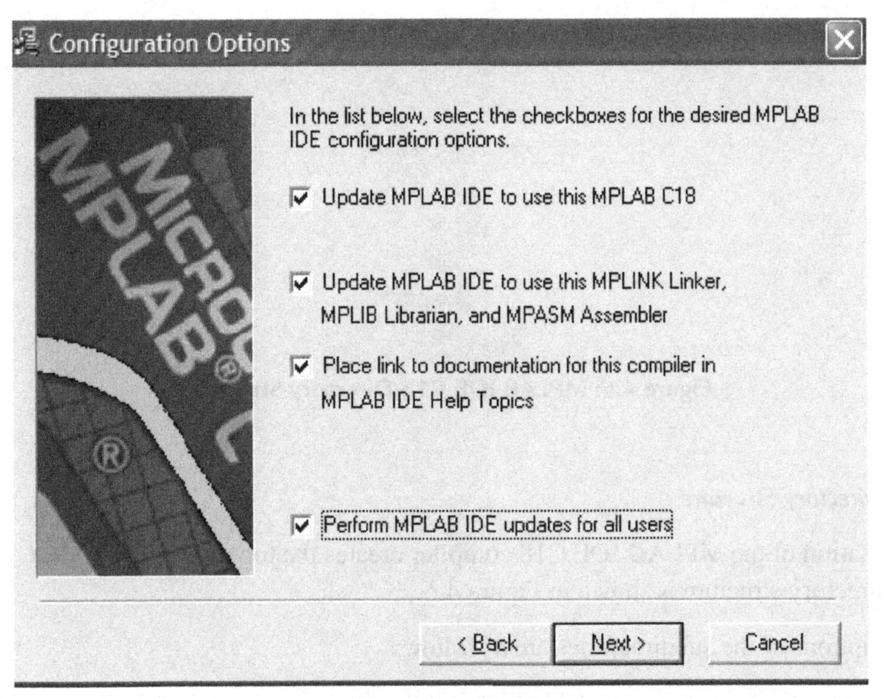

Figure 4.4: MPLAB IDE Options

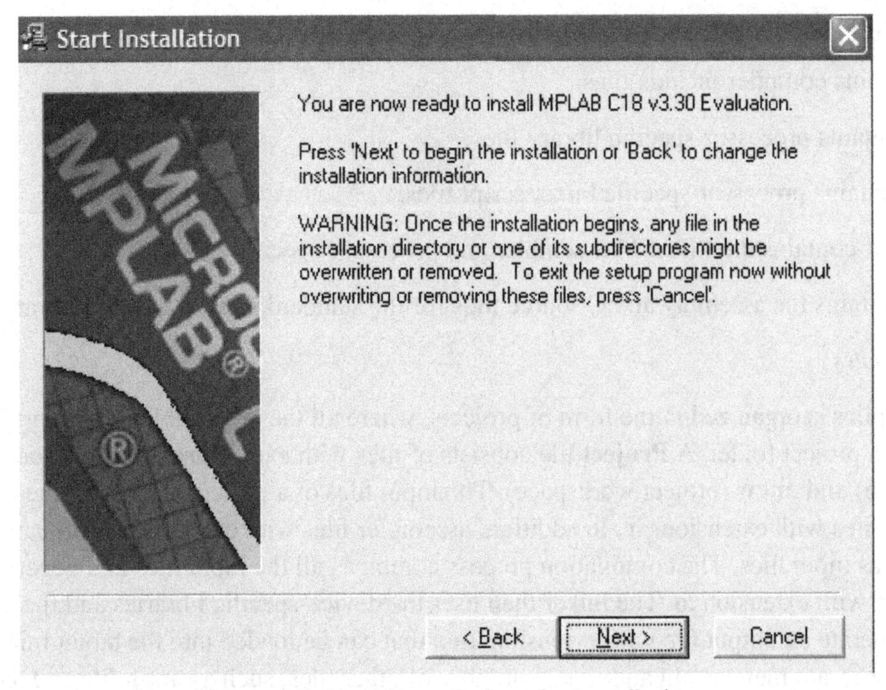

Figure 4.5: Final Form Before the Installation

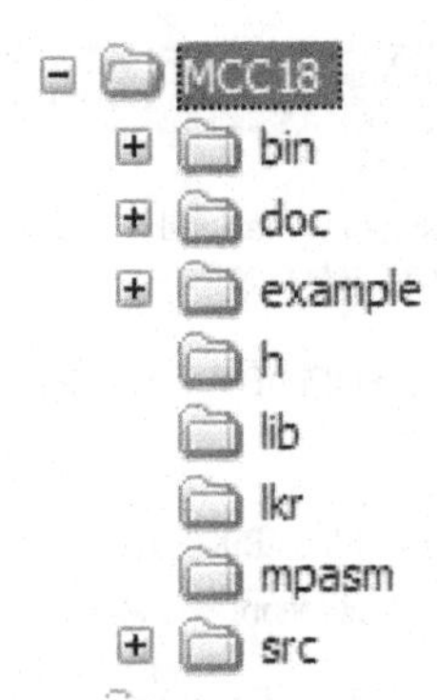

Figure 4.6: MPLAB IDE C18 Directory Structure

MPLAB Directory Structure

The installation of the MPLAB IDE C18 compiler creates the top directory **C:\MCC18**, and the directory structure is shown in Figure 4.6.

The descriptions of the subdirectories are as follows:

bin: It contains the compiler executable files and the linker script file subdirectory.

doc: It contains the compiler documentation and help files.

example: It contains sample C programs and documents.

h: It contains compiler include files.

lib: It contains processor-specific library files.

lkr: It contains processor-specific linker script files.

mpasm: It contains the MPASM assembler and processor-specific assembly files.

src: It contains the assembly and C source files for the standard and extended C library.

Compiler Files

The compiler is organized in the form of projects, where all the required files for a project are stored in a project folder. A **Project** file consists of files with extensions **.mcp** (project information file) and **.mcw** (project workspace). The input files of a project consist of one or more C source files with extensions **.c**. In addition, assembler files with extensions **.asm** can be included as input files. The compilation process combines all the input files and generates an object file with extension **.o**. The linker then uses the device-specific libraries and the object file to generate an output file with extension **.hex** that can be loaded into the target microcontroller's program memory. In addition, a number of other files, such as **.map** file and **.cof** file, are also generated by the compiler.

4.3 An Example Program

An example program and its simulation are given in this section to demonstrate how a C program can be created and then simulated. This program displays the message **PIC MICROCONTROLLERS**. Do not worry if you do not understand the program, because the C18 programming details will be given in the next chapter.

4.3.1 Building the Project

The step-by-step solution of the example is given below:

- Create a folder to store the project files. In this example, the project folder is given the name **MYC** and is under the top directory, i.e., **C:\MYC**
- Start MPLAB IDE as in Figure 4.7.
- Select **File->New** and write the following C program (see Figure 4.8):

```
#include <stdio.h>
#pragma config WDT = OFF

void main (void)
{
        printf ("PIC MICROCONTROLLERS\n");
        while (1);
}
```

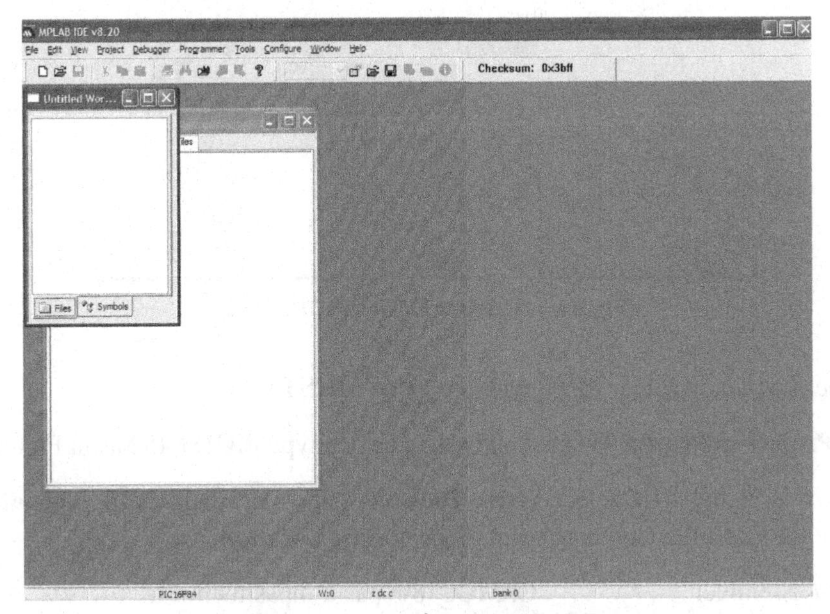

Figure 4.7: Starting MPLAB IDE

```
#include <stdio.h>

#pragma config WDT = OFF

void main (void)
{
        printf ("PIC MICROCONTROLLERS\n");

        while (1);

}
```

Figure 4.8: Example C Program

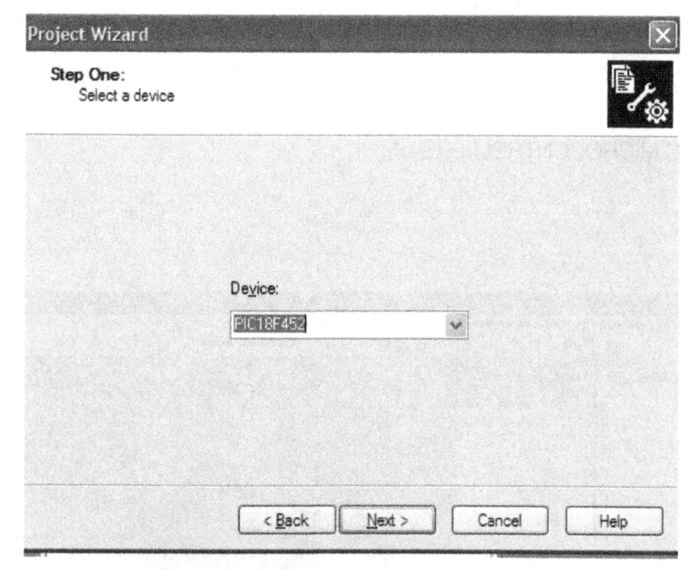

Figure 4.9: Select Device PIC18F452

- Save the program in folder MYC with the name **FIRST.C**.
- Select **Project -> Project Wizard** and select device type PIC18F452 as in Figure 4.9.
- As shown in Figure 4.10, select **Active Toolsuite** to be **Microchip C18 Toolsuite**. Make sure that the Toolsuite Contents point to the correct **Location**:

MPASM assembler	–	C:\MCC18\mpasm\mpasmwin.exe
MPLINK object	–	C:\MCC18\bin\mplink.exe

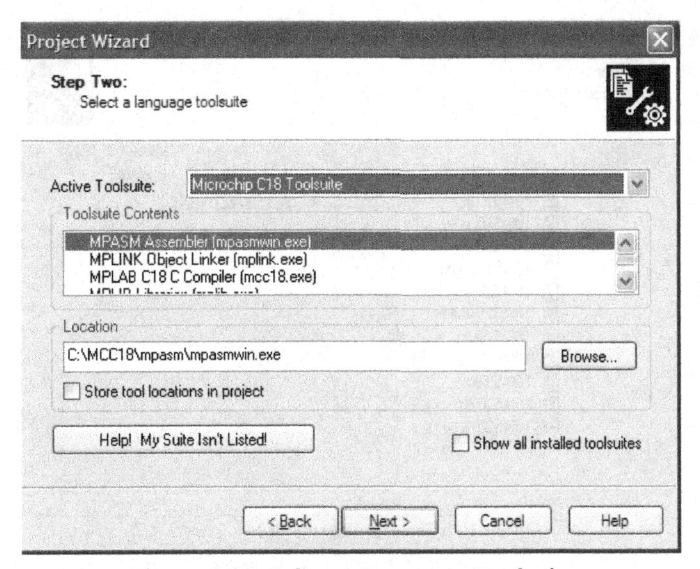

Figure 4.10: Select a Language Toolsuite

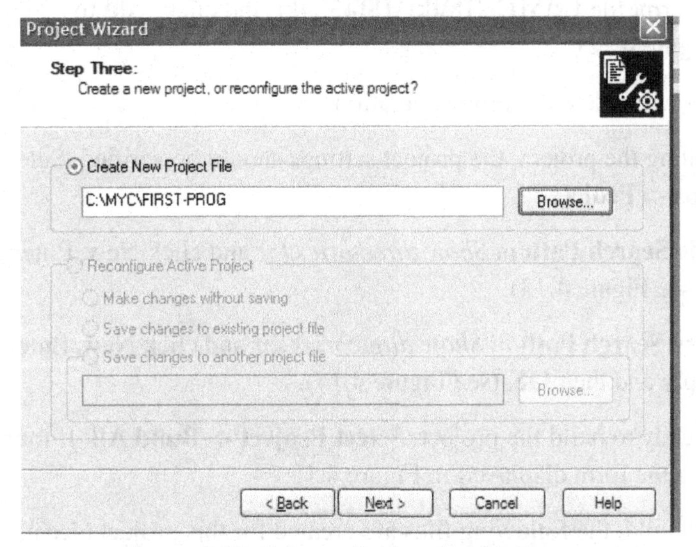

Figure 4.11: Select a Project Folder Name

MLAB C18 C compiler	–	C:\MCC18\bin\mcc18.exe
MPLIB librarian	–	C:\MCC18\bin\mplib.exe

- Select a folder, as shown in Figure 4.11, to store the project files. In this example, the folder is given the name MYC and is under the top directory, i.e., **C:\MYC**. Also choose a name for the project. In this example, the name **FIRST-PROG** is chosen.
- Select file **C:\MYC\FIRST.C** and click **Add** to add to the project folder.

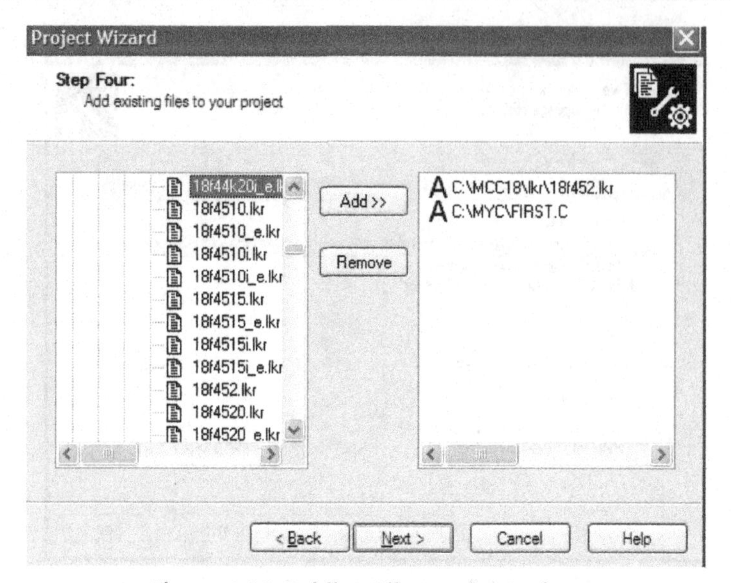

Figure 4.12: Adding Files to the Project

- Select linker script file **C:\MCC18\lkr\18f452.lkr** and click **Add** to add to the project folder (see Figure 4.12).
- Click **Finish** to complete the project creation.
- Before compiling the project, the project settings should be verified. Select **Project -> Build Options -> Project**
- Select **Include Search Path** in *Show directories for* and click **New.** Enter c:\mcc18\h and click **Apply** (see Figure 4.13).
- Select **Library Search Path** in *Show directories for* and click **New.** Enter c:\mcc18\lib and click **Apply** and then **OK** (see Figure 4.14).
- We are now ready to build the project. **Select Project -> Build All**. If there are no errors, you should see the form displayed in Figure 4.15.

After a successful build, the following files are created for this project (see also Figure 4.16) in project folder **C:\MYC**:

First.c	Source C program
First.o	Object program created
FIRST-PROG.cof	Object cof file
FIRST-PROG.hex	Program HEX file
FIRST-PROG.map	Program map file
FIRST-PROG.mcp	Project definition file
FIRST-PROG.mcw	Project work file

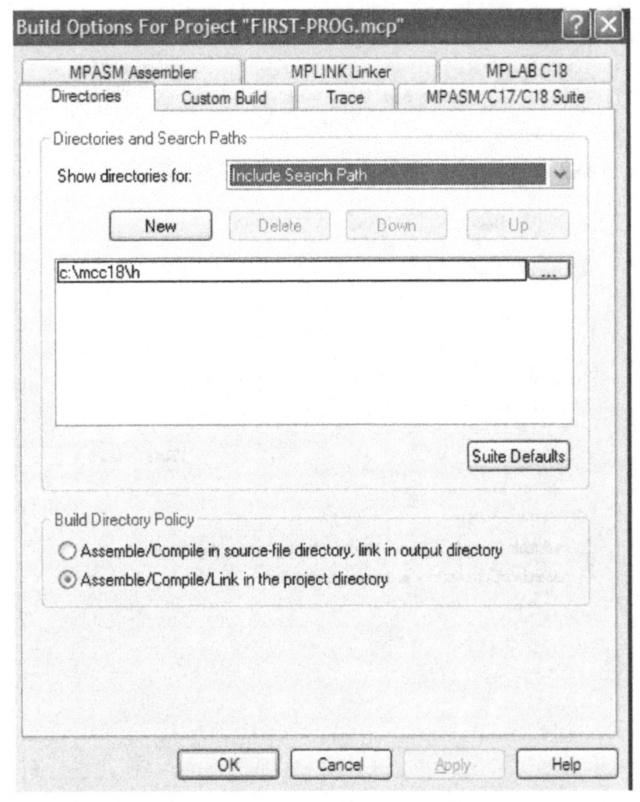

Figure 4.13: The Include Search Path

Program HEX file (**FIRST-PROG.hex**) is the file loaded into the program memory of the target microcontroller.

4.3.2 Simulating the Project

The C program we have written can be tested using the MPLAB simulator. The steps are given below:

- Select the debugger (as shown in Figure 4.17) by clicking **Debugger -> Select Tool -> MPLAB SIM**
- Select **Debugger -> Settings** and click on **Uart IO** tab and **Enable Uart IO**. Set the **Output** to **Window** and click **OK** as shown in Figure 4.18.
- You can see the debug toolbar in Figure 4.19.
- Start the simulator by clicking the **Run** icon in the debug toolbar. The program will run and you can see the output in Figure 4.20.
- Stop the program by clicking the **Halt** button in the debugger menu.

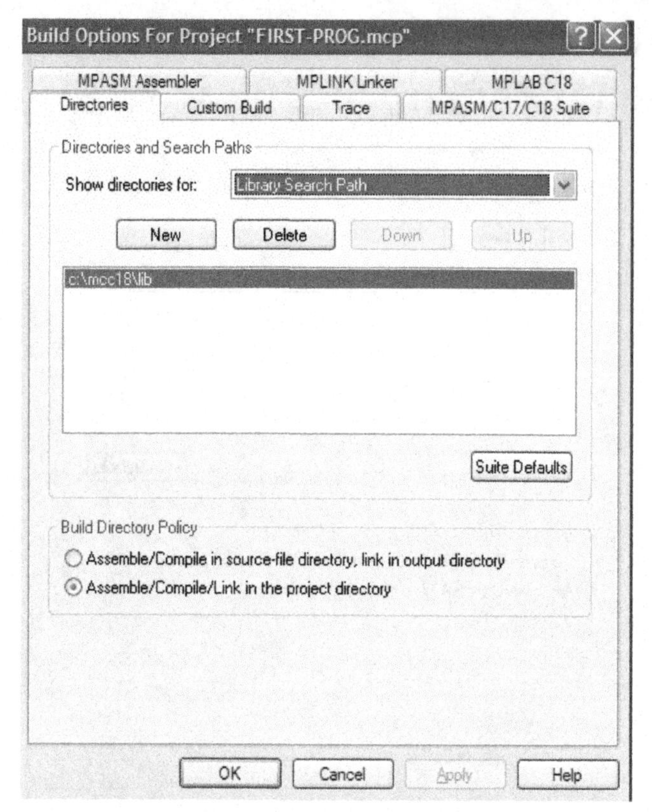

Figure 4.14: The Library Search Path

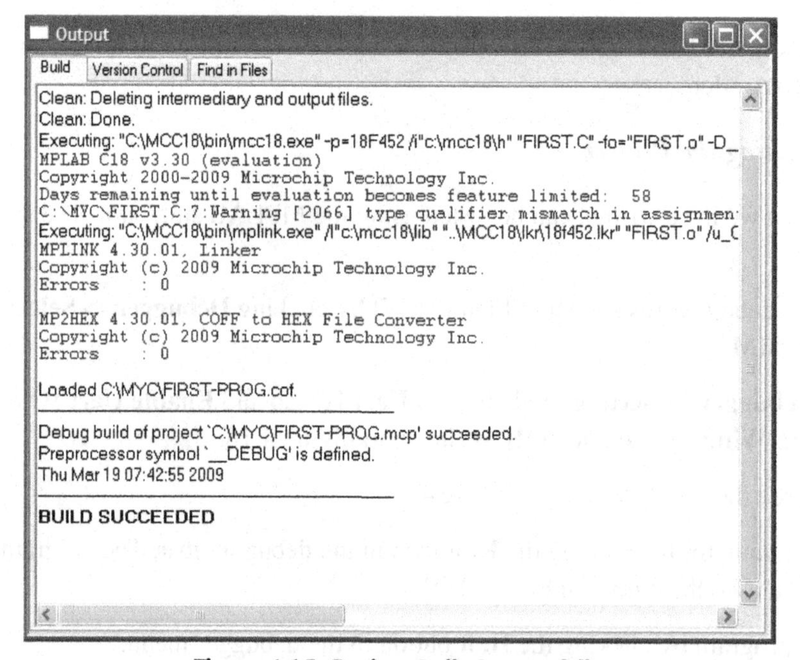

Figure 4.15: Project Built Successfully

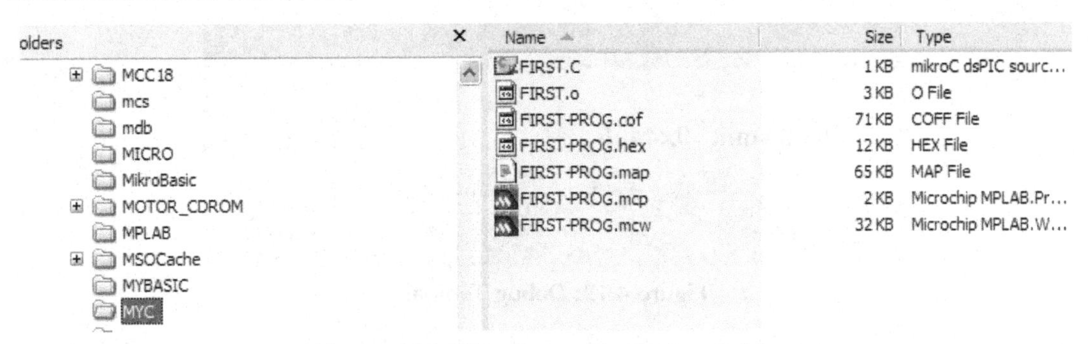

Figure 4.16: Files Created in Project Folder

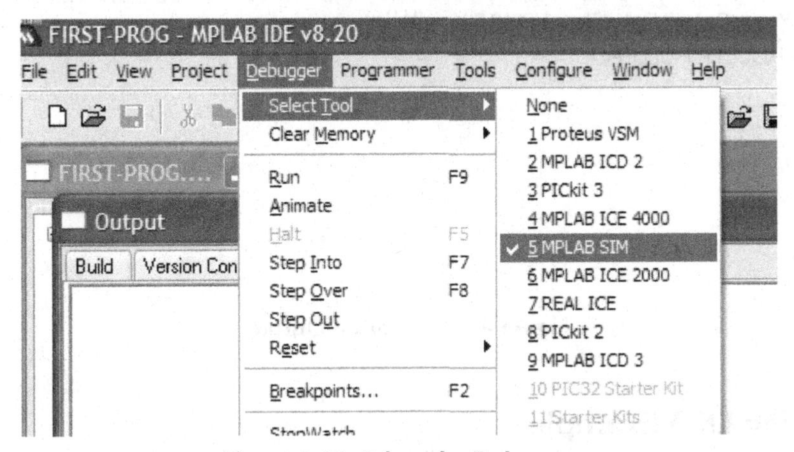

Figure 4.17: Select the Debugger

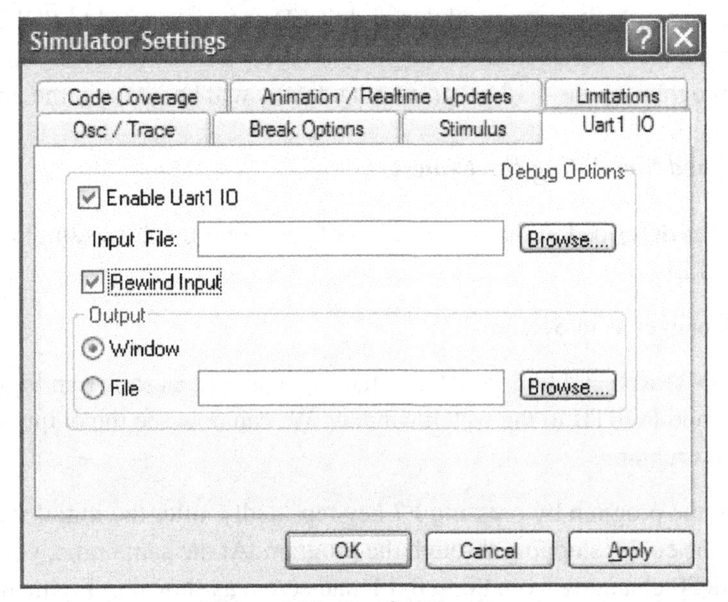

Figure 4.18: Set the Simulator

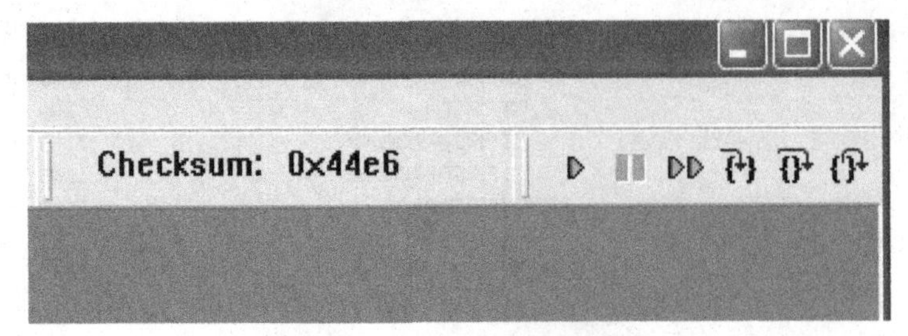

Figure 4.19: Debug Toolbar

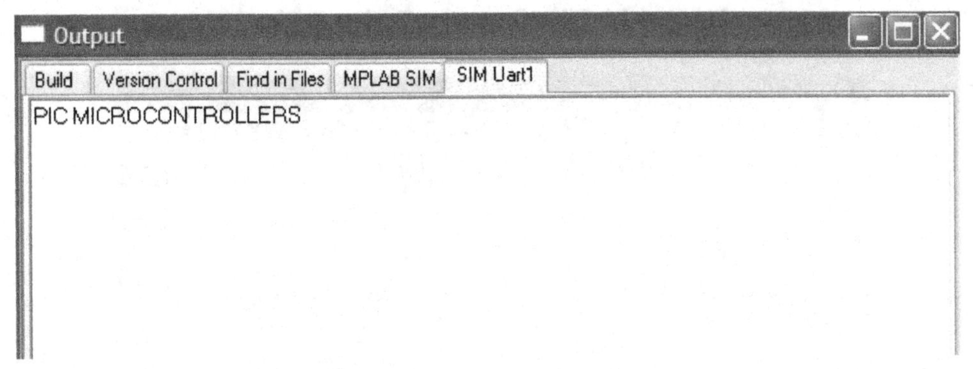

Figure 4.20: Program Output

4.4 Flashing LED Example

Another example is given in this section to show some other aspects of the MPLAB simulator. In this example, it is assumed that eight LEDs are connected to PORTB of the microcontroller and these LEDs flash as the program runs. Do not worry if you do not understand the program, as the C18 programming details will be given in the next section.

4.4.1 Building and Simulating the Project

Build the project as described in Section 4.3. Your C program for this example will be as shown in Figure 4.21.

- Select the debugger as in Section 4.3.
- Click **View -> Watch** and find PORTB in the first list-box, as shown in Figure 4.22. Click **Add SFR** to add PORTB to the watch window. We can now see the output of PORTB as the program is running.
- Step through the program by pressing **F7** key repeatedly; after the initial start-up code, you can see the cursor stepping through the program. At the same time, you can see the value of PORTB changing from 00 to 0xFF and so on, as shown in Figure 4.23.

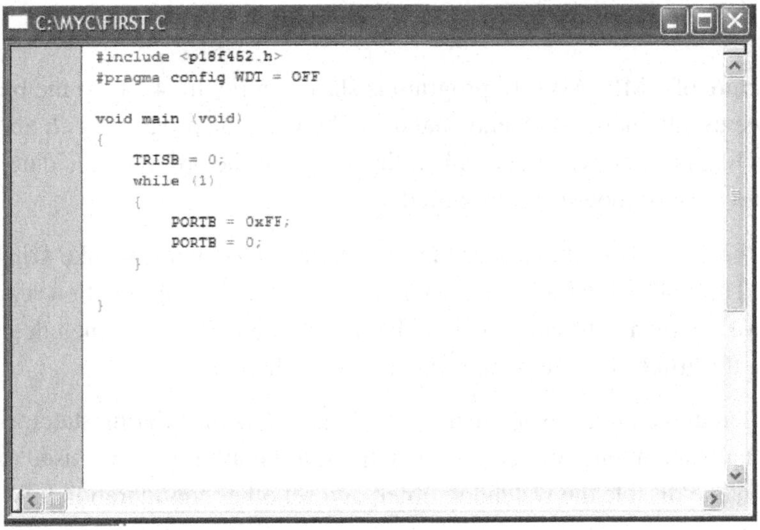

C:\MYC\FIRST.C

```
#include <p18f452.h>
#pragma config WDT = OFF

void main (void)
{
    TRISB = 0;
    while (1)
    {
        PORTB = 0xFF;
        PORTB = 0;
    }

}
```

Figure 4.21: The C Program of the Example

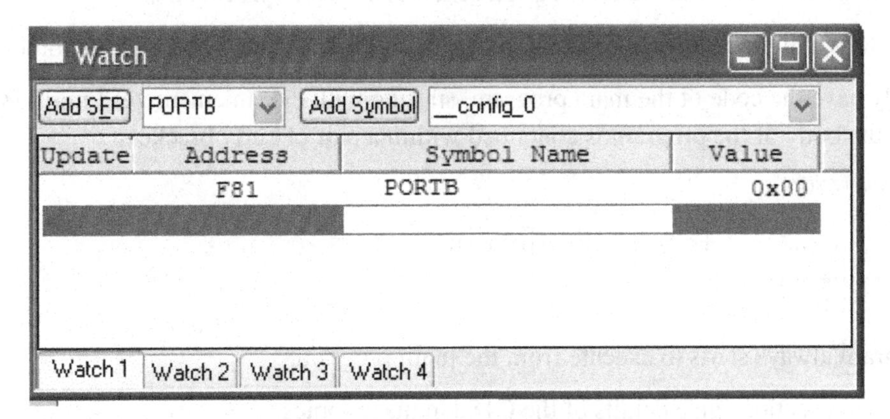

Figure 4.22: Set the Watch Window

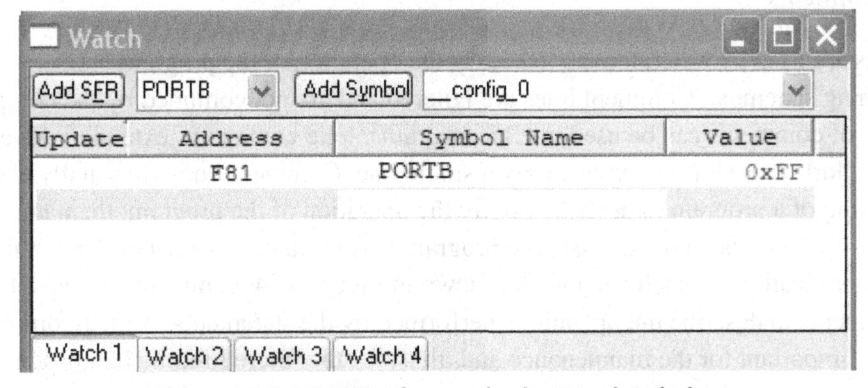

Figure 4.23: PORTB Changes in the Watch Window

4.5 Structure of the MPLAB C18 Compiler

The basic structure of a MPLAB C18 program is shown in Figure 4.24. At the beginning of the program, we usually have comments that describe what the program is all about, the name of the program file, program version number, the author of the program, the date the program was created, and a list of modifications with dates.

Then we have the device-specific header files declared using statements like **#include <p18xxxx.h>**. The header file includes the names of various special function registers used in the program. We can then optionally declare the other source files to be included in our main program, using the **#include** statement, followed by the filenames.

The symbols used in the program can then be declared using the **#define** statements. The next statement is usually a compiler directive called **#pragma config**, which is used to select the clock source, enable/disable the watchdog timer, and set other configuration bits.

We can then define any global variable used in the program. These global variables can be accessed from any part of the main program or its functions or procedures.

The functions and procedures used in the program are then declared one after the other.

We finally have the code of the main program, starting with the statement **void main (void)** and then the body of the program is contained within a pair of curly brackets:

```
void main (void)
{
        .............        Body of the main program
        .............
}
```

The program always starts to execute from the main program.

The following sections give details of the C18 language topics.

4.5.1 Comments

Comments are used by programmers to clarify the operation of the program or a programming statement. Comment lines are ignored and are not compiled by the compiler. Two types of comments can be used in C18 programs: long comments, extending several lines, and short comments occupying only a single line. Comment lines are usually used at the beginning of a program to describe briefly the operation of the program, the name of the author, the program filename, the date the program was written, and a list of version numbers with the modifications in each version. As shown in Figure 4.24, comments can also be used after statements to describe the operations performed by the statements. A well-commented program is important for the maintenance and, thus, for the future lifetime of a program.

```
/* This is a comment line which describes the program

   File:        Example.c
   Version:     1.0
   Author:      Dogan Ibrahim
   Date:        March, 2009

   List of Modifications:

*/

/* Include Files */
#include <p18xxxxx.h>
#include ...
.....

/* Symbol Definitions */
#define MAX 100
#define ...
....

#pragma config WDTEN = OFF

/* Global Variables */
int x,y,z,.....
char a, b, c,......
.....

/* Function Declarations */
int Func(char r)
{
          .............
          .............
          return ...
}

....
....


/* Start of MAIN Program */
void main(void)
{
          ..............
          ..............
          ..............
          ..............
}
```

Figure 4.24: Structure of a C18 Program

In general, any programmer will find it easier to modify and/or update a well-commented program.

As shown in Figure 4.24, long comments start with characters "/*" and terminate with characters "*/". Similarly, short comments start with characters "//"; they can only occupy one line and there is no need to terminate short comments.

4.5.2 Terminating Program Statements

In C language, all program statements must be terminated with the semicolon (";") character, otherwise a compiler error will be generated:

```
j = 5;          // correct
j = 5           // error
```

4.5.3 White Spaces

White spaces are spaces, blanks, tabs, and new-line characters. All white spaces are ignored by the C compiler. Thus, the following three sequences are all identical:

```
int i;          char j;

or

int i;
char j;

or

int i;
      char j;
```

Similarly, the following sequences are identical:

```
i = j + 2;

or

i = j
   + 2;
```

4.5.4 Case Sensitivity

The C language is case sensitive and variables with lowercase names are different from those with uppercase names. Thus, the following variables are all different and they represent different locations in memory:

total TOTAL Total ToTal total totaL

4.5.5 Variable Names

In C language, variable names can begin with an alphabetical character or with an underscore. In essence, variable names can be any of the characters a–z and A–Z, the digits 0–9, and the underscore character "_". Each variable name should be unique within the first 31 characters of its name. Variable names can contain uppercase and lowercase characters, and numeric characters can be used inside a variable name. Examples of valid variable names are as follows:

Sum count sum100 counter il UserName _myName

Some names are reserved for the compiler itself, and they cannot be used as variable names in our programs. Some reserved names are as follows:

int	char	float	static	for	while	switch
long	do	if	else	struct	union	break

4.5.6 Variable Types

C18 language supports the variable types shown in Table 4.1. Examples of variables are given in this section.

unsigned char: These are 8-bit unsigned variables with a range 0–255. In the following example, two 8-bit variables named **total** and **sum** are created and **sum** is assigned a decimal value 150:

```
unsigned char total, sum;
sum = 150;
```

Table 4.1: C18 Variable Types

Types	Size (bits)	Range
char	8	−128 to +127
signed char	8	−128 to +127
unsigned char	8	0 to 255
int	16	−32 768 to +32 767
unsigned int	16	0 to 65 535
short	16	−32 768 to +32 767
unsigned short	16	0 to 65 535
short long	24	−8 388 608 to +8 388 607
unsigned short long	24	0 to 16 777 215
long	32	−2 147 483 648 to +2 147 483 647
unsigned long	32	0 to 4 294 967 295
float	32	±1.17549E-38 to ±6.80565E38
double	32	±1.17549E-38 to ±6.80565E38

Variables can be assigned values during their declaration. Thus, the above statements can also be written as:

```
unsigned char total, sum = 150;
```

char or **signed char:** These are 8-bit signed character variables ranging from –128 to +127. In the following example, a signed 8-bit variable named **counter** is created with a value of –50:

```
signed char counter = –50;
```

or,

```
char counter = –50;
```

or,

```
unsigned char counter;
counter = –50;
```

int or **short:** These are 16-bit signed variables ranging from –32 768 to +32 767. In the following example, a signed integer named **Big** is created and assigned the value 32 000:

```
int Big = 32000;
```

or

```
short Big;
Big = 32000;
```

unsigned int or **unsigned short:** These are 16-bit unsigned variables with a range 0–65 535. In the following example, an unsigned 16-bit variable named **count** is created and is assigned value 12 000:

```
unsigned int count = 12000;
```

or

```
unsigned short count = 12000;
```

short long: These variables are signed and 24 bits long ranging from –8 388 608 to +8 388 607. An example is given below:

```
short long LargeNumber;
short long Big = –50000;
```

or

```
short long BigNo;
BigNo = –25000;
```

unsigned short long: These are 24-bit unsigned variables having the range 0–16 777 215. An example is given below:

```
unsigned short long VeryLargeNumber;
```

long: These are 32-bit signed numbers used in arithmetic operations where large integer numbers are required. The range of long numbers is –2 147 483 648 to +2 147 483 647. In the following example, variable Count is assigned the value 2 000 250 500

```
long Count = 2000250500;
```

unsigned long: These integer numbers are 32 bits wide and unsigned. The range is 0 to +4 294 967 295. In the following example, variable Total is assigned a value 3 250 900 290.

```
unsigned long Total = 3250900290;
```

float or **double:** These are floating point variables implemented using Microchip AN575 32-bit format, which is IEEE 754 compliant. Floating point numbers range from ±1.17549E-38 to ±6.80565E38. In the following example, a floating point variable named **area** is created and is assigned a value 12.235

```
float area;
area = 12.235;
```

4.5.7 Constants

Constants represent fixed values (numeric or character) in programs that cannot be changed. Constants are stored in the flash program memory of the PIC microcontroller; thus, the valuable and limited RAM is not wasted. In C18, constants can be integer, floating point, character, string, or enumerated types. It is important that values must be assigned to constants when they are declared.

Integer Constants

Integer constants can be decimal, hexadecimal, octal, or binary. The data type of a constant is derived by the compiler from its value. However, suffixes can be used to change the type of a constant.

From Table 4.1, we can see that integer variables can be 8, 16, 24, or 32 bit wide. In C18 language, Constants are declared using the key word **const rom**, and they are stored in the flash *program memory* of the PIC microcontroller, thus not wasting any valuable RAM space (it is important to note that in most C languages constants are declared using the key word **const** only). In the following example, constant integer **MAX** is declared as 100 and is stored in the flash program memory of the PIC microcontroller:

```
const rom int MAX = 100;
```

Hexadecimal constants start with the characters 0x and may contain numeric data 0–9 and hexadecimal characters A–F. In the following example, constant **TOTAL** is given the hexadecimal value FF:

```
const rom int TOTAL = 0xFF;
```

Octal constants have a zero at the beginning of the number and may contain numeric data 0–7. In the following example, constant **CNT** is given an octal value 17:

```
const rom int CNT = 017;
```

Binary constant numbers start with 0b and may contain only 0 or 1. In the following example, a constant named **Min** is declared having the binary value "11110000":

```
const rom int Min = 0b11110000
```

Floating Point Constants

Floating point constant numbers have integer parts, a dot, fractional part, and an optional e or E followed by a signed integer exponent. In the following example, a constant named **TEMP** is declared having the fractional value 37.50:

```
const rom TEMP = 37.50
```

or

```
const rom TEMP = 3.750E1
```

Character Constants

A character constant is a character enclosed in a single quote. In the following example, a constant named **First_Alpha** is declared having the character value "A":

```
const rom char First_Alpha = 'A';
```

Character Array Constants

A character array consists of a number of characters stored sequentially in a variable. In the following example, a character array named Product is declared storing the characters COMPUTER:

```
const rom char Product[ ] = {'C', 'O', 'M', 'P', 'U', 'T', 'E', 'R'};
```

Note that it is not necessary to declare the size of the array as this is automatically calculated by the compiler. The above statement could also be written as follows to show that the character array consists of eight characters:

```
const rom char Product[8] = {'C', 'O', 'M', 'P', 'U', 'T', 'E', 'R'};
```

String Constants

String constants are fixed sequences of characters stored in the flash memory of the microcontroller. The string must begin with a double quote character (") and also terminate with a double quote character. The compiler automatically inserts a null character as a terminator. An example string constant is

```
const rom char x = "This is an example string constant";
```

A string constant can be extended across a line boundary using a backslash character ("\"):

```
const rom char x = "This is first part \
  and this is second part"
```

The above string constant declaration is same as

```
const rom char x = "This is first part and this is second part"
```

Enumerated Constants

Enumeration constants are integer type, and they are used to make a program easier to follow. In the following example, the **colors** constant stores the names of colors. The first element is given the value 0 by default but it can be changed if desired:

```
const rom enum colors {black, brown, red, orange, yellow, green, blue, gray, white};
```

where black is assigned 0, brown is assigned 1, red is assigned 2, and so on.

or

```
const rom enum days {Monday = 1, Tuesday, Wednesday, Thursday, Friday, Saturday};
```

where Monday is assigned 1, Tuesday is assigned 2, Wednesday is assigned 3, and so on.

4.5.8 Escape Sequences

Escapes sequences are used to represent nonprintable ASCII characters. Table 4.2 shows some of the commonly used escape sequences and their representation in C language. For example, the character combination "\n" represents the new-line character. An ASCII character can also be represented by specifying its hexadecimal code after a backslash. For example, the new-line character can also be represented as "\x0A."

Table 4.2: Some of the Commonly Used Escape Sequences

Escape Sequences	Hex Values	Characters
\a	0x07	BEL (bell)
\b	0x08	BS (backspace)
\t	0x09	HT (horizontal tab)
\n	0x0A	LF (line feed)
\v	0x0B	VT (vertical feed)
\f	0x0C	FF (form feed)
\r	0x0D	CR (carriage return)
\xH		String of hex digits

4.5.9 Static Variables

Static variables are local variables usually used in functions when it is required to preserve the last value of a variable between successive calls to the function. As shown below, static variables are declared using the key word static:

```
static unsigned char count;
```

4.5.10 External Variables

Using the key word **extern** before a variable name declares that variable as external. This key word indicates to the compiler that the variable is declared elsewhere in a separate source code module. In the following example, variables sum1 and sum2 are declared as external unsigned integer:

```
extern int sum1, sum2;
```

4.5.11 Volatile Variables

Volatile variables are important, especially in interrupt-based programs and in input–output (I/O) routines. Using the key word **volatile** indicates that the value of a variable may change during the lifetime of the program, independent from the normal flow of the program. Variables declared as volatile are not optimized by the compiler since their values can change at any unexpected time. In the following example, variable Led is declared as volatile unsigned char:

```
volatile unsigned char Led;
```

4.5.12 Enumerated Variables

Enumerated variables are used to make a program more readable. In an enumerated variable, a list of items is specified and the value of the first item is set to 0, the next item is set to 1, and so on. In the following example, type Week is declared as an enumerated list and MON = 0, TUE = 1, WED = 2, and so on:

```
enum Week {MON, TUE, WED, THU, FRI, SAT, SUN};
```

It is possible to change the values of the elements in an enumerated list. In the following example, black = 2, blue = 3, red = 4, and so on.

```
enum colors {black = 2, blue, red, white, gray};
```

Similarly, in the following example, black = 2, blue = 3, red = 8, and gray = 9:

```
enum colors {black = 2, blue, red = 8, gray};
```

Variables of type enumeration can be declared by specifying them after the list of items. For example, to declare variable My_Week of enumerated type Week, use the following statement:

```
enum Week {MON, TUE, WED, THU, FRI, SAT, SUN} My_Week;
```

Now, we can use variable My_Week in our programs:

```
My_Week = WED          // assign 2 to My_Week
```

or

```
My_Week = 2            // same as above
```

After defining the enumeration type Week, we can declare variables This_Week and Next_Week of type Week as

```
enum Week This_Week, Next_Week;
```

4.5.13 Arrays

Arrays are used to store related items together and sequentially in the same block of memory and under a specified name. An array is declared by specifying its type, name, and the number of elements it will store. For example,

```
unsigned int Total[5];
```

creates an array of type unsigned int, with name Total and has five elements. The first element of an array is indexed with 0. Thus, in the above example, Total[0] refers to the first element of this array and Total[4] refers to the last element. The array total is stored in memory in five consecutive locations as follows:

Total[0]
Total[1]
Total[2]
Total[3]
Total[4]

Data can be stored in the array by specifying the array name and index. For example, to store 25 in the second element of the array, we have to write

```
Total[1] = 25;
```

Similarly, the contents of an array can be read by specifying the array name and its index. For example, to copy the third array element to a variable called temp, we have to write

```
Temp = Total[2];
```

The contents of an array can be initialized during the declaration of the array by assigning a sequence of comma delimited values to the array. An example is given below where array months has 12 elements and months[0] = 31, months[1] = 28, and so on.:

```
unsigned char months[12] = {31,28,31,30,31,30,31,31,30,31,30,31};
```

The above array can also be declared without specifying the size of the array:

```
unsigned char months[ ] = {31,28,31,30,31,30,31,31,30,31,30,31};
```

Character arrays can be declared similarly. In the following example, a character array named Hex_Letters is declared with six elements:

```
unsigned char Hex_Letters[ ] = {'A', 'B', 'C', 'D', 'E', 'F'};
```

Strings are character arrays with a null terminator. Strings can either be declared by enclosing the string in double quotes or by specifying each character of the array within single quotes and can then be terminated with a null character. In the following example, the two string declarations are identical and both occupy five locations in memory:

```
unsigned char Mystring[ ] = "COMP";
```

and

```
unsigned char Mystring[ ] = {'C', 'O', 'M', 'P', '\0'};
```

In C programming language, we can also declare arrays with multiple dimensions. One-dimensional arrays are usually called vectors and two-dimensional arrays are called matrices. A two-dimensional array is declared by specifying the data type of the array, array name, and the size of each dimension. In the following example, a two-dimensional array named P having three rows and four columns is created. Altogether the array has 12 elements. The first element of the array is P[0][0] and the last element is P[2][3]. The structure of this array is shown below:

P[0][0]	P[0][1]	P[0][2]	P[0][3]
P[1][0]	P[1][1]	P[1][2]	P[1][3]
P[2][0]	P[2][1]	P[2][2]	P[2][3]

Elements of a multidimensional array can be specified during the declaration of the array. In the following example, two-dimensional array Q has two rows and two columns and its diagonal elements are set to one and nondiagonal elements are cleared to zero:

```
unsigned char Q[2][2] = { {1,0}, {0,1} };
```

4.5.14 Pointers

Pointers are an important part of the C language and they hold the memory addresses of variables. Pointers are declared same as any other variables but with the character ("*") before the variable name. In general, pointers can be created to point to (or hold the addresses of) character variables, integer variables, long variables, floating point variables, or they can point to functions.

In the following example, an unsigned character pointer named pnt is declared:

```
unsigned char *pnt;
```

When a new pointer is created, its content is initially unspecified and it does not hold the address of any variable. We can assign the address of a variable to a pointer using the ("&") character:

```
pnt = &Count;
```

Now pnt holds the address of variable Count. Variable Count can be set to a value using the character ("*") before its pointer. For example, Count can be set to 10 using its pointer:

```
*pnt = 10;                  // Count = 10
```

which is same as

```
Count = 10;                 // Count = 10
```

or the value of Count can be copied to variable Cnt using its pointer:

```
Cnt = *pnt;                 // Cnt = Count
```

Array Pointers

In C language, the name of an array is also a pointer to the array. Thus, for the array

```
unsigned int Total[10];
```

The name Total is also a pointer to this array and it holds the address of the first element of the array. Thus, the following two statements are equal:

```
Total[2] = 0;
```

and

```
*(Total + 2) = 0;
```

Also, the following statement is true:

```
&Total[j] = Total + j
```

In C language, we can perform pointer arithmetic that may involve the following:

- Comparing two pointers
- Adding or subtracting a pointer and an integer value
- Subtracting two pointers
- Assigning one pointer to another one
- Comparing a pointer to null

For example, let us assume that pointer P is set to hold the address of array element Z[2]

```
P = &Z[2];
```

We can now clear elements 2 and 3 of array Z as in the following two examples. The two examples are identical except that in the first example pointer P holds the address of Z[3] at the end of the statements and it holds the address of Z[2] at the end of the second set of statements:

```
*P = 0;            // Z[2] = 0
 P = P + 1;        // P now points to element 3 of Z
*P = 0;            // Z[3] = 0
```

or

```
*P = 0;            // Z[2] = 0
*(P + 1) = 0;      // Z[3] = 0
```

A pointer can be assigned to another pointer. An example is given below where both variables Cnt and Tot are set to 10 using two different pointers:

```
unsigned int *i, *j;         // declare 2 pointers
unsigned int Cnt, Tot;       // declare two variables
i = &Cnt;                    // i points to Cnt
*i = 10;                     // Cnt = 10
j = i;                       // copy pointer i to pointer j
Tot = *j;                    // Tot = 10
```

Incrementing Pointers

It is important to realize that when a pointer is incremented, it is scaled by the size of the object it points to. For example, if the pointer is of type **long**, then incrementing the pointer will increment its value by 4 since a **long** is 4 bytes long. An example is given below where a long array called w is declared with three elements and then each element is cleared to zero using pointers. Notice that the statement $p = p + 1$ increments the value of p by 4 and not by 1:

```
long w[5] = {1, 2, 3};       // Declare array w
long p;                      // Declare a long pointer

p = w;                       // Point to array w
*p = 0;                      // Clear first element to 0
p = p + 1;                   // Point to next element
*p = 0;                      // Clear second element to 0
p = p + 1;                   // Point to next element
*p = 0;                      // Clear third element to zero
```

4.5.15 Structures

Structures can be used to collect related items as single objects. Unlike arrays, the members of structures can be a mixture of any data type. For example, a structure can be created to store the personal details (name, surname, age, date of birth, etc.) of a student.

A structure is created using the key word **struct**, followed by a structure name, and a list of member declarations. Optionally, variables of the same type as the structure can be declared at the end of the structure.

The following example declares a structure named Person:

```
struct Person
{
    unsigned char name[20];
    unsigned char surname[20];
    unsigned char nationality[20];
    unsigned char age;
}
```

Declaring a structure does not occupy any space in memory, but the compiler creates a template describing the names and types of the data objects or member elements that will eventually be stored within such a structure variable. It is only when variables of the same type as the structure are created, then these variables occupy space in memory. We can declare variables of the same type as the structure by giving the name of the structure and the name of the variable. For example, two variables **Me** and **You** of type Person can be created by the statement:

```
struct Person Me, You;
```

Variables of type Person can also be created during the declaration of the structure as shown below:

```
struct Person
{
    unsigned char name[20];
    unsigned char surname[20];
    unsigned char nationality[20];
    unsigned char age;
} Me, You;
```

We can assign values to members of a structure by specifying the name of the structure, followed by a dot ("."), and the name of the member. In the following example, the **age** of structure variable **Me** is set to 25 and variable *M* is assigned to the value of **age** in structure variable **You**.

```
Me.age = 25;
M = You.age;
```

Structure members can be initialized during the declaration of the structure. In the following example, the radius and height of structure Cylinder are initialized to 1.2 and 2.5, respectively.

```
struct Cylinder
{
    float radius;
    float height;
} MyCylinder = {1.2, 2.5};
```

Values can also be set to members of a structure using pointers by defining the variable types as pointers. For example, if **TheCylinder** is defined as a pointer to structure Cylinder, then we can write

```
struct Cylinder
{
    float radius;
    float height;
} *TheCylinder;

TheCylinder -> radius = 1.2;
TheCylinder -> height = 2.5;
```

The size of a structure is the number of bytes contained within the structure. We can use the **sizeof** operator to get the size of a structure. Considering the above example,

```
sizeof(MyCylinder)
```

returns 8, since each float variable occupies 4 bytes in memory.

Bit fields can be defined using structures. With bit fields, we can assign identifiers to bits of a variable. For example, to identify bits 0, 1, 2, and 3 of a variable as **LowNibble** and to identify the remaining 4 bits as **HighNibble**, we can write

```
struct
{
    LowNibble: 4;
    HighNibble: 4;
} MyVariable;
```

We can then access the nibbles of variable MyVariable as

```
MyVariable.LowNibble = 12;
MyVariable.HighNibble = 8;
```

In C language, we can use the **typedef** statements to create new types of variables. For example, a new structure data type named **Reg** can be created as follows:

```
typedef struct
{
    unsigned char name[20];
    unsigned char surname[20];
    unsigned age;
} Reg;
```

Variables of type **Reg** can then be created in exactly the same way as creating any other types of variables. In the following example, variables **MyReg**, **Reg1**, and **Reg2** are created from data type **Reg**:

```
Reg MyReg, Reg1, Reg2;
```

The contents of one structure can be copied to another structure provided that both structures have been derived from the same template. In the following example, two structure variables *P*1 and *P*2 of same type have been created and *P*2 is copied to *P*1:

```
struct Person
{
    unsigned char name[20];
    unsigned char surname[20];
    unsigned int age;
    unsigned int height;
    unsigned weight;
}

struct Person P1, P2;
........................
........................
P2 = P1;
```

4.5.16 Unions

Unions are used to overlay variables. A union is similar to a structure, and it is even defined in a similar manner; both are based on templates and their members are accessed using the "." or "->" operators. The difference of a union is that all variables in a union occupy the same memory area. In other words, all member elements of a union share the same common storage. An example for union declaration is given below:

```
union flags
{
    unsigned char x;
    unsigned int y;
} P;
```

In this example, variables *x* and *y* occupy the same memory area, and the size of this union is 2 bytes long, which is the size of the biggest member of the union. When variable *y* is loaded with a 2-byte value, variable *x* will have the same value as the low byte of *y*. In the following example, *y* is loaded with 16-bit hexadecimal value 0xAEFA and *x* is loaded with 0xFA:

```
P.y = 0xAEFA;
```

The size of a union is the size (number of bytes) of its largest member. Thus, the statement

```
sizeof(P)
```

returns 2.

The above union could also have been declared as

```
union flags
{
    unsigned char x;
    unsigned int y;
}

union flags P;
```

4.5.17 Operators in C

Operators are applied to variables and other objects in expressions, and they cause some conditions or some computations to occur.

C18 language supports the following operators:

- Arithmetic operators
- Logical operators
- Bitwise operators
- Conditional operators
- Assignment operators
- Relational operators
- Preprocessor operators

Arithmetic Operators

Arithmetic operators are used in arithmetic computations. Arithmetic operators associate from left to right, and they return numerical results. A list of the C18 arithmetic operators is given in Table 4.3.

Table 4.3: C18 Arithmetic Operators

Operators	Operations
+	Addition
−	Subtraction
*	Multiplication
/	Division
%	Remainder (integer division)
++	Autoincrement
−−	Autodecrement

Example for use of arithmetic operators is given below:

```
/* Adding two integers */
5 + 12                              // equals 17

/* Subtracting two integers */
120 – 5                             // equals 115
10 – 15                             // equals –5

/* Dividing two integers */
5 / 3                               // equals 1
12 / 3                              // equals 4

/* Multiplying two integers */
3 * 12                              // equals 36

/* Adding two floating point numbers */
3.1 + 2.4                           // equals 5.5

/* Multiplying two floating point numbers */
2.5 * 5.0                           // equals 12.5

/* Dividing two floating point numbers */
25.0 / 4.0                          // equals 6.25

/* Remainder (not for float) */
7 % 3                               // equals 1

/* Post-increment operator */
j = 4;
k = j++;                            // k = 4, j = 5

/* Pre-increment operator */
j = 4;
k = ++j;                            // k = 5, j = 5

/* Post-decrement operator */
j = 12;
k = j––;                            // k = 12, j = 11

/* Pre-decrement operator */
j = 12;
k = ––j;                            // k = 11, j = 11
```

Relational Operators

Relational operators are used in comparisons. If the expression evaluates to TRUE, 1 is returned otherwise 0 is returned.

All relational operators associate from left to right, and a list of mikroC relational operators is given in Table 4.4.

Example for use of relational operators is given below:

```
x = 10
x > 8        // returns 1
x = = 10     // returns 1
x < 100      // returns 1
x > 20       // returns 0
x != 10      // returns 0
x >= 10      // returns 1
x <= 10      // returns 1
```

Logical Operators

Logical operators are used in logical and arithmetic comparisons, and they return TRUE (i.e., logical 1) if the expression evaluates to nonzero or FALSE (i.e., logical 0) if the expression evaluates to zero. If more than one logical operator is used in a statement and if the first condition evaluates to false, the second expression is not evaluated.

A list of the C18 logical operators is given in Table 4.5.

Example for use of logical operators is given below:

```
x = 7;

x > 0 && x < 10       // returns 1
x > 0 || x < 10       // returns 1
```

Table 4.4: C18 Relational Operators

Operators	Operations
= =	Equal to
!=	Not equal to
>	Greater than
<	Less than
>=	Greater than or equal to
<=	Less than or equal to

Table 4.5: C18 Logical Operators

Operators	Operations
&&	AND
||	OR
!	NOT

```
x >=0 && x <=10                 // return 1
x >=0 && x < 5                  // returns 0

a = 10; b = 20; c = 30; d = 40;

a > b && c > d                  // returns 0
b > a && d > c                  // returns 1
a > b || d > c                  // returns 1
```

Bitwise Operators

Bitwise operators are used to modify the bits of a variable. A list of the C18 bitwise operators is given in Table 4.6.

Bitwise AND returns 1 if both bits are 1, else it returns 0.

Bitwise OR returns 0 if both bits are 0, otherwise it returns 1.

Bitwise XOR returns 1 if both bits are complementary, otherwise it returns 0.

Bitwise complement inverts each bit.

Bitwise shift left and shift right move the bits to the left or right, respectively.

Example for use of bitwise operators is given below:

i. 0xFA & 0xEE returns 0xEA

```
0xFA: 1111 1010
0xEE: 1110 1110
---------------------
0xEA: 1110 1010
```

ii. 0x01 | 0xFE returns 0xFF

```
0x08: 0000 0001
0xFE: 1111 1110
---------------------
0xFE: 1111 1111
```

Table 4.6: C18 bitwise operators

Operators	Operations
&	Bitwise AND
|	Bitwise OR
^	Bitwise EXOR
~	Bitwise complement
<<	Shift left
>>	Shift right

iii. 0xAA ^ 0x1F returns 0xB5

```
0xAA: 1010 1010
0x1F:  0001 1111
--------------------
0xB5: 1011 0101
```

iv. ~0xAA returns 0x55

```
0xAA: 1010 1010
~     : 0101 0101
-------------------
0x55: 0101 0101
```

v. 0x14 >> 1 returns 0x08 (shift 0x14 right by 1 digit)

```
0x14: 0001 0100
>>1 : 0000 1010
--------------------
0x0A: 0000 1010
```

vi. 0x14 >> 2 returns 0x05 (shift 0x14 right by 2 digits)

```
0x14: 0001 0100
>> 2: 0000 0101
--------------------
0x05: 0000 0101
```

vii. 0x235A << 1 returns 0x46B4 (shift left 0x235A left by 1 digit)

```
0x235A: 0010 0011 0101 1010
<<1     : 0100 0110 1011 0100
-----------------------------------
0x46B4 : 0100 0110 1011 0100
```

viii. 0x1A << 3 returns 0xD0 (shift left 0x1A by 3 digits)

```
0x1A: 0001 1010
<<3 : 1101 0000
--------------------
0xD0: 1101 0000
```

Assignment Operators

In C language, there are two types of assignments: simple assignments and compound assignments. In simple assignments, an expression is simply assigned to another expression or an operation is performed using an expression and the result is assigned to another expression:

```
Expression1 = Expression2
```

or

```
Result = Expression1 operation Expression2
```

Examples of simple assignments are

```
Temp = 10;
Cnt = Cnt + Temp;
```

Compound assignments have the general format:

```
Result operation = Expression1
```

Here, the specified operation is performed on Expression1 and the result is stored in **Result**. For example,

$$j += k; \quad \text{is same as} \quad j = j + k;$$

also,

$$p *= m; \quad \text{is same as} \quad p = p * m;$$

The following compound operators can be used in C18 programs:

+=	−=	*=	/=	%=
&=	|=	^=	>>=	<<=

Conditional Operator

The syntax of the conditional operator is

Result = Expression1 ? Expression2: Expression3

Expression1 is evaluated first and if its value is true, Expression2 is assigned to Result, otherwise Expression3 is assigned to Result. In the following example, maximum of x and y is found, where x is compared with y, and if $x > y$, then max $= x$, otherwise max $= y$

```
max=(x>y)?x:y;
```

In the following example, lowercase characters are converted to uppercase. If the character is lowercase (between "a" and "z"), then by subtracting 32 from the character, we obtain the equivalent uppercase character:

```
c=(c>='a'&&c<='z')?(c-32):c;
```

Preprocessor Operators

The preprocessor allows a programmer to

- Compile a program conditionally such that parts of the code are not compiled
- Replace symbols with other symbols or values
- Insert text files into a program

The preprocessor operator is the ("#") character, and any line of code with a leading ("#") is assumed to be a preprocessor command. Semicolon character (";") is not needed to terminate a preprocessor command.

C compiler supports the following preprocessor commands:

```
#define     #undef
#if         #elif       #endif
#ifdef      #ifndef
#error
#line
#pragma
```

#define, #undef, #ifdef, #ifndef

#define preprocessor command provides Macro expansion where every occurrence of an identifier in the program is replaced with the value of the identifier. For example, to replace every occurrence of MAX with value 100, we can write

```
#define MAX 100
```

An identifier that has already been defined cannot be defined again unless both definitions have the same values. One way to get round this problem is to remove the Macro definition:

```
#undef MAX
```

or the existence of a Macro definition can be checked. In the following example, if **MAX** has not already been defined, then it is given value 100, otherwise the **#define** line is skipped:

```
#ifndef MAX
   #define MAX 100
#endif
```

Note that the **#define** preprocessor command does not occupy any space in memory.

We can pass parameters to a Macro definition by specifying the parameters in parenthesis after the Macro name. For example, consider the Macro definition

```
#define ADD(a, b) (a + b)
```

When this Macro is used in a program, (a,b) will be replaced with (a + b) as shown below:

```
p = ADD(x, y)    will be transformed into p = (x + y)
```

Similarly, we can define a Macro to calculate the square of two numbers:

```
#define SQUARE(a) (a * a)
```

When we now use this Macro in a program,

p = SQUARE(x) will be transformed into p = (x * x)

#include

The preprocessor directive **#include** is used to include a source file in our program. Usually, header files with extension ".h" are used with **#include**. There are two formats for using the **#include**:

```
#include <file>
```

and

```
#include "file"
```

In the first option, the file is searched in the C18 installation directory first and then in the user search paths. In the second option, the specified file is searched in the C18 project folder, then in the C18 installation folder, and then in the user search paths. It is also possible to specify a complete directory path as

```
#include "C:\temp\last.h"
```

The file is then searched only in the specified directory path.

#if, #elif, #else, #endif

The above preprocessor commands are used for conditional compilation where parts of the source code can be compiled only if certain conditions are met. In the following example, the code section where variables *A* and *B* are cleared to zero is compiled if *M* has a nonzero value, otherwise the code section where *A and B* are both set to 1 is compiled. Notice that the **#if** must be terminated with **#endif**:

```
#if M
    A = 0;
    B = 0;
#else
    A = 1;
    B = 1;
#endif
```

We can also use the **#elif** condition that tests for a new condition if the earlier condition was false:

```
#if M
    A = 0;
    B = 0;
#elif N
    A = 1;
    B = 1;
```

```
#else
    A = 2;
    B = 2;
#endif
```

In the above example, if M has a nonzero value code section, then $A = 0$; $B = 0$; is compiled. If N has a nonzero value, then code section $A = 1$; $B = 1$; is compiled. Finally, if both M and N are zero, then code section $A = 2$; $B = 2$; is compiled. Notice that only one code section is compiled between **#if** and **#endif** and a code section can contain any number of statements.

#pragma

The **#pragma** directive is used to define device-specific constructs. This directive is used with the key words as shown below:

- #pragma code
- #pragma romdata
- #pragma idata
- #pragma config
- #pragma interrupt
- #pragma interruptflow
- #pragma varlocate

#pragma code is used to instruct the compiler to compile all subsequent instructions into the program memory section of the target processor.

#pragma romdata is used to instruct the compiler to compile the subsequent static data into the program memory of the target processor.

#pragma udata is used to locate the uninitialized user variables in data memory.

#pragma idata is used to locate the initialized user variables in data memory.

#pragma interrupt is used to instruct the compiler to compile the code from the named C function as a high-priority interrupt service routine.

#pragma interruptflow is used to instruct the compiler to compile the named C function as a low-priority interrupt service routine.

#pragma varlocate is used to specify where the variables will be located so that the compiler will not generate extraneous instructions to set the bank when accessing the variables.

#pragma config is used to define the processor Configuration bits. Document "*PIC18 Configuration Settings Addendum*" of Microchip Inc. gives tables of all the configuration bits available for any member of the PIC18 family. Some of the widely used configuration bit definitions for the popular PIC18F452 microcontroller are shown in Table 4.7. As an example, to disable the watchdog and set the oscillator to XT type crystal operation, we can use the following statement:

```
#pragma config WDT = OFF
#pragma config OSC = XT
```

It is permissible to combine the various settings on a single line, i.e.,

```
#pragma config WDT = OFF, OSC = XT
```

Table 4.7: Widely Used PIC18F452 Microcontroller Configuration Bits

Oscillator Selection	
OSC = LP	LP
OSC = XT	XT
OSC = HS	HS
OSC = RC	RC
OSC = EC	EC and OSC2 as clock out
OSC = ECIO	EC and OSC2 as port RA6
OSC = HSPLL	HS-PLL enabled
OSC = RCIO	RC and OSC2 as port RA6
Power-up Timer	
PWRT = ON	Timer enabled
PWRT = OFF	Timer disabled
Watchdog Timer	
WDT = OFF	Watchdog disabled
WDT = ON	Watchdog enabled
Watchdog Postscaler	
WDTPS = 1	1:1
WDTPS = 2	1:2
WDTPS = 4	1:4
WDTPS = 8	1:8
WDTPS = 16	1:16
WDTPS = 32	1:32
WDTPS = 64	1:64
WDTPS = 128	1:128

4.5.18 Modifying the Flow of Control

Statements are normally executed sequentially from the beginning to the end of a program. We can use control statements to modify the normal sequential flow of control in a C program. The following control statements are available in C18 programs:

- Selection statements
- Unconditional modification of flow
- Iteration statements

Selection Statements

There are two selection statements: **If** and **switch.**

If Statement

The general format of the **if** statement is

```
if(expression)
        Statement1;
else
        Statement2;
```

or,

```
if(expression)Statement1; else Statement2;
```

If the expression evaluates to TRUE, Statement1 is executed, otherwise Statement2 is executed. The **else** key word is optional and may be omitted if not required. In the following example, if the value of *x* is greater than **MAX**, then variable *P* is incremented by 1, otherwise it is decremented by 1:

```
if(x > MAX)
    P++;
else
    P--;
```

We can have more than one statement by enclosing the statements within curly brackets. For example,

```
if(x > MAX)
{
    P++;
    Cnt = P;
    Sum = Sum + Cnt;
}
else
    P--;
```

In the above example, if *x* is greater than **MAX**, then the three statements within the curly brackets are executed, otherwise the statement *P--* is executed.

Another example using the **if** statement is given below:

```
if(x > 0 && x < 10)
{
    Total += Sum;
    Sum++;
}
else
{
    Total = 0;
    Sum = 0;
}
```

switch Statement

The **switch** statement is used when there are a number of conditions, and different operations are performed when a condition is true. The syntax of the **switch** statement is

```
switch (condition)
{
    case condition1:
        Statements;
        break;
    case condition2:
        Statements;
        break;
    .....................
    .....................
    case conditionn:
        Statements;
        break;
    default:
        Statements;
}
```

The **switch** statement functions as follows: First, the **condition** is evaluated. The **condition** is then compared to **condition1**, and if a match is found, statements in that case block are evaluated and control jumps outside the **switch** statement when the **break** key word is encountered. If a match is not found, **condition** is compared to **condition2**, and if a match is found, statements in that case block are evaluated and control jumps outside the switch statements and so on. The **default** is optional and statements following **default** are evaluated if the **condition** does not match to any of the conditions specified after the **case** key words.

In the following example, the value of variable **Cnt** is evaluated. If **Cnt** = 1, *A* is set to 1, if **Cnt** = 10, *B* is set to 1, and if **Cnt** = 100, *C* is set to 1. If **Cnt** is not equal to 1, 10, or 100, then *D* is set to 1:

```
switch (Cnt)
{
```

```
        case 1:
            A = 1;
            break;
        case 10:
            B = 1;
            break;
        case 100:
            C = 1;
            break;
        default:
            D = 1;
    }
```

Because white spaces are ignored in C language, we could also write the above code as

```
    switch (Cnt)
    {
        case 1:         A = 1; break;
        case 10:        B = 1; break;
        case 100:       C = 1; break;
        default:        D = 1;
    }
```

Example 4.1

In an experiment, the relationship between *X* and *Y* values are found to be

X	*Y*
1	3.2
2	2.5
3	8.9
4	1.2
5	12.9

Write a switch statement that will return the *Y* value, given the *X* value.

Solution

The required switch statement is

```
switch (X)
{
    case 1:
        Y = 3.2;
        break;
    case 2:
        Y = 2.5;
```

```
        break;
    case 3:
        Y = 8.9;
        break;
    case 4:
        Y = 1.2;
        break;
    case 5:
        Y = 12.9;
}
```

■

4.5.19 Iteration Statements

Iteration statements enable us to perform loops in our programs, where part of a code is repeated required number of times. In C18 language, there are four ways that iteration can be performed, and we will look at each one with examples:

- Using **for** statement
- Using **while** statement
- Using **do** statement
- Using **goto** statement

for *Statement*

The syntax of the **for** statement is

```
for(initial expression; condition expression; increment expression)
{
    Statements;
}
```

The **initial expression** sets the starting variable of the loop, and this variable is compared against the **condition expression** before an entry to the loop. Statements inside the loop are executed repeatedly, and after each iteration, the value of **increment expression** is incremented. The iteration continues until the **condition expression** becomes false. An endless loop is formed if the **condition expression** is always true.

The following example shows how a loop can be set up to execute 10 times. In this example, variable *i* starts from 0 and increments by 1 at the end of each iteration. The loop terminates when $i = 10$ in which case the condition $i < 10$ becomes false. On exit from the loop, the value of *i* is 10:

```
for(i = 0; i < 10; i ++)
{
    statements;
}
```

The above loop could also be formed by starting the **initial expression** with a nonzero value. Here, *i* starts with 1 and the loop terminates when $i = 11$. Thus, on exit from the loop, the value of *i* is 11:

```
for(i = 1; i <= 10; i++)
{
    Statements;
}
```

The parameters of a **for** loop are all optional and can be omitted. If the **condition expression** is left out, it is assumed to be true. In the following example, an endless loop is formed where the **condition expression** is always true and the value of *i* starts with 0 and is incremented after each iteration:

```
/* Endless loop with incrementing i */
for(i = 0; ; i++)
{
    Statements;
}
```

Another example of an endless loop is given below where all the parameters are omitted:

```
/* Example of endless loop */
for(; ;)
{
    Statements;
}
```

In the following endless loop, *i* starts with 1 and is not incremented inside the loop:

```
/* Endless loop with i = 1 */
for(i = 1; ;)
{
    Statements;
}
```

If there is only one statement inside the **for** loop, we can omit the curly brackets as shown in the following example:

```
for(k = 0; k < 10; k++)Total = Total + Sum;
```

Nested for loops can be used in programs. In a nested for loop, the inner loop is executed for each iteration of the outer loop. An example is given below where the inner loop is executed five times and the outer loop is executed 10 times. The total iteration count is 50:

```
/* Example of nested for loops */
for(i = 0; i < 10; i++)
{
    for(j = 0; j < 5; j++)
    {
```

```
        Statements;
    }
}
```

In the following example, the sum of all the elements of a 3 × 4 matrix M is calculated and stored in a variable called Sum:

```
/* Add all elements of a 3x4 matrix */
Sum = 0;
for(i = 0; i < 3; i++)
{
    for(j = 0; j < 4; j++)
    {
        Sum = Sum + M[i][j];
    }
}
```

Because there is only one statement to be executed, the above example could also be written as

```
/* Add all elements of a 3x4 matrix */
Sum = 0;
for(i = 0; i < 3; i++)
{
    for(j = 0; j < 4; j++) Sum = Sum + M[i][j];
}
```

while *Statement*

This is another statement that can be used to create iteration in programs. The syntax of the **while** statement is

```
while (condition)
{
    Statements;
}
```

Here, the statements are executed repeatedly until the **condition** becomes false or the statements are executed repeatedly as long as the **condition** is true. If the **condition** is false on entry into the loop, then the loop will not be executed and the program will continue from the end of the **while** loop. It is important that the **condition** is changed inside the loop; otherwise an endless loop will be formed.

The following code shows how to set up a loop to execute 10 times using the **while** statement:

```
/* A loop that executes 10 times */
k = 0;
while (k < 10)
```

```
{
    Statements;
    k++;
}
```

At the beginning of the code, variable *k* is 0. Because *k* is less than 10, the **while** loop starts. Inside the loop, the value of *k* is incremented by 1 after each iteration. The loop repeats as long as $k < 10$ and is terminated when $k = 10$. At the end of the loop, the value of *k* is 10.

Notice that an endless loop will be formed if *k* is not incremented inside the loop:

```
/* An endless loop */
k = 0;
while (k < 10)
{
    Statements;
}
```

An endless loop can also be formed by setting the **condition** to be always true:

```
/* An endless loop */
while (k = k)
{
    Statements;
}
```

Here is an example of calculating the sum of numbers from 1 to 10 and storing the result in variable called **sum**:

```
/* Calculate the sum of numbers from 1 to 10 */
unsigned int k, sum;
k = 1;
sum = 0;
while(k <= 10)
{
    sum = sum + k;
    k++;
}
```

It is possible to have a **while** statement with no body. Such a statement is useful, for example, if we are waiting for an input port to change its value. An example is given below where the program will wait as long as bit 0 of PORTB (RB0) is at logic 0. The program will continue when the port pin changes to logic 1:

```
while(PORTBbits.RB0 == 0);          // Wait until RB0 becomes 1
```

or

```
while(PORTBbits.RB0);
```

It is possible to have nested **while** statements.

As we shall see later, the bits of a port can be accessed using the key word **bits** after the port name, followed by the port bit name to be accessed.

do *Statement*

The **do** statement is similar to the **while** statement, but here, the loop executes until the **condition** becomes false or the loop executes as long as the **condition** is true. The **condition** is tested at the end of the loop. The syntax of the **do** statement is

```
do
{
    Statements;
} while (condition);
```

The first iteration is always performed whether the **condition** is true or false, and this is the main difference between the **while** statement and the **do** statement.

The following code shows how to setup a loop to execute 10 times using the **do** statement:

```
/* Execute 10 times */
k = 0;
do
{
    Statements;
    k++;
} while (k < 10);
```

The loop starts with $k = 0$, and the value of k is incremented inside the loop after each iteration. k is tested at the end of the loop, and if k is not less than 10, the loop terminates. In this example, because $k = 0$ at the beginning of the loop, the value of k is 10 at the end of the loop.

An endless loop will be formed if the condition is not modified inside the loop as shown in the following example. Here, k is always less than 10:

```
/* An endless loop */
k = 0;
do
{
    Statements;
} while (k < 10);
```

An endless loop can also be created if the condition is set to be true all the time:

```
/* An endless loop */
do
{
    Statements;
} while (k = k);
```

It is possible to have nested **do** statements.

goto *Statement*

The **goto** statement can be used to alter the normal flow of control in a program. This statement causes the program to jump to a specified label. A label can be any alphanumeric character set starting with a letter and terminating with the colon (":") character.

Although not recommended, the **goto** statement can be used together with the **if** statement to create iterations in a program. The following example shows how to setup a loop to execute 10 times using the **goto** and **if** statements:

```
    /* Execute 10 times */
    /k = 0;
Loop:
    /Statements;
    /k++;
    /if(k < 10)goto Loop;
```

The loop starts with label **Loop** and variable $k = 0$ at the beginning of the loop. Inside the loop, the statements are executed and k is incremented by 1. The value of k is then compared with 10 and the program jumps back to label **Loop** if $k < 10$. Thus, the loop is executed 10 times until the condition at the end becomes false. At the end of the loop, the value of k is 10.

continue *and* break *Statements*

The **continue** and **break** statements can be used inside iterations to modify the flow of control. The **continue** statement is usually used with **if** statement and causes the loop to skip an iteration. An example is given below, which calculates the sum of numbers from 1 to 10 except number 5:

```
/* Calculate sum of numbers 1,2,3,4,6,7,8,9,10 */
Sum = 0;
i = 1;
for(i = 1; i <= 10; i++)
{
    if(i == 5) continue;  // Skip number 5
    Sum = Sum + i;
}
```

Similarly, the **break** statement can be used to terminate a loop from inside the loop. In the following example, the sum of numbers from 1 to 5 are calculated even though the loop parameters are set to iterate 10 times:

```
/* Calculate sum of numbers 1,2,3,4,5 */
Sum = 0;
i = 1;
for(i = 1; i <= 10; i++)
{
```

```
    if(i > 5) break;                    // Stop loop if i > 5
    Sum = Sum + i;
}
```

4.5.20 Mixing C18 with Assembly Language Statements

It sometimes becomes necessary to mix PIC microcontroller assembly language statements with the C18 language statements in a program. For example, very accurate program delays can be generated using assembly language statements. Use of the assembly language is beyond the scope of this book, but the techniques for including assembly language instructions in C18 programs will be discussed in this section for those readers who are familiar with using the PIC microcontroller assembly languages.

Assembly language instructions can be included in a C18 program by starting the code, using the key word **_asm** and terminating with **_endasm**. This process is known as **inline assembly** and it differs from full assembly (e.g., using **MPASM** assembler) as follows:

- Comments must be in C18 format
- Directives are not allowed
- All operands must be specified (no defaults allowed)
- Literals are specified using C radix notation (e.g., 0x12 and not H12)
- Default radix is decimal
- Indexed addressing (e.g., [ ]) is not supported
- Labels must end with a colon

An example inline assembly program is given below:

```
_asm
    /* This assembly code introduces delay to the program*/
    MOVLW 6                         // Load W with 6
    .............
    .............
done:
    .............
    .............
    GOTO done
_endasm
```

User declared C variables can be used in assembly language routines. For example, C variable **Temp** can be initialized and then loaded to the **W** register as

```
unsigned char Temp = 10;
_asm
```

```
    MOVLW Temp                    // W = Temp = 10
    ..................
    ..................
_endasm
```

Global symbols, such as the predefined port names and register names, can be used in assembly language routines without having to initialize them:

```
_asm
    MOVWF PORTB, 1
    ......................
    ......................
_endasm
```

4.6 PIC Microcontroller I/O Port Programming

Depending on the type of microcontroller used, PIC microcontroller I/O ports are named as PORTA, PORTB, PORTC, and so on. Port pins can be in analog or digital mode. In analog mode, ports are input only and a built-in analog to digital converter and multiplexer circuits are used. In digital mode, a port pin can be configured as either input or output. The TRIS registers control the port directions and there are TRIS registers for each port, namely TRISA, TRISB, TRISC, and so on. Clearing a TRIS register bit to 0 sets the corresponding port bit to output mode. Similarly, setting a TRIS register bit to 1 sets the corresponding port bit to input mode.

Ports can be accessed either as a single 8-bit register or as individual bits. In the following example, PORTB is configured as an output port and all its bits are set to 1:

```
TRISB = 0;                 // Set PORT B as output
PORTB = 0xFF;              // Set PORTB bits to 1
```

Similarly, the following example shows how four upper bits of PORTC can be set as input and how upper 4 bits of PORTC can be set as output:

```
TRISC = 0xF0;
```

Bits of an I/O port can be accessed using the key word **bits** and then specifying the required port bit name. In the following example, variable *P2* is loaded with bit 2 of PORTB:

```
P2 = PORTBbits.RB2;
```

All the bits of a port can be complemented by the statement:

```
PORTB = ~PORTB;
```

4.7 Programming Examples

In this section, some simple programming examples are given to make the reader familiar with programming in C18 language. In all the following examples, the processor header file must be included at the beginning of the program (e.g., **#include <p18f452.h>**)

Example 4.2

Write a program to set all eight port pins of PORTB to logic 1.

Solution

The required program is given below. PORTB is configured as an output port and then all port pins are set to logic 1 by sending hexadecimal number 0xFF:

```
void main(void)
{
    TRISB = 0;                  // Configure PORT B as output
    PORTB = 0xFF;               // Set all port pins to logic 1
}
```

■

Example 4.3

Write a program to set the odd numbered (bits 1, 3, 5, and 7) PORTB pins to logic 1.

Solution

Odd numbered port pins can be set to logic 1 by sending the bit pattern "10101010" to the port. This bit pattern is the hexadecimal number 0xAA and the required program is

```
void main(void)
{
    TRISB = 0;                  // Configure PORTB as output
    PORTB = 0xAA;               // Turn on odd numbered port pins
}
```

■

Example 4.4

It is required to write a program to continuously count up in binary and send this data to PORTB. Thus, PORTB is required to have the binary data:

00000000
00000001
00000010
00000011
....
....
11111110
11111111
00000000
....

Solution

A **for** loop can be used to create an endless loop, and inside this loop, the value of a variable can be incremented and then sent to PORTB. The required program is:

```
void main(void)
{
    unsigned char Cnt = 0;

    for(;;)                             //Endless loop
    {
        PORTB = Cnt;                    // Send Cnt to PORT B
        Cnt++;                          // Increment Cnt
    }
}
```

■

■ Example 4.5

Write a program to set all bits of PORTB to logic 1 and then to logic 0. Repeat this process 10 times.

Solution

The **for** statement can be used to create a loop and repeat the required operation 10 times:

```
void main(void)
{
    unsigned char j;

    for(j = 0; j < 10; j++)             // Repeat 10 times
    {
        PORTB = 0xFF;                   // Set PORT B pins to 1
        PORTB = 0;                      // Clear PORT B pins
    }
}
```

■

Example 4.6

The radius and height of a cylinder are 2.5 and 10 cm, respectively. Write a program to calculate the volume of this cylinder.

Solution

The required program is

```
void main(void)
{
    float Radius = 2.5, Height = 10;
    float Volume;

    Volume = PI *Radius*Radius*Height;
}
```

Example 4.7

Write a program to find the largest element of an integer array having 10 elements.

Solution

The program is given below. At the beginning, variable *m* is set to the first element of the array. A loop is then formed and the largest element of the array is found:

```
void main (void)
{
    unsigned char j;
    int m, A[10];

    m = A[0];                          //First element of array
    for(j = 1; j < 10; j++)
    {
        if(A[j] > m)m = A[j];
    }
}
```

Example 4.8

Write a program using the **while** statement to clear all 10 elements of an integer array M.

Solution

As shown in the program listing below, **NUM** is defined to be 10 and variable *j* is used as the loop counter:

```
#define NUM 10
void main(void)
{
    int M[NUM];
    unsigned char j = 0;

    while (j < NUM)
    {
        M[j] = 0;
        j++;
    }
}
```

■

■ Example 4.9

Write a program to convert the temperature from °C to °F starting from 0°C, in steps of 1°C up to and including 100°C, and store the results in an array called F.

Solution

Given the temperature in °C, the equivalent in °F is calculated using the formula:

$$F = (C - 32.0)/1.8$$

The required program listing is given below. A for loop is used to calculate the temperature in °F and store in array F:

```
void main(void)
{
    float F[100];
    unsigned char C;

    for (C =0; C <= 100; C++)
    {
        F[C] = (C – 32.0) / 1.8;
    }
}
```

■

4.8 Functions

A function is a self-contained section of code written to perform a well-defined action. Functions are usually created when it is required to perform an operation at several different parts of a main program. In addition, it is a good programming practice to divide a large program into a number of smaller independent functions. The statements within a function may be executed by calling (or invoking) the function.

The syntax of a general function definition is as shown in Figure 4.25. The data type indicates the type of data returned by the function. This is followed by the name of the function, and then a set of brackets are used where any comma separated arguments can be declared inside the brackets. Then the body of the function that includes the operational code of the function is written inside a set of opening and closing curly brackets.

An example of function definition is shown below. This function, named **Mult**, receives two integer arguments **a** and **b** and returns their product. Note that the use of brackets in a return statement is optional:

```
int Mult(int a, int b)
{
    return (a*b);
}
```

When a function is called, it generally expects to be given the number of arguments expressed in the function's argument list. For example, the above function can be called as

```
z = Mult(x,y);
```

where variable *z* has the data type **int**. Note that the arguments declared in the function header and the arguments passed when the function is called are independent of each other, even if they may have the same name. In the above example, when the function is called, variable *x* is copied to *a* and variable *y* is copied to *b* on entry to function **Mult**.

Some functions do not return any data and the data type of such functions must be declared as **void**. An example is given below:

```
void LED(unsigned char D)
{
    PORTB = D;
}
```

```
type name (parameter1, parameter2,.....)
{
            .............
            function body
            .............
}
```

Figure 4.25: General Syntax of a Function Definition

void functions can be called without any assignment statements, but the brackets must be used to tell the compiler that a function call is made:

```
LED( );
```

Also, some functions do not have any arguments. In the following example, the function, named **Compl**, complements PORTC of the microcontroller and it returns no data and has no arguments:

```
void Compl( )
{
    PORTC = ~PORTC;
}
```

The above function can be called as

```
Compl( );
```

Functions are normally defined before the start of the main program.

Some example function definitions and their usage in main programs are illustrated in the following examples.

Example 4.10

Write a function called **Circle_Area** to calculate the area of a circle where the radius is to be used as an argument. Use this function in a main program to calculate the area of a circle whose radius is 2.5 cm. Store the area in a variable called **Circ**.

Solution

The required function definition is given below. The data type of the function is declared as **float**. The area is calculated by the formula:

$$\text{Area} = \pi r^2,$$

where r is the radius of the circle. The area is calculated and stored in a local variable called *s*, which is then returned from the function:

```
float Circle_Area(float radius)
{
    float s;

    s = PI * radius * radius;
    return s;
}
```

Figure 4.26 shows how function **Circle_Area** can be used in a main program to calculate the area of a circle whose radius is 2.5 cm. The function is defined before the main program. Inside the main program, the function is called to calculate and store the area in variable **Circ**.

Example 4.11

Write a function called **Area** and a function called **Volume** to calculate the area and volume of a cylinder, respectively. Then, write a main program to calculate the area and the volume of cylinder whose radius is 2.0 cm and height is 5.0 cm. Store the area in variable **cyl_area** and the volume in variable **cyl_volume**.

Solution

The area of a cylinder is calculated by the formula:

$$\text{Area} = 2\pi rh,$$

where r and h are the radius and the height of the cylinder, respectively. Similarly, the volume of a cylinder is given by the formula:

$$\text{Volume} = \pi r^2 h$$

Figure 4.27 gives the listing of the functions that calculate the area and volume of a cylinder.

The main program that calculates the area and volume of a cylinder whose radius = 2.0 cm and height = 5.0 cm is shown in Figure 4.28.

■

Example 4.12

Write a function called **LowerToUpper** to convert a lowercase character to uppercase.

Solution

The ASCII value of the first uppercase character ("A") is 0x41. Similarly, the ASCII value of the first lowercase character ("a") is 0x61. An uppercase character can be converted into its equivalent lowercase by subtracting 0x20 from the character. The required function listing is shown in Figure 4.29.

■

```
/*****************************************************************************************
                              AREA OF A CIRCLE
                              ================

This program calls to function Circle_Area to calculate the area of a circle.

Programmer:   Dogan Ibrahim
File:         CIRCLE.C
Version:      1.0
Date:         March, 2009
*****************************************************************************************/

/* This function calculates the area of a circle given the radius */
float Circle_Area(float radius)
{
    float s;

    s = 3.14 * radius * radius;
    return s;
}

/* Start of main program. Calculate the area of a circle where radius = 2.5 */
void main(void)
{
    float r, Circ;

    r = 2.5;
    Circ = Circle_Area(r);
}
```

Figure 4.26: Program to Calculate the Area of a Circle

```
float Area(float radius, float height)
{
    float s;

    s = 2.0*3.14 * radius*height;
    return s;
}


float Volume(float radius, float height)
{
    float s;

    s = 3.14 *radius*radius*height;
    return s;
}
```

Figure 4.27: Functions to Calculate Cylinder Area and Volume

```
/*******************************************************************************
                    AREA AND VOLUME OF A CYLINDER
                    =============================

This program calculates the area and volume of a cylinder whose
radius is 2.0 cm  And height is 5.0 cm.

Programmer:   Dogan Ibrahim
File:         CYLINDER.C
Version:      1.0
Date:         March, 2009
*******************************************************************************/

/* Function to calculate the area of a cylinder */
float Area(float radius, float height)
{
    float s;

    s = 2.0*3.14 * radius*height;
    return s;
}


/* Function to calculate the volume of a cylinder */
float Volume(float radius, float height)
{
    float s;

    s = 3.14 *radius*radius*height;
    return s;
}

/* Start of the main program */
void main(void)
{
    float r = 2.0, h = 5.0;
    float cyl_area, cyl_volume;

    cyl_area = Area(r, h);
    cyl_volume(r, h);
}
```

Figure 4.28: Program That Calculates the Area and Volume of a Cylinder

```
unsigned char LowerToUpper(unsigned char c)
{
    if(c >= 'a' && c <= 'z')
        return (c – 0x20);
    else
        return c;
}
```

Figure 4.29: Function to Convert Lowercase to Uppercase

```
/*******************************************************************************************
                        LOWER CASE TO UPPER CASE
                        ========================

This program converts the lower case character in variable Lc to upper case
and stores in variable Uc.

Programmer:  Dogan Ibrahim
File:        LTOUPPER.C
Version:     1.0
Date:        March, 2009
*******************************************************************************************/

/* Function to convert a lower case character to upper case */
unsigned char LowerToUpper(unsigned char c)
{
    if(c >= 'a' && c <= 'z')
        return (c – 0x20);
    else
        return c;
}


/* Start of main program */
void main(void)
{
    unsigned char Lc, Uc;

    Lc = 'r';
    Uc = LowerToUpper(Lc);
}
```

Figure 4.30: Program Calling Function LowerToUpper

Example 4.13

Use the function you have created in Example 4.12 in a main program to convert letter "r" to uppercase.

Solution

The required program is shown in Figure 4.30. Function **LowerToUpper** is called to convert the lowercase character in variable **Lc** to uppercase and store in **Uc**.

■

4.8.1 Function Prototypes

If a function is not defined before it is called, then the compiler will generate an error message. One way to round this problem is to create a function prototype. A function prototype is easily constructed by making a copy of the function's header and appending a semicolon to it. If the function has parameters, it is not compulsory to give names to these parameters but the data type of the parameters must be defined. An example is given below, which declares a function prototype called **Area** and the function is expected to have a floating point type parameter:

```
float Area(float radius);
```

This function prototype could also be declared as

```
float Area(float);
```

Function prototypes should be declared at the beginning of a program. Function definitions and function calls can then be made at any point in the program.

Example 4.14

Repeat Example 4.13 but declare **LowerToUpper** as a function prototype.

Solution

Figure 4.31 shows the program where function **LowerToUpper** is declared as a function prototype at the beginning of the program. In this example, the actual function definition is written after the main program.

4.8.2 Passing Arrays to Functions

There are many applications where we may want to pass arrays to functions. Passing a single array element is straightforward as we simply specify the index of the array element to be passed as in the following function call that passes the second element (index = 1) of array **A** to function **Calc**. It is important to realize that an individual array element is passed *by value*; i.e., a copy of the array element is passed to the function:

```
x = Calc(A[1]);
```

In some applications, we may want to pass complete arrays to functions. An array name can be used as an argument to a function, thus permitting the entire array to be passed to a function. To pass a complete array to a function, the array name must appear by itself with brackets. The size of the array is not specified within the formal argument declaration. In the function header, the array name must be specified with a pair of empty brackets. It is important to realize that when a complete array is passed to a function, what is actually passed is not a copy of the array but the address of the first element of the array; i.e., the array elements are passed *by reference*, which means that the original array elements can be modified inside the function.

```
/******************************************************************************
                    LOWER CASE TO UPPER CASE
                    ========================

This program converts the lower case character in variable Lc to
upper case and stores in variable Uc.


Programmer:  Dogan Ibrahim
File:        LTOUPPER2.C
Version:     1.0
Date:        March, 2009
******************************************************************************/

unsigned char LowerToUpper(unsigned char);


/* Start of main program */
void main(void)
{
    unsigned char Lc, Uc;

    Lc = 'r';
    Uc = LowerToUpper(Lc);
}


/* Function to convert a lower case character to upper case */
unsigned char LowerToUpper(unsigned char c)
{
    if(c >= 'a' && c <= 'z')
        return (c – 0x20);
    else
        return c;
}
```

Figure 4.31: Program Using Function Prototype

Some examples are given below to illustrate the passing of a complete array to a function.

Example 4.15

Write a program to store the numbers 1–10 in an array called **Numbers**. Then, call a function named **Average** to calculate the average of these numbers.

Solution

The required program listing is shown in Figure 4.32. Function **Average** receives the elements of array **Numbers** and calculates the average of the array elements.

```
/*****************************************************************************************
                        PASSING AN ARRAY TO A FUNCTION
                        ==============================

This program stores numbers 1 to 10 in an array called Numbers. Function
Average is then called to calculate the average of these numbers.


Programmer:   Dogan Ibrahim
File:         AVERAGE.C
Version:      1.0
Date:         March, 2009
*****************************************************************************************/

/* Function to calculate the average */
float Average(int A[ ])
{
    float Sum = 0.0, k;
    unsigned char j;

    for (j=0; j<10; j++)
    {
        Sum = Sum + A[j];
    }
    k = Sum / 10.0;
    return k;
}


/* Start of the main program */
void main(void)
{
    unsigned char j;
    float Avrg;
    int Numbers[10];

    for(j=0; j<10; j++)Numbers[j] = j+1;
    Avrg = Average(Numbers);
}
```

Figure 4.32: Program Passing an Array to a Function

Example 4.16

Repeat Example 4.15, but this time define the array size at the beginning of the program and then pass the array size to the function.

Solution

The required program listing is shown in Figure 4.33.

```
/**************************************************************************************
                PASSING AN ARRAY TO A FUNCTION
                ==============================

This program stores numbers 1 to N in an array called Numbers
where N is defined at the beginning of the program. Function Average
is then called to calculate the average of these numbers.


Programmer:  Dogan Ibrahim
File:        AVERAGE2.C
Version:     1.0
Date:        March, 2009
**************************************************************************************/

#define Array_Size 20

/* Function to calculate the average */
float Average(int A[ ], int N)
{
    float Sum = 0.0, k;
    unsigned char j;

    for(j=0; j<N; j++)
    {
        Sum = Sum + A[j];
    }
    k = Sum / N;
    return k;
}


/* Start of the main program */
void main(void)
{
    unsigned char j;
    float Avrg;
    int Numbers[Array_Size];

    for(j=0; j<Array_Size; j++)Numbers[j] = j+1;
    Avrg = Average(Numbers, Array_Size);
}
```

Figure 4.33: Program Passing an Array to a Function

It is also possible to pass a complete array to a function using pointers. Here, the address of the first element of the array is passed to the function and the function can then manipulate the array as required using pointer operations. An example is given below.

■

```
/*****************************************************************************************
                    PASSING AN ARRAY TO A FUNCTION
                    ==============================

This program stores numbers 1 to 10 in an array called Numbers. Function
Average is then called to calculate the average of these numbers.

Programmer:   Dogan Ibrahim
File:         AVERAGE3.C
Version:      1.0
Date:         March, 2009
*****************************************************************************************/

/* Function to calculate the average */
float Average(int *A)
{
    float Sum = 0.0, k;
    unsigned char j;

    for(j=0; j<10; j++)
    {
        Sum = Sum + *(A + j);
    }
    k = Sum / 10.0;
    return k;
}


/* Start of the main program */
void main(void)
{
    unsigned char j;
    float Avrg;
    int Numbers[10];

    for(j=0; j<10; j++)Numbers[j] = j+1;
    Avrg = Average(&Numbers[0]);
}
```

Figure 4.34: Program Passing an Array Using Pointers

Example 4.17

Repeat Example 4.15, but this time, use a pointer to pass the array elements to the function.

Solution

The required program listing is given in Figure 4.34. Here, an integer pointer is used to pass the array elements to the function and the function elements are manipulated using pointer operations. Notice that the address of the first element of the array is passed as an integer with the statement: &Numbers[0]

■

4.8.3 Passing Variables by Reference to Functions

By default, arguments to functions are passed **by value**. Although this method has many distinct advantages, there are occasions when it is more appropriate and also more efficient to pass the address of the arguments instead, that is, pass the argument **by reference**. When the address of an argument is passed, the original value of that argument can be modified by the function, and thus, the function does not have to return any variables. An example is given below, which illustrates how the address of arguments can be passed to a function and how the values of these arguments can be modified inside the function.

■ Example 4.18

Write a function named **Swap** to accept two integer arguments and then to swap the values of these arguments. Use this function in a main program to swap the values of two variables.

Solution

The required program listing is shown in Figure 4.35. Function **Swap** is defined as **void** since it does not return any value and it has two arguments **a** and **b**, and in the function header, two integer pointers are used to pass the addresses of these variables. Inside the function body, the value of an argument is accessed by inserting "*" character before the argument. Inside the main program, the address of the variables are passed to the function using the "&" characters before the variable names. At the end of the program, variables *p* and *q* are set to 20 and 10, respectively.

■

4.8.4 Static Function Variables

Normally, variables declared at the beginning of a program, before the main program, are global, and their values can be accessed and modified by all parts of the program. Declaring a variable used in a global function will ensure that its value is retained from one call of the function to another, but this would undermine the variable's privacy and reduce the portability of the function to other applications. A better approach is to declare such variables as **static**. Static variables are mainly used in function definitions. When a variable is declared as static,

```
/*********************************************************************************************
                    PASSING VARIABLES BY REFERENCE
                    ==============================

This program shows how the address of variables can be passed to functions.
The function in this program swaps the values of two integer variables.

Programmer:   Dogan Ibrahim
File:         SWAP.C
Version:      1.0
Date:         March, 2009
*********************************************************************************************/

/* Function to swap two integers */
void Swap(int *a, int *b)
{
    int temp;

    temp = *a;                          // Store a in temp
    *a = *b;                            // Copy b to a
    *b = temp;                          // Copy temp to b
}


/* Start of the main program */
void main(void)
{
    int p, q;

    p = 10;                             // Set p = 10
    q = 20;                             // Set q = 20
    swap(&p, &q);                       // Swap p and q (p=20, q=10)
}
```

Figure 4.35: Passing Variables by Reference to a Function

its value is retained from one call of the function to another. In the example code given below, variable *k* is declared as static and is initialized to zero. This variable is then incremented before exiting from the function, and the value of *k* remains in existence and holds its last value on the next call to the function; i.e., on the second call to the function, the value of *k* will be 1 and not 0:

```
void Cnt(void)
{
    static int k = 0;                   // Declare k as static
    ..................
    ..................
    k++;                                // increment k
}
```

4.9 MPLAB C18 Library Functions

C18 compiler provides a large set of library functions that can be used in our programs. These library functions can be called from anywhere in a program, and they require the correct header files to be included at the beginning of the program. C18 manual *MPLAB C18 C Compiler Libraries* by Microchip Inc. gives detailed description of each library function.

C18 library functions are given as in the following headings:

- Hardware Peripheral Functions Library
- Software Peripheral Functions Library
- General Software Library
- Math Library

Hardware Peripheral Functions Library consists of functions for the following peripherals:

- Analog-to-Digital (A/D)
- Input capture
- Integrated Interconnect (I^2C) bus
- I/O port
- Microwire bus
- Pulse width modulation (PWM)
- Serial peripheral interface (SPI) bus
- Timer
- Universal synchronous-asynchronous receiver-transmitter (USART)

Software Peripheral Functions Library consists of functions for the following peripherals:

- LCD
- CAN bus
- I^2C bus
- SPI bus
- UART

General Software Library consists of functions for the following general-purpose operations:

- Character classification
- Data conversion
- Memory and string manipulation
- Delay
- Reset
- Character output

Math Library consists of functions for the following:

- A 32-bit floating point math library
- C Standard library math functions

In this section, the description of some commonly used library functions are given with examples.

4.9.1 Delay Functions

These functions are used to create delays in programs. Table 4.8 gives a list of the available C18 delay functions. The header file "delays.h" must be included at the beginning of the program when any of these functions are used. The arguments of the functions must be an 8-bit unsigned character, i.e., the maximum allowed number in an argument is 255.

The delays are specified in terms of instruction cycle times. For example, when using a 4-MHz clock, the instruction cycle time is 1 μs. Thus, calling function

```
Delay100TCYx(5)
```

will cause $100 \times 1 \times 5 = 500$ μs of delay in the program.

Similarly, to generate a 1-s delay when using a 4-MHz clock, we can call the function

```
Delay10KTCYx(100)
```

Table 4.8: C18 Delay Functions

Functions	Descriptions
Delay1TCY	Delay in one instruction cycle
Delay10TCYx	Delay in multiples of 10 instruction cycles
Delay100TCYx	Delay in multiples of 100 instruction cycles
Delay1KTCYx	Delay in multiples of 1000 instruction cycles
Delay10KTCYx	Delay in multiples of 10 000 instruction cycles

With an 8-MHz clock, the required function to generate a 1-s delay is

```
Delay10KTCYx(200)
```

Some examples given below show the use of delay functions in programs.

Example 4.19

An LED is connected to bit 0 of PORTB (pin RB0) of a PIC18F452 microcontroller through a current limiting resistor as shown in Figure 4.36. Choose a suitable value for the resistor and write a program that will flash the LED ON and OFF continuously with 1-s intervals.

Solution

LEDs can be connected to a microcontroller in two modes: current sinking mode and current sourcing mode. In current sinking mode (see Figure 4.37), one leg of the LED is connected to +5 V and the other leg is connected to the microcontroller output port pin through a current limiting resistor *R*.

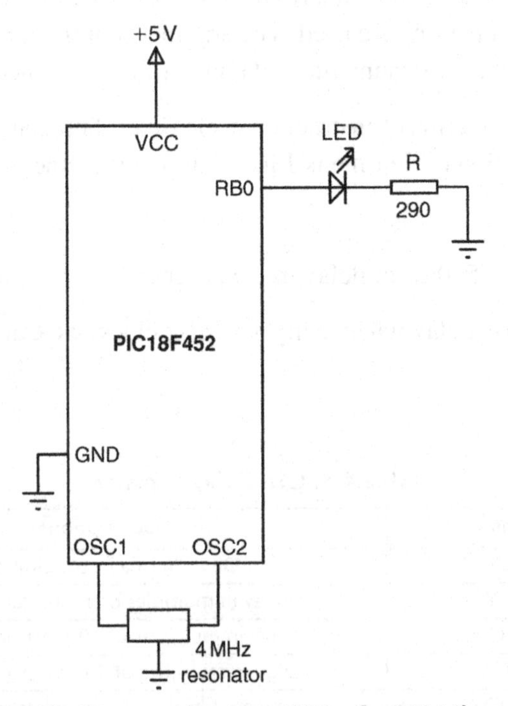

Figure 4.36: LED Connected to Port RB0 of a PIC Microcontroller

Under normal working conditions, the voltage across an LED is approximately 2 V, and the current through the LED is approximately 10 mA (some low-power LEDs can operate at as low as 1 mA current). The maximum current that can be sourced or sinked at the output port of a PIC microcontroller is 25 mA.

The value of the current limiting resistor R can be calculated as follows. In current sinking mode, the LED will be turned ON when the output port of the microcontroller is at logic 0, i.e., at approximately 0 V. The required resistor is then

$$R = \frac{5\,\text{V} - 2\,\text{V}}{10\,\text{mA}} = 0.3\,\text{K}$$

The nearest resistor to choose is 290 Ω(a slightly higher resistor can be chosen for a lower current and less brightness).

In current sourcing mode (see Figure 4.38), one leg of the LED is connected to the output port of the microcontroller and the other leg is connected to ground through a current limiting resistor. The LED will be turned ON when the output port of the

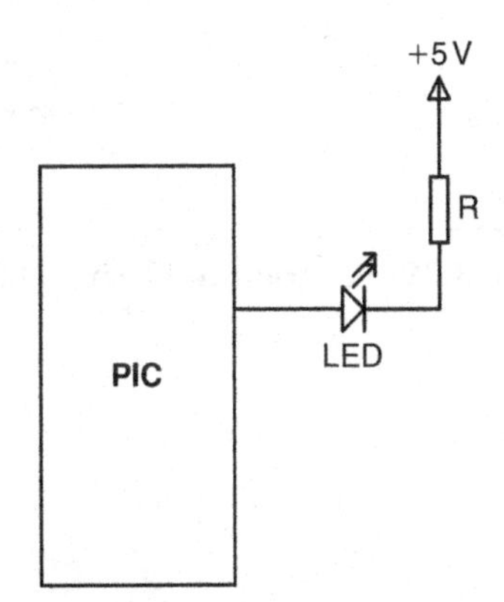

Figure 4.37: Connecting the LED in Current Sinking Mode

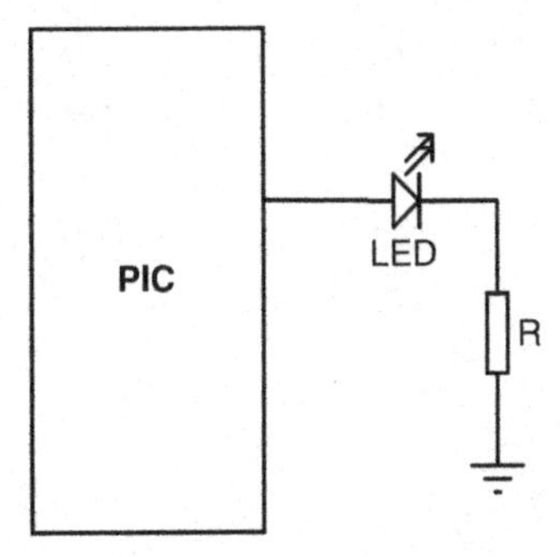

Figure 4.38: Connecting the LED in Current Sourcing Mode

```
/*****************************************************************************************
                              FLASHING AN LED
                              ==============

This program flashes an LED connected to port RB0 of a microcontroller with one
second intervals. C18 library function Delay10KTCYx is used to create a 1 second
delay between the flashes.

A 4MHz clock is used in the project.

Programmer:  Dogan Ibrahim
File:        FLASH.C
Version:     1.0
Date:        March, 2009
*****************************************************************************************/
#include <p18f452.h>
#include <delays.h>

void main(void)
{
    TRISB = 0;                              // Configure PORTB as output
    for(; ;)                                // Endless loop
    {
        PORTB = 1;                          // Turn ON LED
        Delay10KTCYx(100);                  // 1 second delay
        PORTB = 0;                          // Turn OFF LED
        Delay10KTCYx(100);                  // 1 second delay
    }
}
```

Figure 4.39: Program to Flash an LED

microcontroller is at logic 1, i.e., at approximately 5 V. Both in current sinking and in current sourcing modes, we can use the same value resistor.

The required program listing is given in Figure 4.39 (program FLASH.C). At the beginning of the program, device-specific header file "p18f452.h" and delay header file "delays.h" are included. Then, PORTB is configured as output using the TRISB = 0 statement. An endless loop is then formed with the **for** statement, and inside this loop, the LED is turned ON and OFF with 1-s delay between each output. The microcontroller is operated with a 4-MHz clock and thus the instruction cycle time is 1 μs. Function Delay10KTCYx is called as described earlier with an argument of 100 to generate a 1-s delay.

The program given in Figure 4.39 can be made more user friendly and easier to follow using defined statements as shown in Figure 4.40 (program FLASH2.C).

■

```
/**********************************************************************************
                        FLASHING AN LED
                        ===============

This program flashes an LED connected to port RB0 of a microcontroller
with one second intervals. C18 library function Delay10KTCYx is used to
create a 1 second delay between the flashes.

Programmer:   Dogan Ibrahim
File:         FLASH2.C
Version:      1.0
Date:         March, 2009
**********************************************************************************/
#include <p18f452.h>
#include <delays.h>

#define LED PORTBbits.RB0
#define ON  1
#define OFF 0
#define One_Second_Delay Delay10KTCYx(100)

void main(void)
{
    TRISB = 0;                          // Configure PORT B as output

    for(; ;)                            // Endless loop
    {
        LED = ON;                       // Turn ON LED
        One_Second_Delay;               // 1 second delay
        LED = OFF;                      // Turn OFF LED
        One_Second_Delay;               // 1 second delay
    }
}
```

Figure 4.40: Another Program to Flash an LED

Table 4.9: Some Character Classification Functions

Functions	Descriptions
isalnum	Determines if a character is alphanumeric
isalpha	Determines if a character is alphabetic
isdigit	Determines if a character is a decimal digit
islower	Determines if a character is lowercase
isupper	Determines if a character is uppercase

4.9.2 Character Classification Functions

These functions can be used to test the nature of characters. Table 4.9 shows some of the functions in this library. The header file "ctype.h" must be included at the beginning of the program whenever one of these functions is used.

An example is given below showing the use of character classification functions in programs.

Example 4.20

An LED is connected to bit 0 of PORTB (pin RB0) of a PIC18F452 microcontroller through a current limiting resistor as shown in Figure 4.36. Write a program that will turn the LED ON if variable called **mode** contains an uppercase character, otherwise turn the LED OFF.

Solution

The program listing is shown in Figure 4.41. In this example, variable mode is set to a lowercase character and therefore the LED will be in OFF state when the program is run. The program uses library function **isupper** to determine whether or not the character in variable **mode** is uppercase.

```
/***********************************************************************************************
                                        LED CONTROL
                                        ============

This program turns the LED ON if the character in variable mode is an upper case
character, otherwise the LED is turned OFF.

The LED is assumed to be connected to port RB0 of the microcontroller and A
4MHz clock is used in the project.

fC18 library function isupper is used to determine whether or not the character is
upper case.

Programmer:  Dogan Ibrahim
File:        LED.C
Version:     1.0
Date:        March, 2009
***********************************************************************************************/
#include <p18f452.h>
#include <ctype.h>

void main(void)
{
    unsigned char mode = 'a';           // mode lower case

    TRISB = 0;                          // Configure PORT B as output

    if(isupper(mode))
        PORTB = 1;                      // Turn ON LED
    else
        PORTB = 0;                      // Turn OFF LED

    while(1);                           // wait here forever
}
```

Figure 4.41: Program for the Example 4.20

Table 4.10: Some Data Conversion Functions

Functions	Descriptions
atoi	Converts a string into a 16-bit signed integer
atol	Converts a string into a long integer
itoa	Converts a 16-bit signed integer into a string
tolower	Converts a character to lowercase
toupper	Converts a character to uppercase

4.9.3 Data Conversion Functions

These functions can be used to convert one type of data into another type. Table 4.10 shows some of the functions in this library. The header file "stdlib.h" or "ctype.h" must be included at the beginning of the program whenever one of these functions is used.

An example is given below showing the use of data conversion functions in programs.

Example 4.21

A character array named **my_str** contains five lowercase ASCII characters. Write a program that will convert all the characters in this array into uppercase.

Solution

The program listing is shown in Figure 4.42. In this example, data conversion function **toupper** is used to convert the characters to uppercase. Notice that the header file "ctype.h" is included at the beginning of the program.

■

4.9.4 Memory and String Manipulation Functions

These functions can be used to search for a given value in the memory, to compare two arrays, to compare two strings, initialize an array, determine the length of a string, and so on. Table 4.11 shows some of the functions in this library. The header file "string.h" must be included at the beginning of the program whenever one of these functions is used.

An example is given below showing the use of memory and string manipulation functions in programs.

```
/*****************************************************************************
                              UPPER CASE
                              ==========

This program converts the 5 characters in array my_str into upper case.

C18 library function toupper is used in the program.

Programmer:   Dogan Ibrahim
File:         UPPER.C
Version:      1.0
Date:         March, 2009
*****************************************************************************/
#include <p18f452.h>
#include <ctype.h>

void main(void)
{
    unsigned char my_str[5] = {'a','s','e','r','w'};
    unsigned char k;

    for(k = 0; k < 5; k++)my_str[k] = toupper(my_str[k]);

    while(1);                               // wait here forever
}
```

Figure 4.42: Program for Example 4.21

Table 4.11: Some Data Conversion Functions

Functions	Descriptions
memcmp	Compares the contents of two arrays
memset	Initializes an array with a repeated value
strchr	Locates the first occurrence of a value in a string
strlen	Determines the length of a string
strstr	Locates the first occurrence of a string inside another string
strupr	Coverts all characters in a string to uppercase

Example 4.22

A string named **my_str** contains various lowercase and uppercase ASCII characters. Write a program that will convert all the characters in this array into uppercase.

Solution

The program listing is shown in Figure 4.43. In this example, string manipulation function **strupr** is used to convert the characters to uppercase. Notice that the header file "string.h" is included at the beginning of the program.

```
/*******************************************************************************************
                                    UPPER CASE
                                    ===========

This program converts the characters in string my_str into upper case.

C18 library function strupr is used in the program.


Programmer:   Dogan Ibrahim
File:         UPPER.C
Version:      1.0
Date:         March, 2009
*******************************************************************************************/

#include <p18f452.h>
#include <string.h>

void main(void)
{
    char my_str[ ] = "MyString";                    // Declare string my_str

    strupr(my_str);                                 // my_str = "MYSTRING"

    while(1);                                       // Wait here forever
}
```

Figure 4.43: Program for the Example 4.22

Example 4.23

Write a program to find the length of a string called **my_str** and store it in a character variable called **slen**.

Solution

The program listing is shown in Figure 4.44. In this example, function **strlen** is used to find the length of the string. The string is "MyString" and thus **slen** is assigned number 8.

■

Example 4.24

Assume that a string called **First** contains data "PIC" and another string called **Second** contains data "Microcontroller." Write a program to append string **Second** to string **First**. Thus, at the end of the operation, string **First** will have data "PIC Microcontroller" and string **Second** will not change, i.e., it will have the data "Microcontroller."

```
/*****************************************************************************
                              STRING LENGTH
                              =============

This program finds the length of string my_str and stores in variable slen.
String my_str is loaded with "MyString", having 8 characters and thus
Variable slen will be set to 8.

C18 library function strlen is used in the program.


Programmer:  Dogan Ibrahim
File:        LEN.C
Version:     1.0
Date:        March, 2009
*****************************************************************************/
#include <p18f452.h>
#include <string.h>

void main(void)
{
    char my_str[ ] = "MyString";                    // Declare string my_str
    unsigned char slen;

    slen = strlen(my_str);                          // Length of the string

    while(1);                                       // Wait here forever
}
```

Figure 4.44: Program for the Example 4.23

Solution

The program listing is shown in Figure 4.45. In this example, function **strcat** is used to append one string to another string. You should make sure that the destination string (First) is large enough to hold the source string (Second).

■

4.9.5 Reset Functions

These functions are used to determine the cause of reset or wake up in a program. Table 4.12 shows some of the functions in this library. The header file "reset.h" must be included at the beginning of the program whenever one of these functions is used.

An example is given below showing the use of reset functions in programs.

```
/*****************************************************************************************
                              STRING APPEND
                              =============

String First is loaded with "PIC" and string Second is loaded with
"Microcontroller". The program appends string Second to string First so that First
contains the characters "PIC Microcontroller". String Second does not change.

C18 library function strcat is used in the program.


Programmer:   Dogan Ibrahim
File:         StrLen.C
Version:      1.0
Date:         March, 2009
*****************************************************************************************/
#include <p18f452.h>
#include <string.h>

void main(void)
{
    char First[20] = "PIC ";                        // String First
    char Second[ ] = "Microcontroller";             // String Second

    strcat(First, Second);                          // Append strings

    while(1);                                       // Wait here forever
}
```

Figure 4.45: Program for the Example 4.24

Table 4.12: Some Reset Functions

Functions	Descriptions
isMCLR	If the cause of reset was MCLR
isPOR	If the cause of reset was power-on reset
isWDTTO	If the cause of reset was the watchdog timeout

Example 4.25

Write a program to determine the cause of reset in a program. If the cause of reset is Master Clear (MCLR) input or power-on reset, then load variable **Rst** with 1, otherwise load **Rst** with 0.

Solution

The program listing is shown in Figure 4.46.

```
/*****************************************************************************************
                                    RESET
                                    =====

This program determines the cause of reset in a program.

If the cause of Reset is MCLR or Power-on-reset then load variable Rst with 1,
otherwise load Rst with 0.

C18 library functions is MCLR and isPOR are used in the program.


Programmer:   Dogan Ibrahim
File:         Reset.C
Version:      1.0
Date:         March, 2009
*****************************************************************************************/
#include <p18f452.h>
#include <reset.h>

void main(void)
{
    unsigned char Rst;

    if(isMCLR || isPOR)                    // Cause MCLR or Power-on-reset ?
        Rst = 1;                           // Load Rst with 1
    else
        Rst = 0;                           // Load Rst with 0

    while(1);                              // Wait here forever
}
```

Figure 4.46: Program for the Example 4.25

Table 4.13: Some Character Output Functions

Functions	Descriptions
printf	Formatted output to *stdout*
putc	Character output to a string
puts	String output to *stdout*
sprintf	Formatted string output to a data memory buffer
_usart_putc	Single character output to USART

4.9.6 Character Output Functions

These functions are used to process the output to various peripheral devices. Table 4.13 shows some of the functions in this library. The header file "stdio.h" must be included at the beginning of the program whenever one of these functions is used. Note that the data is sent to the device defined by the standard output *stdout*. The *stdout* is defined, by default, as the device _H_USART, which is the USART.

Some examples are given below to show the use of character output functions.

Example 4.26

Write a program to send the message "My Microcontroller" to the USART device of a PIC18F452 microcontroller.

Solution

The program listing is shown in Figure 4.47. Function **printf** is used to send the string to USART. This function can have a large number of arguments to format the data. Some of the commonly used formatting arguments are as follows:

%c	character
%d	integer number
%o	octal number
%u	unsigned integer number
%b	binary number
%x	hexadecimal number
%s	string
%f	floating point number

Example 4.27

Write a program to count up in decimal from 0 to 10 and send the output to the USART in the following format:

Number = nn,

where nn is the number 0–10

Solution

The program listing is shown in Figure 4.48. Function **printf** is used to send the string to USART with the argument "%d." Note that a new-line character "\n" is used at the end of each output.

```
/*****************************************************************************
                              Character Output
                              ============

This program sends the message "My Microcontroller" to the standard
output device of the microcontroller, which is the USART.

C18 library function printf is used in the program.


Programmer:  Dogan Ibrahim
File:        Cout.C
Version:     1.0
Date:        March, 2009
*****************************************************************************/
#include <p18f452.h>
#include <stdio.h>

void main(void)
{
    printf("My Microcontroller");

    while(1);                                          // Wait here forever
}
```

Figure 4.47: Program for Example 4.26

```
/*************************************************************************
                          Character Output
                          =============

This program counts up from 00 to 10 and sends the result to
USART in the Following format:
                          Number = nn
C18 library function printf is used in the program.


Programmer:  Dogan Ibrahim
File:        Count.C
Version:     1.0
Date:        March, 2009
*************************************************************************/
#include <p18f452.h>
#include <stdio.h>

void main(void)
{
    unsigned char cnt;

    for(cnt=0; cnt <= 10; cnt++) printf("Number = %d\n", cnt);

    while(1);                                    // Wait here forever
}
```

Figure 4.48: Program for the Example 4.27

If the debugger UART window is enabled (**Debugger -> Settings -> Uart IO -> Enable Uart IO -> Window**), then the following text will be displayed in the I/O panel:

```
Number = 0
Number = 1
Number = 2
Number = 3
Number = 4
Number = 5
Number = 6
Number = 7
Number = 8
Number = 9
Number = 10
```

Example 4.28

Write a program to count up in decimal from 0 to 20 and send the output to USART in the following hexadecimal format:

```
Number = nn,
```

where nn is the number 0–0x14

Solution

The program listing is shown in Figure 4.49. Function **printf** is used to send the string to USART with the argument "%d." Note that a new-line character "\n" is used at the end of each output.

■

If the debugger UART window is enabled then the following text will be displayed in the I/O panel:

```
Number = 0
Number = 1
Number = 2
Number = 3
Number = 4
Number = 5
Number = 6
Number = 7
```

Number = 8
Number = 9
Number = a
Number = b
Number = c
Number = d
Number = e
Number = f

4.9.7 Math Library Functions

These are trigonometric, logarithmic, and power functions. Table 4.14 shows some of the functions in this library. The header file "math.h" must be included at the beginning of every program using these functions.

```
/******************************************************************************
                          Character Output
                          =============

This program counts up from 0 to 120 and sends the result to
USART in the following hexadecimal format:

                          Number = nn

Where nn is 0 to 0x14.

C18 library function printf is used in the program.


Programmer:  Dogan Ibrahim
File:        HexCount.C
Version:     1.0
Date:        March, 2009
******************************************************************************/
#include <p18f452.h>
#include <stdio.h>

void main(void)
{
    unsigned char cnt;

    for(cnt=0; cnt <= 20; cnt++) printf("Number = %x\n", cnt);

    while(1);                          // Wait here forever
}
```

Figure 4.49: Program for Example 4.28

Table 4.14: Some Math Library Functions

Functions	Descriptions
sin	Trigonometric sine
cos	Trigonometric cosine
tan	Trigonometric tangent
asin	Trigonometric inverse of sine
acos	Trigonometric inverse of cosine
cosh	Hyperbolic cosh
log	Natural logarithm
exp	Exponential factor

Some examples are given below to show how these functions can be used in programs.

Example 4.29

Write a program to calculate the trigonometric sine of 30° and store the result in a variable called **angle**.

Solution

The program listing is shown in Figure 4.50. It is important to note that the angles must be represented in radians and not in degrees. To convert degrees into radians, we can multiply the angle with π (3.14159) and divide by 180. The answer 0.5 is stored in variable **angle**.

■

Example 4.30

Write a program to calculate the squares of numbers from 0 to 10 and send the result to USART as a table in the following format:

```
0     0
1     1
2     4
3     9
4     1
...................
...................
...................
10    100
```

```
/*****************************************************************************
                         Trigonometric Sine
                         ==============

This program calculates the trigonometric sine of 30 degrees and stores
the Result in variable called angle.

C18 library function sin is used in the program.


Programmer:  Dogan Ibrahim
File:        Sine.C
Version:     1.0
Date:        March, 2009
*****************************************************************************/
#include <p18f452.h>
#include <math.h>

void main(void)
{
    float angle;
    float Pi = 3.14159;                    // Pi
    float Conv = 3.14159 / 180.0;          // Conversion factor

    angle = sin(30.0 * Conv);              // angle 0.5

    while(1);                              // Wait here forever
}
```

Figure 4.50: Program for the Example 4.29

Solution

The program listing is shown in Figure 4.51. Function **pow** is used in a for loop with arguments (k, 2.0) so that the squares of numbers *k* are calculated. The numbers are sent to USART in a table format using the **printf** function. Notice that both "stdio.h" and "math.h" header files are included at the beginning of the program.

The USART data can be displayed by enabling the USART in the debugger. The data will be displayed as shown below:

```
0    0
1    1
2    4
3    9
4    16
5    25
6    36
```

```
7    49
8    64
9    81
10   100
```

4.9.8 LCD Functions

All microcontrollers lack some kind of video display. A video display would make a microcontroller much more user friendly as it will enable text messages, graphics, and numeric values to be output in a more versatile manner than the seven-segment displays, LEDs, or alphanumeric displays. Standard video displays require complex interfaces and their cost is relatively high. LCDs are alphanumeric (or graphical) displays, which are frequently used in microcontroller-based

```
/*****************************************************************************************
                              Squares of Numbers
                              ================

This program calculates the squares of integer numbers from 0 to 10
and sends the results to USART in a table form.

C18 library function pow and printf are used in the program.


Programmer:  Dogan Ibrahim
File:        Powers.C
Version:     1.0
Date:        March, 2009
*****************************************************************************************/
#include <p18f452.h>
#include <math.h>
#include <stdio.h>

void main(void)
{
    unsigned char k;
    unsigned int Number;

    for(k = 0; k<= 10; k++)
    {
        Number = pow(k , 2.0);
        printf("%d      %d\n", k, Number);
    }

    while(1);                                          // Wait here forever
}
```

Figure 4.51: Program for the Example 4.30

applications. These display devices come in different shapes and sizes. Some LCDs have 40 or more character lengths with the capability to display several lines. Some other LCD displays can be programmed to display graphic images. Some modules offer color displays, while some others incorporate back lighting so that they can be viewed in dimly lit conditions.

There are basically two types of LCDs as far as the interfacing technique is concerned: parallel LCDs and serial LCDs. Parallel LCDs (e.g., Hitachi HD44780 series) are connected to the microcontroller circuitry such that data is transferred to the LCD using more than one line, and usually, four or eight data lines are used. Serial LCD is connected to a microcontroller using one data line only and data is transferred using the RS232 asynchronous data communications protocol. Serial LCDs are generally much easier to use, but they are more costly than the parallel ones. In this book, only the parallel LCDs will be considered, as they are cheaper and are used more commonly in microcontroller-based projects.

Low-level programming of a parallel LCD is usually a complex task and requires a good understanding of the internal operation of the LCD, including an understanding of the timing diagrams. Fortunately, C18 language provides library functions for text-based LCDs, which simplify the use of external LCDs in PIC microcontroller-based projects.

HD44780 controller is commonly used in parallel LCD-based microcontroller applications. A brief description of this controller and information on some commercially available LCD modules is given below.

HD44780 LCD Controller

HD44780 is one of the most popular LCD controllers used in many LCD modules in industrial and commercial applications and also by hobbyists. The module is monochrome and comes in different shapes and sizes. Modules with character lengths of 8, 16, 20, 24, 32, and 40 can be selected. Depending on the model chosen, the display provides a 14-pin or a 16-pin connector to interface to the external world. Table 4.15 shows the pin configuration and pin functions of a typical 14-pin LCD.

V_{SS} is the 0 V supply or ground. V_{DD} pin should be connected to the positive supply. Although the manufacturers specify a 5-V DC supply, the modules will usually work with as low as 3 V or as high as 6 V.

Pin 3 is named as V_{EE}, and this is the contrast control pin. This pin is used to adjust the contrast of the display and it should be connected to a DC supply. A potentiometer is usually connected to the power supply with its wiper arm connected to this pin and the other leg of the potentiometer connected to the ground. This way, the voltage at the V_{EE} pin and hence the contrast of the display can be adjusted as desired.

Pin 4 is the Register Select (RS), and when this pin is LOW, data transferred to the LCD is treated as commands. When RS is HIGH, character data can be transferred to and from the module.

Table 4.15: Pin Configuration of the HD44780 LCD Module

Pin Nos	Names	Functions
1	V_{SS}	Ground
2	V_{DD}	+ve supply
3	V_{EE}	Contrast
4	RS	Register select
5	R/W	Read/write
6	EN	Enable
7	D0	Data bit 0
8	D1	Data bit 1
9	D2	Data bit 2
10	D3	Data bit 3
11	D4	Data bit 4
12	D5	Data bit 5
13	D6	Data bit 6
14	D7	Data bit 7

Pin 5 is the Read/Write (R/W) pin. This pin is pulled LOW to write commands or character data to the LCD module. When this pin is HIGH, character data or status information can be read from the module.

Pin 6 is the Enable (EN) pin that is used to initiate the transfer of commands or data between the module and the microcontroller. When writing to the display, data is transferred only on the HIGH to LOW transition of this pin. When reading from the display, data becomes available after the LOW to HIGH transition of the enable pin, and this data remains valid as long as the enable pin is at logic HIGH.

Pins 7–14 are the eight data bus lines (D0–D7). Data can be transferred between the microcontroller and the LCD module either using a single, 8-bit byte or as two, 4-bit nibbles. In the latter case, only the upper four data lines (D4–D7) are used. The 4-bit mode has the advantage that fewer I/O lines are required to communicate with the LCD.

C18 LCD library provides large number of functions to control text-based LCDs with 4-bit and 8-bit data interface. Four-bit interface-based text LCDs are the most commonly used LCDs, and this section describes the important C18 functions to control and send data to these LCDs. Further information on other LCD functions can be obtained from the manual *MPLAB C18 C Compiler Libraries*.

Table 4.16 gives a list of the commonly used LCD functions available for 4-bit interface text-based LCDs. Note that the header file "xlcd.h" must be included at the beginning of a program when any of these functions are used.

Table 4.16: Commonly Used LCD Functions

Functions	Descriptions
BusyXLCD	Checks if the LCD controller is busy
OpenXLCD	Configures I/O port lines for the LCD and initializes
putcXLCD	Writes a byte of data to the LCD
putsXLCD	Writes a string from data memory to the LCD
putrsXLCD	Writes a string from program memory to the LCD
WriteCmdXLCD	Writes a command to the LCD

The LCD library requires that the following delay functions be defined by the user before using the LCD functions:

DelayFor18TCY Delay for 18 cycles

DelayPORXLCD Delay for 15 ms

DelayXLCD Delay for 5 ms

Assuming a microcontroller clock frequency of 4 MHz, the instruction cycle time is 1 μs. With a clock frequency of 8 MHz, the instruction cycle time is 0.5 μs. Figure 4.52 shows how the above delay functions could approximately be obtained for both 4 and 8-MHz clock frequencies. The 18-cycle delay is obtained using no operation (NOP) statements, where each NOP operation takes one cycle to execute. The end of a function with no "return" statement takes two cycles. When a "return" statement is used, a BRA statement branches to the end of the function where a RETURN 0 is executed to return from the function, thus adding two more cycles. For example, the following function takes four cycles to execute:

```
void test(void)
{
        nop( );                         ;1 cycle
        nop( );                         ;1 cycle
}                                       ; RETURN 0, takes 2 cycles
```

and the following function takes six cycles to execute:

```
void test(void)
{
        nop( );                         ; 1 cycle
        nop( );                         ; 1 cycle
        return;                         ; BRA X, 2 cycles
}                                       ; X: RETURN 0, 2 cycles
```

A brief description of the C18 LCD functions is given below.

BusyXLCD

This function checks to determine whether or not the LCD controller is busy, and data or commands should not be sent to the LCD if the controller is busy. The function returns 1 if

4MHz Clock

```
#include<delays.h>

void DelayFor18TCY(void)
{
        Nop( ); Nop( ); Nop( ); Nop( );              // 18 cycle delay
        Nop( ); Nop( ); Nop( ); Nop( );
        Nop( ); Nop( ); Nop( ); Nop( );
        Nop( ); Nop( );
        return;
}


void DelayPORXLCD(void)                              // 15ms delay
{
        Delay1KTCYx(15);
}


void DelayXLCD(void)
{
        Delay1KTCYx(5);                              // 5ms delay
}
```

8MHz Clock

```
#include <delays.h>

void Delayfor18TCY(void)
{
        Nop( ); Nop( ); Nop( ); Nop( );              // 18 cycle delay
        Nop( ); Nop( ); Nop( ); Nop( );
        Nop( ); Nop( ); Nop( ); Nop( );
        Nop( ); Nop( );
        return;
}


void DelayPORXLCD(void)                              // 15ms delay
{
        Delay1KTCYx(30);
}


void DelayXLCD(void)
{
        Delay1KTCYx(10);                             // 5ms delay
}
```

Figure 4.52: LCD Delay Functions for 4-MHz and 8-MHz Clock

the controller is busy or 0 if it is otherwise. The program can be forced to wait until the LCD controller is ready using the following statement:

```
while(BusyXLCD( ));
```

OpenXLCD

This function is used to configure the interface between the microcontroller I/O ports and the LCD pins. The function requires an argument to specify the interface mode (4 or 8 bit), the LCD character mode, and the number of lines used. A value should be selected and logically AND ed from the following two groups:

FOUR_BIT
EIGHT_BIT

LINE_5 × 7
LINE_5 × 10
LINES_5 × 7

For example, if we are using a four-wire connection with an LCD having a single row with 5 × 7 characters, then the function should be initialized as follows:

```
OpenXLCD(FOUR_BIT & LINE_5 × 7);
```

The actual physical connection between the LCD and microcontroller I/O ports is defined in file "**xlcd.h**," and the default settings use PORTB pins in 4-bit mode where the low 4 bits of the port (RB0-RB3) are connected to the upper data lines (D4–D7) of the LCD (see the manual *MPLAB C18 C Compiler Libraries* for more information on the default connection):

LCD Pins	Microcontroller Pins
E	RB4
RS	RB5
RW	RB6
D4	RB0
D5	RB1
D6	RB2
D7	RB3

Figure 4.53 shows the default connection between a PIC18F452 microcontroller and an LCD.

putcXLCD

This function is used to write a byte to the LCD. The byte is passed as an argument to the function. In the following example, character "A" is displayed on the LCD:

```
unsigned char x = 'A';
putcXLCD(x);
```

putsXLCD

This function writes a string of characters from the data memory to the LCD. The writing stops when a NULL character is detected. An example use of this function is given below:

```
putsXLCD("My Computer");
```

putrsXLCD

This function writes a string of characters from the program memory to the LCD. The writing stops when a NULL character is detected. An example use of this function is given below:

```
char txt[ ] = "My text";
putrsXLCD(txt);
```

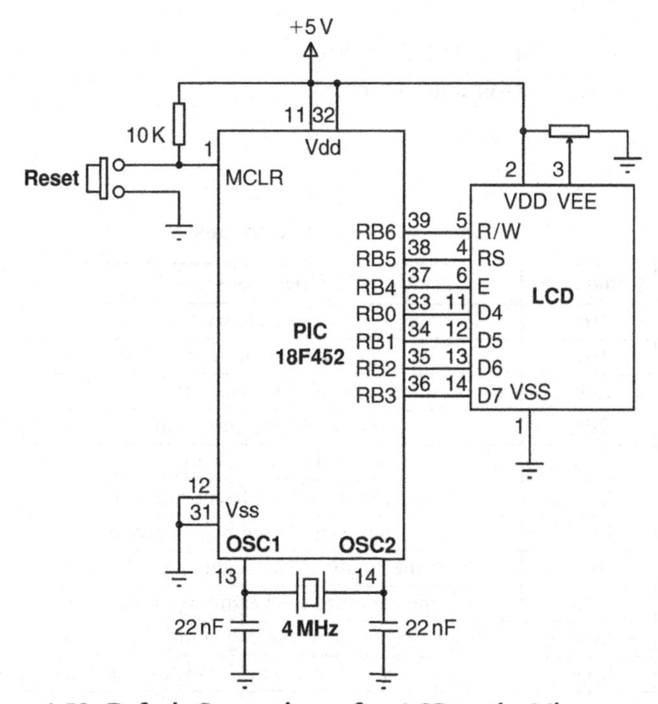

Figure 4.53: Default Connections of an LCD to the Microcontroller

WriteCmdXLCD

This function sends a command to the LCD. The following commands can be specified in the command argument:

DOFF – Turns display off

CURSOR_OFF – Enables display, hide cursor

BLINK_ON – Enables cursor blinking

BLINK_OFF – Disables cursor blinking

SHIFT_CUR_LEFT – Shifts cursor left

SHIFT_CUR_RIGHT – Shifts cursor right

SHIFT_DISP_LEFT – Shifts display to the left

SHIFT_DISP_RIGHT – Shifts display to the right

In addition, the LCD control functions given in Table 4.17 can be specified in the argument to control the LCD.

This command can also be used to set the LCD display characteristics using the following arguments as bitwise AND:

FOUR_BIT – 4-bit data interface
EIGHT_BIT – 8-bit data interface

Table 4.17: LCD Functions

Commands	Operations
0x1	Clears display
0x2	Moves cursor home
0x0C	Turns the cursor off
0x0E	Underlines the cursor on
0x0F	Blinking cursor on
0x10	Moves the cursor left one position
0x14	Moves the cursor right one position
0x80	Moves the cursor to the beginning of first row
0xC0	Moves the cursor to the beginning of second row
0x94	Moves the cursor to the beginning of third row
0xD4	Moves the cursor to the beginning of fourth row

LINE_5 × 7 – 5 × 7 character mode
LINE_5 × 10 – 5 × 10 character mode
LINES_5 × 7 – 5 × 7 multiple line display

It is important that the LCD controller should not be busy (check with function **BusyXLCD**) when commands are sent to it. Some example commands are given below:

```
WriteCmdXL CD(EIGHT_BIT & LINE_5X7);     // 8 bit, 5 × 7 character
WriteCmdXLCD(BLINK_ON);                  // Blink ON
WriteCmdXLCD(1);                         // Clear LCD
```

A complete example is given below, which illustrates how the LCD can be initialized and used.

Example 4.31

A text-based LCD is connected to a PIC18F452 microcontroller in the default mode as shown in Figure 4.53. Write a program to clear the LCD and then send the text "My Computer" to the LCD.

Solution

The required program listing is given in Figure 4.54. The message to be displayed is stored in the character array **msg**. At the beginning of the program, PORTB is configured as output with the TRISB = 0 statement. The LCD is then initialized, display cleared, and the text message "My Computer" is displayed on the LCD. Notice that the LCD is cleared by sending command 1 to the LCD controller.

■

Modifying the Default Configuration

It may sometimes be required to use different ports for the LCD. In this section, an example is given to show how the default port configuration can be modified.

Example 4.32

A text-based LCD is connected to a PIC18F452 microcontroller as shown in Figure 4.55. Show how the default configuration can be modified and write a program to count on the LCD from 0 to 99 with a delay of 1 s between each count. The display should be as follows:

```
NO = nn,
```

where nn is from 0 to 99.

Assume that a PIC18F452 microcontroller is used in the design with a 4-MHz clock.

Solution

The required pin configuration is as follows:

Microcontroller Port	LCD Pin
RC0	D4
RC1	D5
RC2	D6
RC3	D7
RC4	E
RC5	RS
RC6	RW

```
/*****************************************************************************************************

                                    LCD MESSAGE
                                    ============

This program displays the message "My Computer" on the LCD. A PIC18F452 microcontroller is used in
the design and the LCD is connected in the default mode. i.e. the connections between the LCD and the
microcontroller are as follows:

         RB0  - D4
         RB1  - D5
         RB2  - D6
         RB3  - D7
         RB4  - E
         RB5  - RS
         RB6  - RW

The LCD is operated in 4-bit mode with 5x7 character font.

A 4MHz crystal is used in the design.


File:        LCD1.C
Version:     V1.0
Author:      Dogan Ibrahim
Date:        April, 2009

*****************************************************************************************************/
```

Figure 4.54: LCD Program Listing

```
#include <p18f452.h>
#include <xlcd.h>
#include <delays.h>

#pragma config WDT = OFF
#pragma config OSC = XT

//
// Defines
//
#define CLR_LCD     1
#define HOME_LCD 2

//
// LCD Delays
//
void DelayFor18TCY(void)
{
        Nop( ); Nop( ); Nop( ); Nop( );                     // 18 cycle delay
        Nop( ); Nop( ); Nop( ); Nop( );
        Nop( ); Nop( ); Nop( ); Nop( );
        Nop( ); Nop( );
        return;
}


void DelayPORXLCD(void)                                   // 15ms delay
{
        Delay1KTCYx(15);
}


void DelayXLCD(void)
{
        Delay1KTCYx(5);                                     // 5ms delay
}


/*================ Start of MAIN program ================ */

void main(void)
{
        char msg[ ] = "My Computer";

        OpenXLCD(FOUR_BIT & LINE_5X7);                      // 8 bit,5x7 character
        WriteCmdXLCD(CLR_LCD);                              // Clear LCD
        while(BusyXLCD( ));
        WriteCmdXLCD(HOME_LCD);                             // Home cursor
        while(BusyXLCD( ));
        putsXLCD( msg );                                    // Write data

        while(1);
}
```

Figure 4.54: *Cont'd*

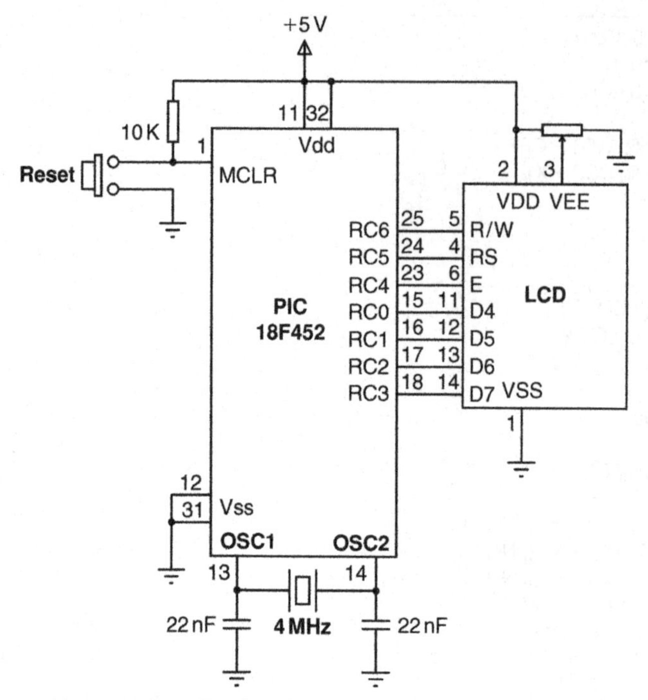

Figure 4.55: Circuit Diagram for the Example 4.32

The first step is to modify the LCD configuration file **xlcd.h**. The steps are given below:

- Copy file **xlcd.h** to **xlcd_default.h** in directory **C:\MCC18\h**. The default LCD library is named as **xlcd_default** and is available if required.
- Modify the following lines in file **xlcd.h** to reflect the required interface between the microcontroller and the LCD:

```
#define DATA_PORT PORTC
#define TRIS_DATA_PORT TRISC
#define RW_PIN LATCbits.LATC6           /* PORT for RW */
#define TRIS_RW TRISCbits.TRISC6        /* TRIS for RW */
#define RS_PIN LATCbits.LATC5           /* PORT for RS */
#define TRIS_RS TRISCbits.TRISC5        /* TRIS for RS */
#define E_PIN LATCbits.LATC4            /* PORT for D */
#define TRIS_E TRISCbits.TRISC4         /* TRIS for E */
```

- Start a command session. **Start -> Run -> Cmd**
- Go to directory **C:\MCC18\src**

- Enter the command **make_one_subsystem_t 18f452 XLCD** to rebuild the LCD library for the PIC18F452 microcontroller
- Wait until the new device library is built incorporating the modifications

The required program listing is shown in Figure 4.56. In addition to the standard LCD delay functions, a delay function called **wait_a_sec** is created to delay for 1 s. Inside the main program, the LCD is initialized and then a **for** loop is entered with variable **cnt** to count from 0 to 99. Inside this **for** loop, the LCD is cleared, cursor is set to home position, and the value of variable **cnt** is converted into a string using function **itoa** and then displayed on the LCD using function **putsXLCD**. The **for** loop is executed 100 times with a 1-s delay between each iteration. Thus, the display counts up from 0 to 99 as shown below:

NO = 0
NO = 1
NO = 2
....
....
NO = 99

■

```
/*****************************************************************************************************

                                        LCD COUNTER
                                        ===========

This program counts up from 0 to 99 and displays on the LCD as:

            NO = nn

Where nn is 0 to 99.

A PIC18F452 microcontroller is used in the design and the LCD is
connected to PORTC of the microcontroller as follows:

            RC0  - D4
            RC1  - D5
            RC2  - D6
            RC3  - D7
            RC4  - E
            RC5  - RS
            RC6  - RW

File XLCD.h is modified to reflect the new connection and then the
PIC18F452 library file is rebuilt for the new changes to take effect.

The LCD is operated in 4-bit mode with 5x7 character font.

A 4MHz crystal is used in the design.
```

Figure 4.56: Program Listing for the Example 4.32

```
File:       LCD2.C
Version:    V1.0
Author:     Dogan Ibrahim
Date:       April, 2009

*****************************************************************************************************/
#include <p18f4520.h>
#include <xlcd.h>
#include <delays.h>
#include <stdlib.h>

#pragma config WDT = OFF
#pragma config OSC = XT

//
// Defines
//
#define CLR_LCD    1
#define HOME_LCD 2

//
// LCD Delays
//
void DelayFor18TCY(void)
{
      Nop( ); Nop( ); Nop( ); Nop( );                        // 18 cycle delay
      Nop( ); Nop( ); Nop( ); Nop( );
      Nop( ); Nop( ); Nop( ); Nop( );
      Nop( ); Nop( );
      return;
}

void DelayPORXLCD(void)                                     // 15ms delay
{
      Delay1KTCYx(15);
}

void DelayXLCD(void)
{
      Delay1KTCYx(5);                                        // 5ms delay
}

void wait_a_sec(void)
{
      Delay10KTCYx(100);                                     // 1 sec delay
}

//
// Start of MAIN program
//
void main(void)
{
```

Figure 4.56: *Cont'd*

```
    char msg[ ] = "No =    ";
    int cnt;

    TRISC = 0;

    OpenXLCD(FOUR_BIT & LINE_5X7);               // 8 bit,5x7 character

    for(cnt = 0; cnt < 100; cnt++)               // Do 100 times
    {
       WriteCmdXLCD(CLR_LCD);                    // Clear LCD
       while(BusyXLCD( ));
       WriteCmdXLCD(HOME_LCD);                   // Home cursor
       while(BusyXLCD( ));
       itoa(cnt, msg+5);                         // Convert to string
       putsXLCD( msg );                          // Display data
       wait_a_sec( );                            // Wait 1 second
    }

    while(1);
}
```

Figure 4.56: *Cont'd*

4.9.9 Software CAN2510 Functions

These functions implement the CAN bus functions for the MCP2510. This is a specialized field and more information can be obtained from the document *MPLAB C18 C Compiler Libraries.*

4.9.10 Software I²C Bus Functions

These functions are used to implement the I²C bus functions. More information can be obtained from the document *MPLAB C18 C Compiler Libraries.*

4.9.11 Software SPI Bus Functions

These functions are used to implement the SPI bus functions. More information can be obtained from the document *MPLAB C18 C Compiler Libraries.*

4.9.12 Software UART Functions

These functions implement RS232-based serial communication using software functions. Because the serial communication is an important topic in the microcontroller field, more details will be given about the serial communication in general and the use of this library.

UART functions are used for RS232-based serial communication between two electronic devices. In serial communication, only two cables (plus a ground cable) are required to transfer data in either direction. Data is sent in serial format over the cable bit by bit. Normally, the receiving device is in idle mode with its transmit (TX) pin at logic 1, also known as MARK. Data transmission starts when this pin goes to logic 0, also known as SPACE. The first bit sent is the **start bit** at logic 0. Following this bit, 7 or 8 **data bits** are sent followed by an optional **parity bit**. The last bit sent is the **stop bit** at logic 1. Serial data is usually sent as a 10-bit frame consisting of a start bit, 8 data bits, a stop bit, and no parity bits. Figure 4.57 shows how character "A" can be sent using serial communication. Character "A" has the ASCII bit pattern "01000001." As shown in the figure, first the start bit is sent, followed by 8 data bits "01000001," and finally the stop bit is sent.

The bit timing is very important in serial communication, and both the transmitting (TX) and receiving (RX) devices must have the same bit timings. The bit timing is measured by the **baud rate**, which specifies the number of bits transmitted or received each second. Typical baud rates are 4800, 9600, 19 200, 38 400, and so on. For example, when operating at a baud rate of 9600 with a frame size of 10 bits, 960 characters are transmitted or received each second. The timing between each bit is then approximately 104 ms.

In RS232-based serial communication, the two devices are connected to each other (see Figure 4.58) using either a 25-way connector or a 9-way connector. Normally, only the TX, RX, and GND pins are required for communication. The required pins for both types of connectors are given in Table 4.18.

The voltage levels specified by the RS232 protocol are ±12 V. A logic HIGH signal is at −12 V and a logic LOW signal is at +12 V. On the other hand, PIC microcontrollers normally operate at 0–5-V voltage levels, and it is required to convert the RS232 signals to 0–5 V when input to a microcontroller. Similarly, the output of the microcontroller must be converted to ±12-V voltage level before sending to the receiving RS232 device. The voltage conversion

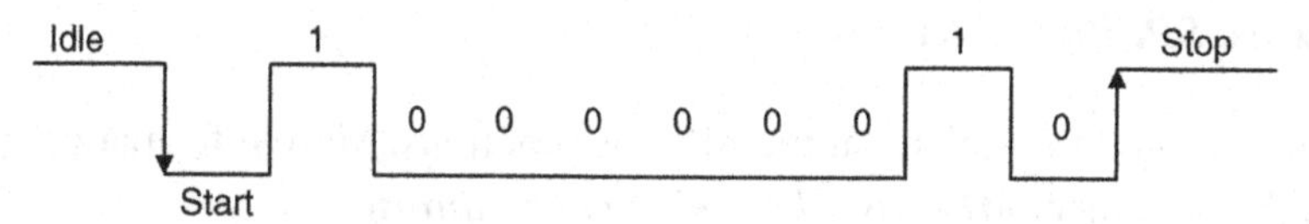

Figure 4.57: Sending Character "A" in Serial Communication

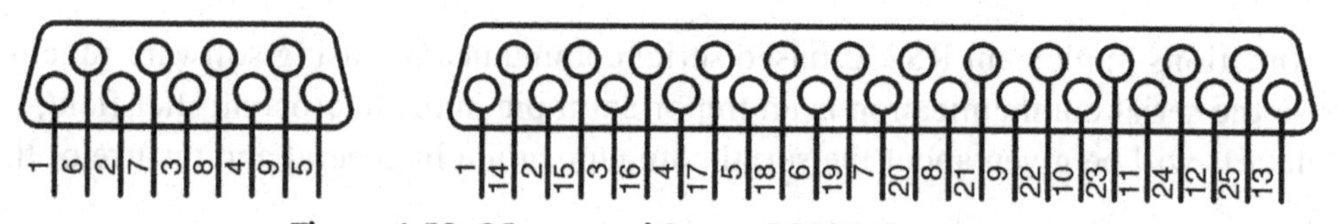

Figure 4.58: 25-way and 9-way RS232 Connectors

Table 4.18: Pins Required for Serial Communication

Pins	9-Way Connectors	25-Way Connectors
TX	2	2
RX	3	3
GND	5	7

Table 4.19: C18 Software UART Functions

Functions	Descriptions
OpenUART	Configures the UART I/O pins
ReadUART	Reads a byte from software UART
WriteUART	Writes a byte to software UART
putsUART	Writes a string to software UART

is usually carried out using RS232 converter chips, such as the MAX232, manufactured by Maxim Inc.

Serial communication is either implemented in the hardware using a specific pin of a microcontroller or the required signals can be generated in the software from any required pin of a microcontroller. Hardware implementation requires either an on-chip UART (or USART) circuit or an external UART chip to be connected to the microcontroller. On the other hand, software-based UART is more commonly used and it does not require any special circuits. Serial data is generated by delay loops in the software-based UART applications. In this section, only the software-based UART functions will be described.

C18 compiler supports the software UART functions shown in Table 4.19. The header file "sw_uart.h" must be included at the beginning of a program using the software UART functions.

A brief description of the C18 software UART functions is given below.

OpenUART

This function configures the I/O pins for the software UART. The default pin configuration is as follows:

- TX pin – port pin RB4
- RX pin – port pin RB5

The above UART pin configurations can be modified by redefining the "equ" statements in files "writuart.asm2," "readuart.asm," and "openuart.asm" found in directory "c:\MCC18\src\traditional\pmc_common\sw_uart."

It is required that the following functions be defined by the user to provide the appropriate delay functions for the software UART library:

DelayTXBitUART – delay for $[(2 * f)/(4 * baud) + 1]/2 - 12$ cycles

DelayRXHalfBitUART – delay for $[(2 * f)/(8 * baud) + 1]/2 - 9$ cycles

DelayRXBitUART – delay for $[(2 * f)/(4 * baud) + 1]/2 - 14$ cycles

As an example, using a clock frequency of 4 MHz and assuming the required baud rate to be 2400, the needed delays are as follows:

$$DelayTXBitUART = [(2 * 4 \times 10^6)/(4 * 2400) + 1]/2 - 12 = 405 \text{ cycles}$$

$$DelayRXHalfBitUART = [(2 * 4 \times 10^6)/(8 * 2400) + 1]/2 - 9 = 199 \text{ cycles}$$

$$DelayRXBitUART = [(2 * 4 \times 10^6)/(4 * 2400) + 1]/2 - 14 = 403 \text{ cycles}$$

Figure 4.59 shows how the required delays can be obtained for the above example.

ReadUART

This function reads a byte from the software UART. An example is given below:

```
char z;
z = ReadUART( );            // Read a byte from UART
```

WriteUART

This function sends a byte to the software UART. An example is given below:

```
char z == 'A';
WriteUART(z);               // Send a byte to UART
```

```
void DelayTXBitUART(void)

        Delay10TCYx(40);                        // 405 cycle delay
        Nop( ); Nop( ); Nop( );
}


void delayRXHalfBitUART(void)
{
        Delay10TCYx(19);                        // 199 cycles
        Nop( ); Nop( ); Nop( ); Nop( );
        Nop( ); Nop( ); Nop( );
}


void DelayRXBitUART(void)
{
        Delay10TCYx(40);                        // 403 cycles
        Nop( );
}
```

Figure 4.59: Delay Functions for 2400 baud with a 4-MHz Clock

putsUART

This function sends a string of characters to software UART. An example is given below:

```
char buff[ ] = "Hello World";
putsUART(buff);              // Send a string to UART
```

Example 4.33

A PIC18F452 microcontroller is connected to a PC using a MAX232 type level converter chip using the default connections as shown in Figure 4.60. Write a program to receive a character from the PC, then increment this character, and send it back to the PC. Assume that the required baud rate is 2400 and a PIC18F452 microcontroller is used with a 4-MHz crystal.

Solution

The program listing is shown in Figure 4.61. The delay functions are used at the beginning of the program. Then, an indefinite loop is formed and a character is received from the serial line. The character is incremented by one and then sent back to the serial line.

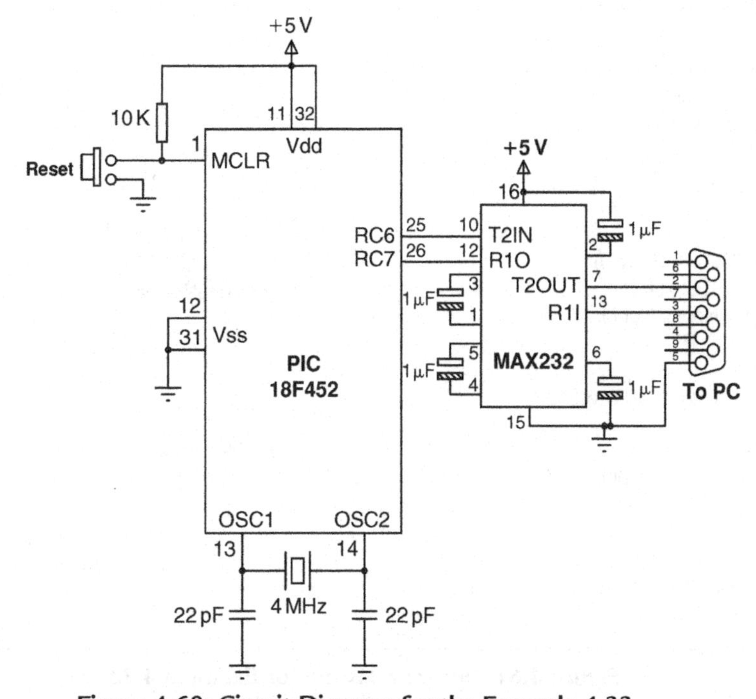

Figure 4.60: Circuit Diagram for the Example 4.33

```
/*******************************************************************************************************

                    READING AND WRITING TO SERIAL PORT
                    ==================================

In this program PORTC pins RC6 and RC7 are configured as serial RX and TX pins respectively. The
communication baud rate is set to 2400 Baud, 8 bits, no parity and with 1 stop bit.
The default UART pins are TX = RB4 and RX = RB5. It is therefore necessary to modify the equ
statements in the following files (found in src/traditional/pmc/sw_uart or src/extended/pmc/sw_uart) so
that pins RC6 and RC7 can be used by the UART routines:

          writuart.asm
          readuart.asm
          openuart.asm


The UART library should then be re-compiled using the provided batch files and then the library included
in the project.

A PIC18F452 type microcontroller, operated with 4MHz is used in the design. The program receives a
character from the serial port of a PC, increments this character by one, and then sends it back to the
PC. Thus, for example, if character "A" is entered on the PC keyboard, character "B" will be sent back
and displayed on the PC screen.


File:               SERIAL.C
Version:            V1.0
Author:             Dogan Ibrahim
Date:               May, 2009
*******************************************************************************************************/

#include <p18f452.h>
#include <sw_uart.h>
#include <delays.h>

#pragma config WDT = OFF
#pragma config OSC = XT

void DelayTXBitUART(void)
     Delay10TCYx(40);                                   // 405 cycle delay
     Nop( ); Nop( ); Nop( );
}

void DelayRXHalfBitUART(void)
{
     Delay10TCYx(19);                                   // 199 cycles
     Nop( ); Nop( ); Nop( ); Nop( );
     Nop( ); Nop( ); Nop( );
}

void DelayRXBitUART(void)
{
```

Figure 4.61: Program Listing for Example 4.33

```
        Delay10TCYx(40);                                    // 403 cycles
        Nop( );
}
//
// Start of MAIN program
//
void main(void)
{
        unsigned char z;

        OpenUART( );                              // Initialize UART

        for(;;)                                   // Endless loop
        {
           z = ReadUART( );                       // Read a character
           z++;                                   // Increment character
           WriteUART(z);                          // Write a character
        }
}
```

Figure 4.61: *Cont'd*

Table 4.20: C18 A/D Functions

A/D Functions	Descriptions
BusyADC	Is A/D converter busy?
CloseADC	Disables the A/D converter
ConvertADC	Starts an A/D conversion
OpenADC	Configures the A/D converter
ReadADC	Reads the conversion result
SetChanADC	Selects A/D channel to be used

4.9.13 Hardware Analog-to-Digital (A/D) Converter Functions

C18 compiler provides the A/D functions shown in Table 4.20. The header file "adc.h" must be included at the beginning of a program using the A/D functions. These functions are described in this section.

BusyADC: This function checks if the A/D converter is currently performing a conversion. A logic "1" is returned when the A/D converter is ready and a logic "0" is returned if the A/D converter is not performing a conversion.

CloseADC: This function disables the A/D converter module.

ConvertADC: This function starts an A/D conversion. The function BusyADC( ) should be used to find out when the conversion is complete.

OpenADC: This function is used to configure the A/D converter module. The function takes two arguments: **config** and **config2**. The values defined can be bitwise AND ed. The type and number of available definitions for **config** and **config2** depend on the type of microcontroller used. For example, for PIC18F452 microcontrollers, the important definitions are given below:

Config

A/D Clock source definitions:

ADC_FOSC_2	FOSC / 2
ADC_FOSC_4	FOSC / 4
ADC_FOSC_8	FOSC / 8
ADC_FOSC_16	FOSC / 16
ADC_FOSC_32	FOSC / 32
ADC_FOSC_64	FOSC / 64
ADC_FOSC_RC	Internal RC oscillator

A/D Result justification:

ADC_RIGHT_JUST	Right justify the result
ADC_LEFT_JUST	Left justify the result

A/D Voltage reference source:

ADC_8ANA_0REF	Vref+ = VDD and Vref- = VSS All analog channels

Config2

Channel:

ADC_CH0	Channel 0
ADC_CH1	Channel 1
ADC_CH2	Channel 2
ADC_CH3	Channel 3
ADC_CH4	Channel 4
ADC_CH5	Channel 5
ADC_CH6	Channel 6
ADC_CH7	Channel 7

A/D Interrupt

ADC_INT_ON	Interrupts enabled
ADC_INT_OFF	Interrupts disabled

An example for the use of OpenADC function is given below:

```
OpenADC( ADC_FOSC_64      &
         ADC_RIGHT_JUST   &
```

```
            ADC_8ANA_0REF       &
            ADC_CH0             &
            ADC_INT_OFF );
```

ReadADC: This function reads the 16-bit result of the A/D conversion. An example is given below to show how A/D result can be read:

```
int result;

OpenADC(    ADC_FOSC_64         &
            ADC_RIGHT_JUST      &
            ADC_8ANA_0REF       &
            ADC_CH0             &
            ADC_INT_OFF );

ConvertADC( );
while(BusyADC( ));          // Wait until conversion is complete
Result = ReadADC( );        // Read the A/D result
CloseADC( );
```

4.9.14 Hardware Input Capture Functions

These functions are used to implement the input capture functions in hardware. More information can be obtained from the document *MPLAB C18 C Compiler Libraries.*

4.9.15 Hardware I²C Functions

These functions are used to implement the I²C bus functions in hardware. More information can be obtained from the document *MPLAB C18 C Compiler Libraries.*

4.9.16 Hardware I/O Port Functions

These functions are used to implement I/O port functions in hardware. More information can be obtained from the document *MPLAB C18 C Compiler Libraries.*

4.9.17 Hardware Microwire Functions

These functions are used to implement the microwire bus functions in hardware. More information can be obtained from the document *MPLAB C18 C Compiler Libraries.*

4.9.18 Hardware Pulse Width Modulation Functions

These functions are used to implement pulse width modulation functions in hardware. More information can be obtained from the document *MPLAB C18 C Compiler Libraries.*

4.9.19 Hardware SPI Functions

These functions are used to implement the SPI bus functions in hardware. More information can be obtained from the document *MPLAB C18 C Compiler Libraries.*

4.9.20 Hardware Timer Functions

These functions are used to configure the microcontroller timers. The header file "timers.h" must be included at the beginning of a program using the timer functions. Table 4.21 gives a list of the available functions.

Only Timer 0 functions are described in this section. Further information on the functions available for other timers can be obtained from the *MPLAB C18 C Compiler libraries.*

CloseTimer 0: This function disables the interrupt and the specified timer.

OpenTimer 0: This function is used to enable a timer. The argument consists of a parameter called **config** that can be made up of a number of bitwise AND ed definitions. The following definitions are available:

Config
Enable Timer 0 interrupt:

TIMER_INT_ON	Interrupt enabled
TIMER_INT_OFF	Interrupt disabled

Timer width:

T0_8BIT	8-bit mode
T0_16BIT	16-bit mode

Clock source:

T0_SOURCE_EXT	External clock source
T0_SOURCE_INT	Internal clock source

External clock trigger

T0_EDGE_FALL	External clock on falling edge
T0_EDGE_RISE	External clock on rising edge

Table 4.21: C18 Timer Functions

Functions	Descriptions
CloseTimerx	Disables timer x
OpenTimerx	Configures and enables timer x
ReadTimerx	Reads the value of timer x
WriteTimerx	Writes a value into timer x

x can be 0, 1, 2, 3, or 4 depending on device type.

Prescale value:

T0_PS_1_1	1:1 prescale
T0_PS_1_2	1:2 prescale
T0_PS_1_4	1:4 prescale
T0_PS_1_8	1:8 prescale
T0_PS_1_16	1:16 prescale
T0_PS_1_32	1:32 prescale
T0_PS_1_64	1:64 prescale
T0_PS_1_128	1:128 prescale
T0_PS_1_256	1:256 prescale

An example for the use of OpenTimer0 function I given below:

```
OpenTimer0( TIMER_INT_OFF    &
            T0_8BIT          &
            T0_SOURCE_INT    &
            T0_PS_1_64);
```

4.9.21 Hardware USART Functions

Hardware USART functions enable RS232 type serial communication to be implemented using the hardware USART module of the microcontroller. In general, hardware-based USART can give faster and more reliable communication. In addition, the processor can carry out other tasks while the USART is handling the serial communication.

C18 compiler provides the USART functions given in Table 4.22 (in microcontrollers with more than one USART a number is added to the end of these functions to identify the USARTs). The header file "usart.h" must be defined at the beginning of a program using these functions.

The definition of these functions is given in this section.

Table 4.22: C18 Hardware USART Functions

Functions	Descriptions
BusyUSART	Checks if the USART is transmitting data
CloseUSART	Disables the USART
DataRdyUSART	Makes data available in USART read buffer
OpenUSART	Configures USART
getcUSART	Reads a byte from the USART
getsUSART	Reads a string from the USART
putcUSART	Writes a byte to USART
putsUSART	Writes a string from data memory to the USART
putrsUSART	Writes a string from program memory to the USART
baudUSART	Sets the baud rate configuration bits for the USART

BusyUSART: This function returns a "1" if the USART transmitter is busy transmitting a character. This function should be checked before sending a new byte to the USART. The function returns a "0" if the USART transmitter is idle.

CloseUSART: This function disables the USART.

DataRdyUSART: This function returns a "1" if data is available in the USART read buffer. A "0" indicates that data is not available in the read buffer.

getcUSART: This function reads a byte from the USART buffer. An example is given below:

```
int result;
result = getcUSART( );
```

getsUSART: This function reads a string of characters from the USART. This function waits and reads a specified number of characters. There is no timeout and the program will wait forever if the specified number of characters are not received. An example is given below to show how this function can be used:

```
char buff[20];
getsUSART(buff, 6);            // Wait to receive 6 characters
```

OpenUSART: This function configures the USART. Two arguments are required: a configuration argument called **config** and an integer called **spbrg**, which specifies the value to be written to the baud rate generator register to determine the baud rate.

config

Interrupt on transmission:

USART_TX_INT_ON	Transmit interrupt ON
USART_TX_INT_OFF	Transmit interrupt OFF

Interrupt on reception:

USART_RX_INT_ON	Receive interrupt ON
USART_RX_INT_OFF	Receive interrupt OFF

USART mode:

USART_ASYNCH_MODE	Asynchronous mode
USART_SYNCH_MODE	Synchronous mode

Transmission width:`

USART_EIGHT_BIT	8-bit transmit/receive
USART_NINE_BIT	9-bit transmit/receive

Slave/Master select

USART_SYNC_SLAVE	Synchronous slave
USART_SYNCH+MASTER	Synchronous master

Reception mode:

USART_SINGLE_RX	Single reception
USART_CONT_RX	Continuous reception

Baud rate:

USART_BRGH_HIGH	High baud rate
USART_BRGH_LOW	Low baud rate

spbrg

This is the value written onto the baud rate generator register to define the baud rate to be used. The formula for the baud rate is as follows:

For High Speed (USART_BRGH_HIGH),

$$\text{Baud} = \text{FOSC}/[16 * (\text{spbrg} + 1)]$$

or

$$\text{spbrg} = \text{FOSC}/(16 * \text{baud}) - 1$$

and

For Low Speed (USART_BRGH_LOW),

$$\text{Baud} = \text{FOSC}/[16 * (\text{spbrg} + 1)]$$

or

$$\text{spbrg} = \text{FOSC}/(64 * \text{baud}) - 1,$$

where FOSC is the microcontroller clock frequency.

For example, assuming that the clock frequency is 4 MHz and the required baud rate is 960, using high-speed setting, the value to be specified as **spbrg** can be calculated as

$$\text{spbrg} = \text{FOSC}/(16 * \text{baud}) - 1 = 4 \times 10^6/(16 \times 9600) - 1 = 25$$

Then, the OpenUSART function can be declared as follows (in this example, it is assumed that asynchronous mode is used with 9600 baud and 8 data bits):

```
OpenUSART(USART_TX_INT_OFF          &
          USART_RX_INT_OFF          &
          USART_ASYNCH_MODE         &
```

```
USART_EIGHT_BIT              &
USART_CONT_RX                &
USART_BRGH_HIGH,
25);
```

putcUSART: This function sends a byte to USART.

putsUSART: This function sends a string of characters to USART from the data memory. An example is given below:

```
putrsUSART("My Computer");
```

putrsUSART: This function sends a string of characters to USART from the program memory.

baudUSART: This function sets the baud rate configuration bits for enhanced USART operation. The valid arguments can be formed from bitwise AND of the following definitions:

Clock idle state:

BAUD_IDLE_CLK_HIGH	Clock idle state is high level
BAUD_IDLE_CLK_LOW	Clock idle state is low level

Baud rate generation:

BAUD_16_BIT_RATE	16-bit baud rate generation
BAUD_8_BIT_RATE	8-bit baud rate generation

RX pin monitoring:

BAUD_WAKEUP_ON	RX pin monitored
BAUD_WAKEUP_OFF	RX pin not monitored

Baud rate measurement:

BAUD_AUTO_ON	Autobaud rate measurement enabled
BAUD_AUTO_OFF	Autobaud rate measurement disabled

4.10 Summary

There are many commercially available C compilers. MPLAB C18 is one of the most popular C compiler used by students and by professional programmers. Student's version of the MPLAB C18 compiler is available free of charge and can be downloaded from the Microchip Web site. The MPLAB C18 compiler has been described in detail in this chapter including the use of built-in library functions with simple examples.

4.11 Exercises

1. Write a C program to set bits 0 and 7 of PORTC to logic 1.
2. Write a C program to count down continuously and send the count to PORTB.
3. Write a C program to multiply each element of a 10-element array with number 2.
4. It is required to write a C program to add two matrices **P** and **Q**. Assume that the dimension of each matrix is 3 × 3 and store the result in another matrix called **W**.
5. Repeat Exercise 4, but this time multiply matrices **P** and **Q** and store the product in matrix **R**.
6. What is meant by the terms "variable" and "constant"?
7. What is meant by program repetition? Describe the operation of **while, do-while,** and **for** loops in C.
8. What is an array? Write example statements to define the following arrays:
   a) An array of 10 integers
   b) An array of 30 float
   c) A two-dimensional array having 6 rows and 10 columns
9. Trace the operation of the following loops. What will be the value of variable *z* at the end of the loops?
   a)
   ```
   unsigned char j = 0, z = 0;
   while(j < 10)
   {
           z++;
           j++;
   }
   ```
   b)
   ```
   unsigned char z = 10;
   for(j = 0; j < 10; j++)z--;
   ```
10. Given the following variable definitions, list the outcome of the following conditional tests in terms of "true" or "false":

    ```
    unsigned int a = 10, b = 2;
    if(a > 10)
    if(b >= 2)
    if(a == 10)
    if(a > 0)
    ```
11. Write a program to calculate whether a number is odd or even.

12. Determine the value of the following bitwise operations using AND, OR, and EXOR operations:

```
Operand 1: 00010001
Operand 2: 11110001
```

13. How many times does each of the following loops iterate and what is the final value of the variable *j* in each case?

    a) **for**(j = 0; j < 5; j++)
    b) **for**(j = 1; j < 10; j++)
    c) **for**(j = 0; j <= 10; j++)
    d) **for**(j = 0; j <= 10; j += 2)
    e) **for**(j = 10; j > 0; j -= 2)

14. Write a program to calculate the sum of all positive integer numbers from 1 to 100.

15. Write a program to calculate the average value of the numbers stored in an array. Assume that the array is called **M** and it has 20 elements.

16. Modify the program in Exercise 15 to find the smallest and the largest values of the array. Store the smallest value in variable called **Sml** and the largest value in variable called **Lrg**.

17. Given that f1 and f2 are both floating point variables, explain why the following test expression controlling the **while** loop may not be safe:

```
do
{
    ...............
    ...............
} while(f1 != f2);
```

Why would the problem not occur if both f1 and f2 were integers? How would you correct the above **while** loop?

18. What can you say about the following **while** loop?

```
k = 0;
Total = 0;
while (k < 10)
{
    Sum++;
    Total += Sum;
}
```

19. What can you say about the following **for** loop:

```
Cnt = 0;
for(;;)
{
   Cnt++;
}
```

20. Write a function to calculate the circumference of a rectangle. The function should receive the two sides of the rectangle as floating point numbers and then return the circumference as a floating point number.

21. Write a main program to use the function you developed in Exercise 20. Find the circumference of a rectangle whose sides are 2.3 and 5.6 cm. Store the result in a floating point number called MyResult.

22. Write a function to convert inches to centimeters. The function should receive inches as a floating point number and then calculate the equivalent centimeters.

23. Write a main program to use the function you developed in Exercise 22. Convert 12.5 inches into centimeters and store the result in a floating point number.

24. An LED is connected to port pin RB0 of a PIC18F452-type microcontroller through a current limiting resistor in current sinking mode. Write a program to flash the LED with 5-s intervals.

25. Eight LEDs are connected to PORTB of a PIC18F452-type microcontroller. Write a program so that the LEDs count up in binary sequence with 1-s delay between each output.

26. An LED is connected to port pin RB7 of a PIC18F452 microcontroller. Write a program to flash the LED such that the ON time is 5 s and the OFF time is 3 s.

27. A text-based LCD is connected to a PIC18F452 type microcontroller in 4-bit data mode. Write a program that will display a count from 0 to 255 on the LCD with 1-s interval between each count.

28. Write a program to configure port pin RB2 of a PIC18F452 microcontroller as the RS232 serial output port. Send character "X" to this port at 4800 baud.

29. Port RB0 of a PIC18F452 microcontroller is configured as the RS232 serial output port. Write a program to send out string "SERIAL" at 9600 baud using software USART functions.

30. Repeat Exercise 29 but use the hardware USART available on the microcontroller chip.

CHAPTER 5

PIC18 Microcontroller Development Tools

The development of a microcontroller-based system is a complex process. Development tools are hardware and software tools, which help the programmers to develop and test systems in a relatively short time.

Developing software and hardware for microcontroller-based systems involves the use of editors, assemblers, compilers, debuggers, simulators, emulators, and device programmers. A typical development cycle starts with writing the application program using a text editor. The program is then translated into the executable code using an assembler or a compiler. If the program consists of several modules, then these are combined together into a single application program using a linker. At this stage, any syntax errors are detected by the assembler or the compiler and have to be corrected before an executable code can be generated. In the next stage of the development cycle, a simulator can be used to test the application program without the actual hardware. Simulators can be useful to test the correctness of an algorithm or a program with limited or no input–outputs. Most of the programming and algorithmic errors can be detected and removed during the simulation. If the programmer is happy and the program seems to be working, the next stage of the development cycle is to load the executable code to the target microcontroller chip using a device programmer and then to test the overall hardware and software system. During this cycle, software and hardware tools, such as in-circuit debuggers (ICDs) or in-circuit emulators (ICEs), can be used to analyze the operation of the program and to display the variables and registers in real time with the help of breakpoints set in the program.

5.1 Software Development Tools

Software development tools are computer programs, and they usually run on personal computers, helping the programmer (or system developer) to create and/or modify or test application programs. Some common software development tools are

- Text editors
- Assemblers/compilers
- Simulators

D.O.I.: 10.1016/B978-1-85617-719-1.00009-9

- High-level language simulators
- Integrated development environments (IDEs)

5.1.1 Text Editors

A text editor is a program that allows us to create or edit programs and text files. Windows operating system has a text editor program called *Notepad*. Using Notepad, we can create a new program file, modify an existing file, or display or print the contents of a file. It is important to realize that programs used for word processing, such as the *Microsoft Word*, cannot be used as a text editor. This is because word processing programs are not true text editors, because they embed word formatting characters, such as bold, italic, underline, and so on, inside the text.

Most assemblers and compilers come with built-in text editors. Using these editors, we can create a program and then assemble or compile it without having to exit from the editor. These editors also provide additional features, such as automatic key word highlighting, syntax checking, parenthesis matching, comment-line identification, and so on. Different parts of a program can be shown in different colors to make the program more readable. For example, comments can be shown in one color, key words in another color, etc. Such features help to eliminate syntax errors during the programming stage, thus speeding up the overall development process.

5.1.2 Assemblers and Compilers

Assemblers generate executable code from assembly language programs, and that generated code can then be loaded into the flash program memory of a PIC18-based microcontroller. Similarly, compilers generate executable code from high-level language programs. Some of the commonly used compilers for the PIC18 microcontrollers are BASIC, C, and Pascal.

Assembly language is used in applications where the processing speed is very critical and the microcontroller is required to respond to external and internal events in the shortest possible time. The main disadvantage of assembly language is that it is difficult to develop complex programs using it. In addition, assembly language programs cannot be maintained easily. High-level languages, on the other hand, are easier to learn, and complex programs can be developed and tested in a much shorter time. High-level programs are also maintained more easily than assembly language programs.

Discussions of programming in this book are limited to the C language. Many different C language compilers are available for developing the PIC18 microcontroller-based programs. Some of the popular ones are

- CCS C (http://www.ccsinfo.com)
- Hi-Tech C (http://www.htsoft.com)

- MPLAB C18 C (http://www.microchip.com)
- mikroC C (http://www.mikroe.com)
- Wiz-C C (http://www.fored.co.uk)

Although most C compilers are essentially the same, each one has its own additions or modifications to the standard language. The C compiler used in this book is the MPLAB C18, developed by Microchip Inc.

5.1.3 Simulators

A simulator is a computer program that runs on a PC without any microcontroller hardware, and it simulates the behavior of the target microcontroller by interpreting the user program instructions using the target microcontroller instruction set. Simulators can display the contents of registers, memory, and the status of input–output ports of the target microcontroller as the user program is interpreted. The user can set breakpoints to stop the execution of the program at desired locations and then examine the contents of various registers at the breakpoint. In addition, the user program can be executed in a single-step mode so that the memory and registers can be examined as the program executes one instruction at a time each time a key is pressed.

Some assembler programs also contain built-in simulators to enable programmers to develop and simulate their programs before loading onto a physical microcontroller chip. Some of the popular PIC18 microcontroller tools with built-in simulators are

- MPLAB IDE (http://www.microchip.com)
- Oshon Software PIC18 simulator (http://www.oshonsoft.com)
- Forest Electronics PIC18 assembler (http://www.fored.co.uk)

5.1.4 High-Level Language Simulators

High-level language simulators are also known as source-level debuggers, and like simulators, they are programs that run on a PC and locate errors in high-level programs. We can set breakpoints in high-level statements, execute the program up to the breakpoint, and then display the values of program variables, the contents of registers, and memory locations at that breakpoint.

A source-level debugger can also invoke hardware-based debugging activity using a hardware debugger device. For example, the user program on the target microcontroller can be stopped, and the values of various variables and registers can be examined.

Some high-level language compilers, including the following three, have built-in source-level debuggers:

- MPLAB C18 C
- Hi-Tech PIC18 C
- MikroC C

5.1.5 Integrated Development Environments

IDEs are powerful PC-based programs that have everything; hence, it is possible to edit, assemble, compile, link, simulate, source-level debug, and download the generated executable code to the physical microcontroller chip (using a programmer device). These programs are in the form of graphical user interface (GUI), where the user can select various options from the program without having to exit the program. IDEs can be extremely useful during the development phases of microcontroller-based systems. Most PIC18 high-level language compilers are in the form of an IDE, thus enabling the programmer to do most tasks within a single software development tool.

5.2 Hardware Development Tools

Numerous hardware development tools are available for the PIC18 microcontrollers. Some of these products are manufactured by Microchip Inc. and some by third-party companies. The popular hardware development tools are

- Development boards
- Device programmers
- ICDs
- ICE
- Breadboards

5.2.1 Development Boards

Development boards are invaluable microcontroller development tools. Simple development boards contain just a microcontroller and the necessary clock circuitry. Some sophisticated development boards contain LEDs, LCD, push buttons, serial ports, USB port, power supply circuit, device programming hardware, and so on.

This section is a survey of various commercially available PIC18 microcontroller development boards and their specifications.

LAB-XUSB Experimenter Board

The LAB-XUSB Experimenter board (see Figure 5.1) is manufactured by microEngineering Labs Inc. and can be used in 40-pin PIC18-based project development. The board is available either as an assembled or as a bare board.

The board contains

- 40-pin ZIF socket for PIC microcontroller
- 5-V regulator
- 20-MHz oscillator
- **Reset** button
- 16-switch keypad
- Two potentiometers
- Four LEDs
- Two-line by 20-character LCD module
- Speaker
- RC servo connector

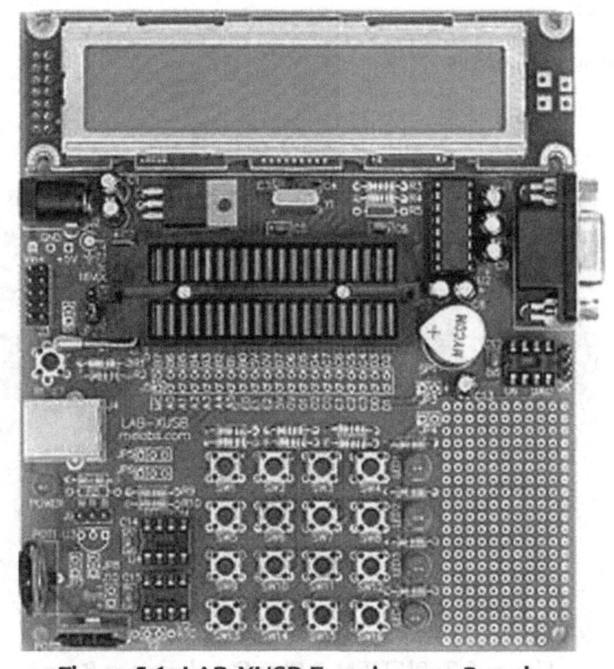

Figure 5.1: LAB-XUSB Experimenter Board

- RS232 interface
- USB connector
- Socket for digital-to-analog converter (device not included)
- Socket for I2C serial EEPROM (device not included)
- Socket for Dallas DS1307 real-time clock (device not included)
- Pads for Dallas DS18S20 temperature sensors (device not included)
- In-circuit programming connector
- Prototyping area for additional circuits

PICDEM 2 Plus

The PICDEM 2 Plus kit (see Figure 5.2) is manufactured by Microchip Inc. and can be used in the development of PIC18 microcontroller-based projects.

The board contains

- 2 × 16 LCD display
- Piezo sounder driven by PWM signal
- Active RS 232 port
- On-board temperature sensor
- Four LEDs

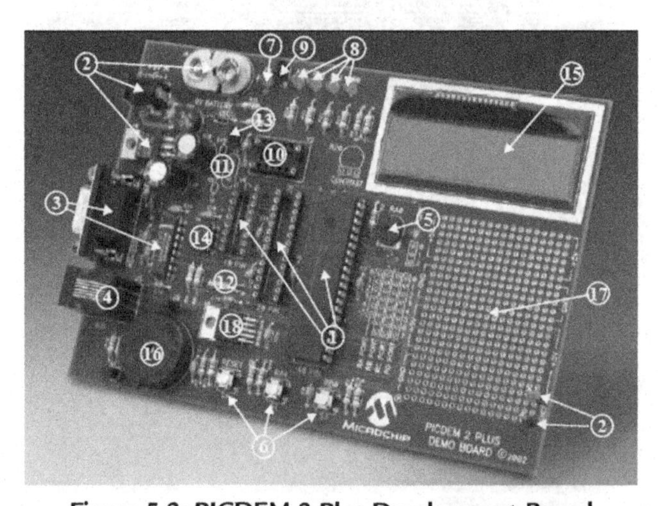

Figure 5.2: PICDEM 2 Plus Development Board

- Two push-button switches and master reset
- Sample PIC18F4520 and PIC16F877A flash microcontrollers
- MPLAB REAL ICE/MPLAB ICD 2 connector
- Source code for all programs
- Demonstration program displaying a real-time clock and ambient temperature
- Generous prototyping area

The board works without a 9-V battery or DC power pack.

PICDEM 4

The PICDEM 4 kit (see Figure 5.3) is manufactured by Microchip Inc. and can be used in the development of PIC18 microcontroller-based projects.

The board contains

- Three different sockets supporting 8-, 14-, and 18-pin DIP devices
- On-board +5-V regulator for direct input from 9-V, 100-mA AC/DC wall adapter
- Active RS-232 port
- Eight LEDs
- 2 × 16 LCD display
- Three push-button switches and master reset
- Generous prototyping area

Figure 5.3: PICDEM 4 Development Board

- I/O expander
- Supercapacitor circuitry
- Area for a LIN transceiver
- Area for a motor driver
- MPLAB ICD 2 connector

PICDEM HPC Explorer Board

The PICDEM HPC Explorer development board (see Figure 5.4) is manufactured by Microchip Inc. and can be used in the development of high-pin-count PIC18 series of microcontroller-based projects.

The main features of this board are

- PIC18F8722, 128 K Flash, 80-pin TQFP microcontroller
- Supports PIC18 J-series devices with Plug-in Modules (PIMs)
- 10-MHz crystal oscillator (to be used with internal PLL to provide 40-MHz operation)
- Power supply connector and programmable voltage regulator, capable of operation from 2.0 to 5.5 V
- Potentiometer (connected to 10-bit A/D, analog input channel)
- Temperature sensor demo included
- Eight LEDs (connected to PORTD with jumper disable)
- RS-232 port (9-pin D-type connector, UART1)
- **Reset** button
- 32-kHz crystal for real-time clock demonstration

Figure 5.4: PICDEM HPC Explorer Board

MK-1 Universal PIC Development Board

The MK-1 Universal PIC development board (see Figure 5.5) is manufactured by Baji Labs and can be used for the development of the PIC microcontroller-based project with up to 40 pins. The board has a key mechanism that allows any peripheral device to be mapped to any pin of the processor, making the board very flexible. In addition, a small breadboard area is provided on the board, enabling users to design and test their own circuits.

The board has the following features

- On-board selectable 3.3 or 5 V
- 16 × 2 LCD character display (8- or 4-bit mode supported)
- Four-digit multiplexed 7-segment display
- 10-LED bar graph (can be used as individual LEDs)
- Eight-position DIP switch
- Socketed oscillator for easy change of oscillators
- Stepper motor driver with integrated driver
- I^2C real-time clock with crystal and battery backup support
- I^2C temperature sensor with 0.5°C precision

Figure 5.5: MK-1 Universal PIC Development Board

- Three potentiometers for direct A/D development
- 16-button telephone keypad wired as 4×4 matrix
- RS232 driver with standard DB9 connector
- Socketed SPI and I²C EEPROM
- RF Xmit and receive sockets
- IR Xmit and receive
- External drive buzzer
- Easy access to pull-up resistors
- AC adapter

SSE452 Development Board

The SSE452 development board (see Figure 5.6), manufactured by Shuan Shizu Electronic Laboratory, can be used for developing the PIC18-based microcontroller projects, especially the PIC18FXX2 series of microcontrollers, and also for programming the microcontrollers.

The main features of this board are as follows:

- One printed circuit board (PCB) suitable for any 28/40-pin PIC18 devices
- Three external interrupt pins
- Two input capture/output compare/pulse width modulation modules (CCP)
- Support for SPI, I²C functions

Figure 5.6: SSE452 Development Board

- 10-bit analog-to-digital (A/D) converter
- One RS-232 connector
- Two debounced push-button switches
- An 8-bit DIP switch for digital input
- A 4 × 4 keypad connector
- A rotary encoder with push button
- TC77 SPI temperature sensor
- An EEPROM (24LC04B)
- A 2 × 20 bus expansion port
- ICD2 connector
- On-board multiple digital signals from 1 Hz to 8 MHz
- Optional devices are 2 × 20 character LCD, 48/28-pin ZIF socket.

SSE8720 Development Board

The SSE8720 development board (see Figure 5.7), manufactured by Shuan Shizu Electronic Laboratory, can be used for the development of PIC18-based microcontroller projects. A large amount of memory and I/O interface are provided, and the board can also be used to program microcontrollers.

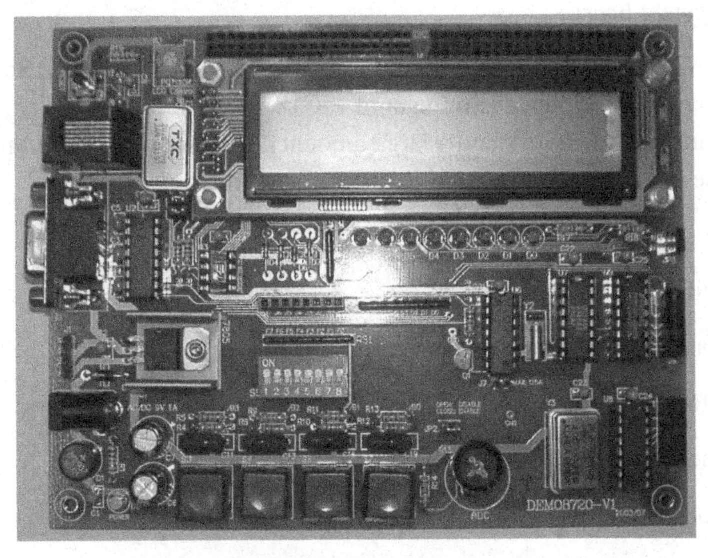

Figure 5.7: SSE8720 Development Board

The main features of this board are as follows:

- 20-MHz oscillator with socket
- One DB9 connector, which provides an EIA232 interface
- ICD connector
- Four debounced switches and one **Reset** switch
- A 4 × 4 keypad connector
- One potentiometer for experiencing A/D conversion
- Eight red LEDs
- One 8-bit DIP switch for digital inputs
- One 2 × 20 character LCD module
- 24 different digital signals from 1 Hz to 16 MHz
- On-board 5-V regulator
- One I^2C EEPROM with socket
- An SPI-compatible digital temperature sensor
- An SPI-compatible real-time clock
- One CCP1 output via a NPN-transistor

SSE8680 Development Board

The SSE8680 development board (see Figure 5.8), manufactured by Shuan Shizu Electronic Laboratory, can be used for the development of PIC18-based microcontroller projects. The board supports CAN network and a large amount of memory and an I/O interface are provided, and the board can also be used to program the microcontrollers.

The main features of this board are as follows:

- 20-MHz oscillator with socket
- One DB9 connector, which provides the EIA232 interface
- ICD connector
- Four debounced switches and one **Reset** switch
- A 4 × 4 keypad connector
- One potentiometer for experiencing A/D conversion

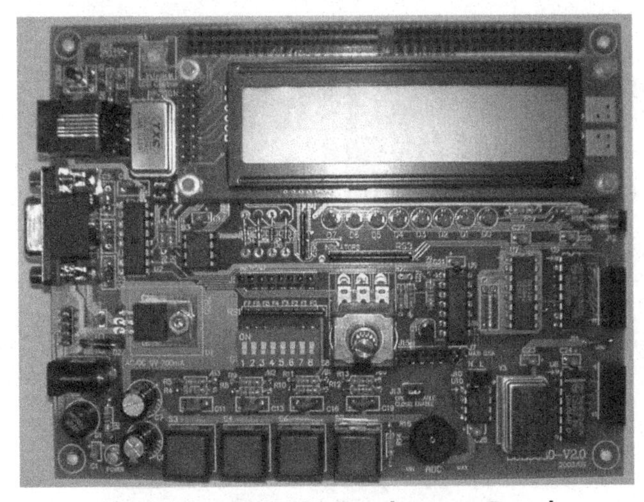

Figure 5.8: SSE8680 Development Board

- Eight red LEDs
- One 8-bit DIP switch for digital inputs
- One 2 × 20 character LCD module
- 24 different digital signals from 1 Hz to 16 MHz
- On-board 5-V regulator
- One I^2C EPROM with socket
- An SPI-compatible digital temperature sensor
- An SPI-compatible real-time clock
- One CCP1 output via an NPN transistor
- A rotary encoder
- CAN transceiver

PIC18F4520 Development Kit

The PIC18F4520 development kit (see Figure 5.9), manufactured by Custom Computer Services Inc., includes a C compiler (PCWH), a prototyping board with the PIC18F4520 microcontroller, an ICD, and a programmer.

The main features of this development kit are

- PCWH compiler
- PIC18F4520 prototyping board

Figure 5.9: PIC18F4520 Development Kit

- Breadboard area
- 93LC56 serial EEPROM chip
- DS1631 digital thermometer chip
- NJU6355 real-time clock IC with attached 32.768-kHz crystal
- Two-digit 7-segment LED module
- ICD/programmer
- DC adapter and cables

Custom Computer Services manufacture a number of other PIC18 microcontroller-based development kits and prototyping boards, such as development kits for CAN, Ethernet, Internet, USB, and serial buses. More information is available at the company's Web site.

BIGPIC4 Development Kit

The BIGPIC4 is a sophisticated development kit (Figure 5.10) that supports the latest 80-pin PIC18 family of microcontrollers. The kit is delivered already assembled, with a PIC18F8520 microcontroller installed and working at 10MHz. The development kit includes an on-board USB port, an on-board programmer, and an ICD. The microcontroller on the board can be replaced easily.

The main features of this development kit are

- 46 buttons
- 46 LEDs

Figure 5.10: BIGPIC4 Development Kit

- USB connector
- External or USB power supply
- Two potentiometers
- Graphics LCD
- 2 × 16 Text LCD
- MMC/SD memory card slot
- Two serial RS232 ports
- ICD
- Programmer
- PS2 connector
- Digital thermometer chip (DS1820)
- Analog inputs
- **Reset** button

A new development board with the name BIGPIC5 is now available from mikroElektronika, offering most functions of the BIGPIC4 at a reduced cost and using 40-pin devices. The BIGPIC5 development board is used in some of the projects in this book.

FUTURLEC PIC18F458 Training Board

The FUTURLEC PIC18F458 training board is a very powerful development kit (see Figure 5.11) based on the PIC18F458 microcontroller and developed by Futurlec (www.futurlec.com). The kit comes already assembled and tested. One of the biggest advantages is its low cost, which is below $45.

The main features are as follows:

- PIC18F458 microcontroller with 10-MHz crystal
- RS232 communication
- Test LED
- Optional real-time clock chip with battery backup
- LCD connection
- Optional RS485/RS422 with optional chip
- CAN and SPI controller
- I^2C expansion
- In-circuit programming
- **Reset** button
- Speaker
- Relay socket
- All port pins available at connectors

Figure 5.11: FUTURLEC PIC18F458 Training Board

PICDEM PIC18 EXPLORER Demonstration Board

The PICDEM PIC18 EXPLORER Demonstration board (Figure 5.12), manufactured by Microchip Inc., is a sophisticated development board that can be used for developing PIC18 microcontroller-based projects. The board comes with a PIC18F8722 microcontroller chip.

The main features of this board are

- PIM for connecting alternate PIC18 microcontroller chips
- 10-MHz crystal
- RS232 communication
- LEDs
- Analog temperature sensor chip
- ICD interface
- **Push-button** switches
- Analog inputs
- LCD display
- SPI I/O expander
- Prototype area for user circuit
- USB connector
- SPI EEPROM

Figure 5.12: PICDEM PIC18 Explorer Demonstration Board

- On-board voltage selection
- PICtail daughter board connector socket

The PICDEM PIC18 Explorer Demonstration board is used in some of the projects in this book.

5.2.2 Device Programmers

After the program has been written and translated into executable code, the resulting HEX file is loaded onto the target microcontroller's program memory with the help of a device programmer. The type of device programmer depends on the type of microcontroller to be programmed. For example, some device programmers can only program PIC16 series, some can program both PIC16 and PIC18 series, and some are designed to program other models of microcontrollers (e.g., Intel 8051 series).

Some microcontroller development kits include on-board device programmers. Hence, the microcontroller chip does not need to be removed and inserted into a separate programming device. This section describes some of the popular device programmers used to program the PIC18 series of microcontrollers.

Forest Electronics USB Programmer

The USB programmer (see Figure 5.13), manufactured by Forest Electronics, can be used to program most PIC microcontrollers with up to 40 pins, including the PIC18 series. The device is connected to the USB port of a PC and receives its power from this port.

Mach X Programmer

The Mach X programmer (Figure 5.14), manufactured by Custom Computer Services Inc., can program microcontrollers of the PIC12, PIC14, PIC16, and PIC18 series ranging from

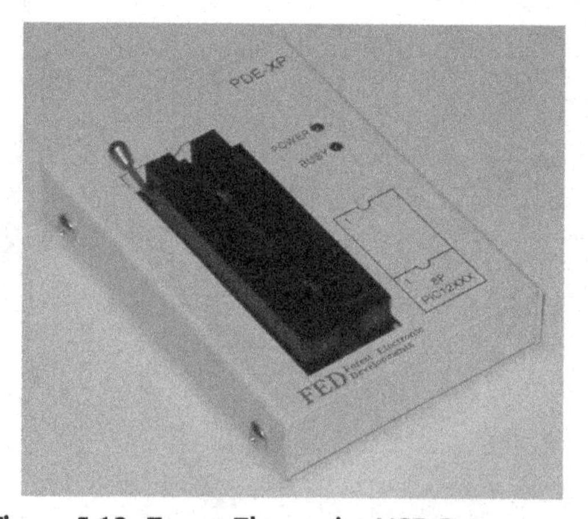

Figure 5.13: Forest Electronics USB Programmer

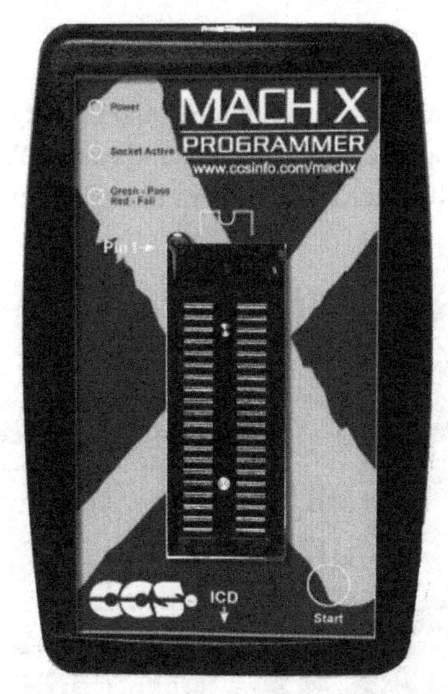

Figure 5.14: Mach X Programmer

8 to 40 pins. This programmer can also read the program inside a microcontroller and then generate a HEX file. In addition, in-circuit debugging is also supported by this programmer.

Melabs U2 Programmer

The Melabs U2 device programmer (see Figure 5.15), manufactured by microEngineering Labs Inc., can be used to program most PIC microcontroller chips from 8 to 40 pins. The device is USB based and receives its power from the USB port of the connected PC.

EasyProg PIC Programmer

The EasyProg PIC is a low-cost programmer (Figure 5.16) that can be used to program PIC16 and PIC18 series of microcontrollers up to 40 pins. The connection to the PC is via a 9-pin serial cable.

PIC Prog Plus Programmer

The PIC Prog Plus (Kanda systems) is another low-cost programmer (Figure 5.17) that can be used to program most PIC microcontrollers. The device is powered from an external 12-V DC supply.

PIC Programmer Module

The PIC Programmer module from Brunning Software (Figure 5.18) can be used to program PIC12, PIC16, and PIC18 series microcontrollers. The module can also be used as a test bed for software and hardware system development.

5.2.3 In-Circuit Debuggers

An ICD is hardware that is connected between a PC and the target microcontroller test system and is used to debug real-time applications quickly and easily. With in-circuit debugging, a monitor program runs in the PIC microcontroller in the test circuit. The programmer can set

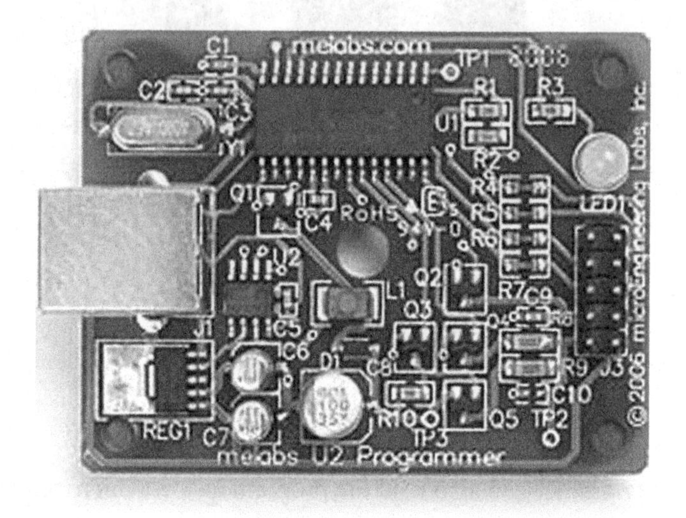

Figure 5.15: Melabs U2 Programmer

Figure 5.16: EasyProg PIC Programmer

breakpoints on the PIC, run code, single-step the program, examine variables and registers on the real device, and change their values if required. An ICD uses some memory and I/O pins of the target PIC microcontroller during the debugging operations. With some ICDs, only the assembly language programs can be debugged. Some more powerful debuggers enable high-level language programs to be debugged.

This section discusses some of the popular ICDs used in PIC18 microcontroller-based system applications.

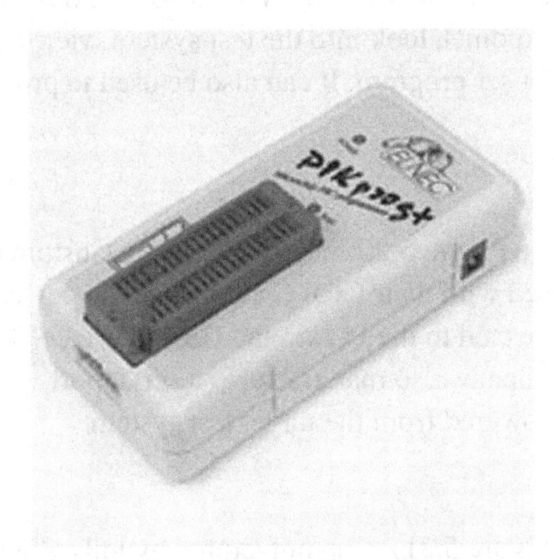

Figure 5.17: PIC Prog Plus Programmer

Figure 5.18: PIC Programmer Module

MPLAB ICD2

The MPLAB ICD2 is a low-cost ICD (see Figure 5.19) manufactured by Microchip Inc. The device can be used for debugging most PIC microcontroller-based systems. With the MPLAB, ICD2 programs are downloaded to a PIC microcontroller chip and executed in real time. This debugger supports both assembly language and C language programs. Breakpoints can be set, the microcontroller can be single-stepped, and registers and variables can be examined or changed if desired.

The MPLAB ICD 2 is connected to the PC using either a serial RS232 interface or via USB. The device acts like an intelligent interface between the PC and the test system, allowing the programmer to set breakpoints, look into the test system, view registers at breakpoints, and single-step through the user program. It can also be used to program the target PIC microcontroller.

ICD-U40

The ICD-U40 is an ICD (see Figure 5.20) manufactured by Custom Computer Services Inc. to debug programs developed with their CCS C compiler. The device operates with a 40-MHz clock frequency and is connected to the PC via the USB interface. The ICD-U40 is powered from the USB port. The company also manufactures a serial port version of this debugger called ICD-S40, which is powered from the target test system.

PICFlash 2

The PICFlash-2 ICD (see Figure 5.21) is manufactured by mikroElektronika and can be used to debug programs developed in mikroBasic, mikroC, or mikroPascal languages. The device is connected to a PC through its USB interface. Power is drawn from the USB port, so the

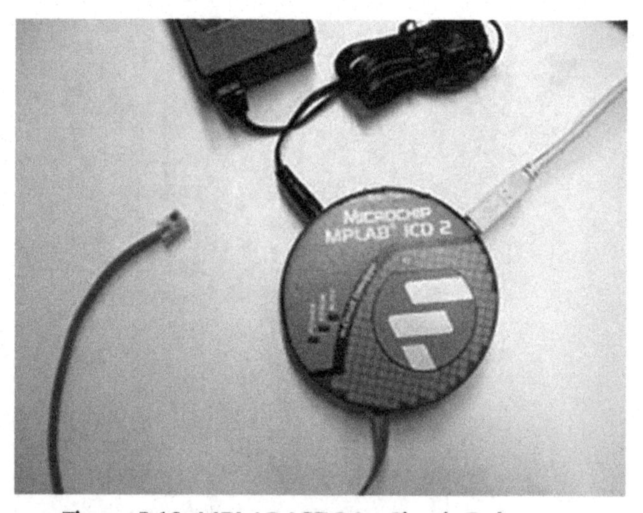

Figure 5.19: MPLAB ICD2 In-Circuit Debugger

Figure 5.20: ICD-U40 In-Circuit Debugger

Figure 5.21: PICFlash 2 In-Circuit Debugger

debugger requires no external power supply. The PICFlash 2 ICD is included in the BIGPIC4 development kit.

MPLAB ICD3

The MPLAB ICD3 is the new ICD device from Microchip Inc. (see Figure 5.22). This is an improved version of MPLAB ICD2, supporting most microcontroller PIC series and dsPIC devices.

The MPLAB ICD3 has the following features:

- Real-time debugging of almost all PIC microcontrollers
- High-speed programming interface

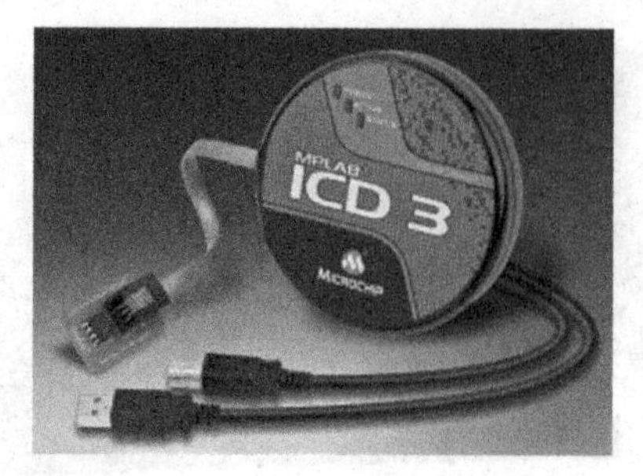

Figure 5.22: MPLAB ICD3 In-Circuit Debugger

- Complex breakpoints and stopwatch
- Simple target microcontroller interface
- Portable, USB interface to a PC
- Standard RJ-11 interface
- Low-voltage emulation
- Low cost

The interface and use of the ICD3 debugger device is described later in this chapter with a detailed example.

5.2.4 In-Circuit Emulators

The ICE is one of the oldest and the most powerful devices for debugging a microcontroller system. It is also the only tool that substitutes its own internal processor for the one in the target system. Like all ICDs, the emulator's primary function is target access – the ability to examine and change the contents of registers, memory, and I/O. As the ICE replaces the CPU, it generally does not require working a CPU on the target system to provide this capability. This makes the ICE by far the best tool for troubleshooting new or defective systems.

In general, each microcontroller family has its own set of ICE. For example, an ICE for the PIC16 microcontrollers cannot be used for the PIC18 microcontrollers. Because of this, to lower the costs, emulator manufacturers provide a multiboard solution to ICE. Usually, a baseboard is provided, which is common to most microcontrollers in the family. For example, the same baseboard can be used by all PIC microcontrollers. Then, probe cards are available for individual microcontrollers. When it is required to emulate a new microcontroller in the same family, it is sufficient to purchase just the probe card for the required microcontroller.

Although ICEs are very powerful debugging tools, their cost is usually very high. Several models of ICEs are available on the market. The following four are some of the more popular ones.

MPLAB ICE 4000

The MPLAB ICE 4000 (Figure 5.23), manufactured by Microchip Inc., can be used to emulate microcontrollers in the PIC18 series. It consists of an emulator pod and the device adapters for the required microcontroller, which are connected using a flex cable. The pod is connected to the PC via its parallel port or using the USB port. Users can insert an unlimited number of breakpoints and examine the register values.

RICE3000

RICE3000 is a powerful ICE (Figure 5.24), manufactured by Smart Communications Ltd, for the PIC16 and PIC18 series of microcontrollers. The device consists of a base unit with different probe cards for the various members of the PIC microcontroller family. The device provides full-speed, real-time emulation up to 40 MHz, supports watching floating-point variables and complex variables, such as arrays and structures, and provides source-level and symbolic debugging in assembly and high-level languages.

ICEPIC 3

The ICEPIC 3 is a modular ICE (see Figure 5.25), manufactured by RF Solutions, for the PIC12/16 and PIC18 series of microcontrollers. The emulator is connected to the PC via its USB port. The device consists of a mother board with additional daughter boards for each microcontroller type. A daughter board is connected to the target system using device adaptors. Additionally, a trace board can be added to the device to capture and analyze execution addresses, opcodes, and external memory read/writes.

Figure 5.23: MPLAB ICE 4000

Figure 5.24: RICE3000 In-Circuit Emulator

Figure 5.25: ICEPIC 3 In-Circuit Emulator

PICE-MC

This is a highly sophisticated emulator (see Figure 5.26), manufactured by Phyton Inc., and supports most PIC microcontrollers. The device consists of a main board, pod, and adapters. The main board contains the emulator logic, and memory, and interface to the PC. The pod contains a slave processor, which emulates the target microcontroller. The adapters are the mechanical parts, which are physically connected to the microcontroller socket of the target

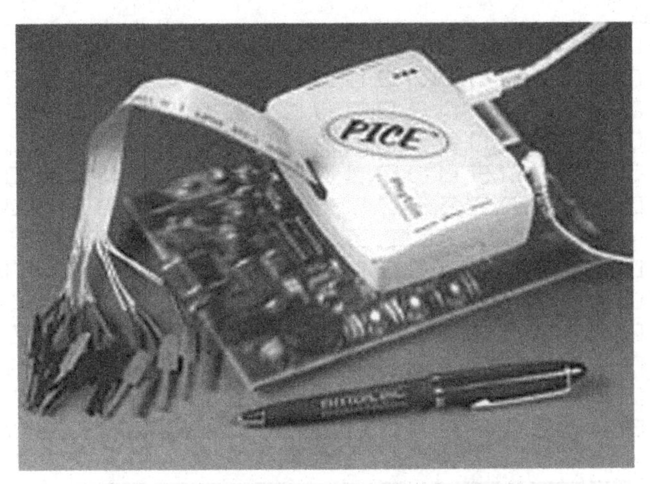

Figure 5.26: PICE-MC In-Circuit Emulator

system. PICE-MC provides source-level debugging of programs written in assembly and high-level languages. A large memory is provided on the system to capture target system data. The user can set up a large number of breakpoints and can access the program and data memories to display or change their contents.

5.2.5 Breadboards

Building an electronic circuit requires connecting the components as shown in the relevant circuit diagram, usually by soldering the components together on a strip board or a PCB. This PCB approach is appropriate for circuits that have been tested and are functioning as desired and also when the circuit is being made permanent. However, making a PCB design for just a few applications – for instance, while still developing the circuit – is not economical.

Instead, while the circuit is still under development, the components are usually assembled on a solderless breadboard. A typical breadboard is shown in Figure 5.27. The board consists of rows and columns of spaced holes so that integrated circuits and other components can be fitted inside them. The holes have spring actions so that the component leads can be held tightly inside the holes. There are various types and sizes of breadboards depending on the complexity of the circuit to be built. The boards can be stacked together to make larger boards for very complex circuits. Figure 5.28 shows the internal connection layout of the breadboard given in Figure 5.27.

The top and bottom half parts of the breadboard are entirely separate with no connection between them. Columns 1–20 in rows A–F are connected to each other on a column basis. Similarly, rows G–L in columns 1–20 are connected to each other on a column basis. Integrated circuits are placed so that the legs on one side are on the top half of the breadboard

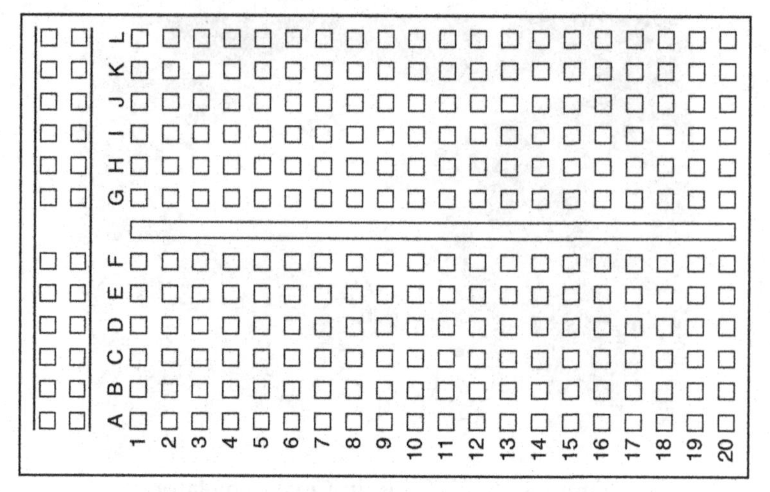

Figure 5.27: A Typical Breadboard Layout

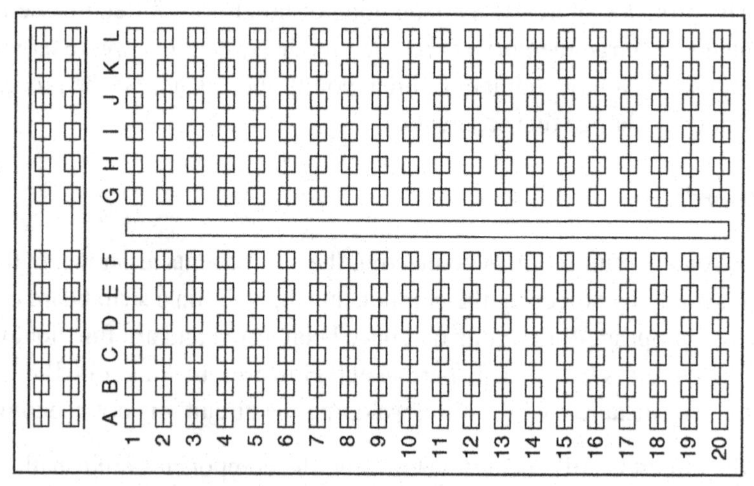

Figure 5.28: Internal Wiring of the Breadboard in Fig. 5.27

and the legs on the other side of the circuit are on the bottom half of the breadboard. The first two columns on the left of the board are usually reserved for power and earth connections. Connections between the components are usually carried out using stranded (or solid) wires plugged inside the holes to be connected.

Figure 5.29 shows picture of a breadboard holding two integrated circuits and a number of resistors and capacitors.

The nice thing about the breadboard design is that the circuit can be modified very easily and quickly and different ideas can be tested without having to solder any components. The

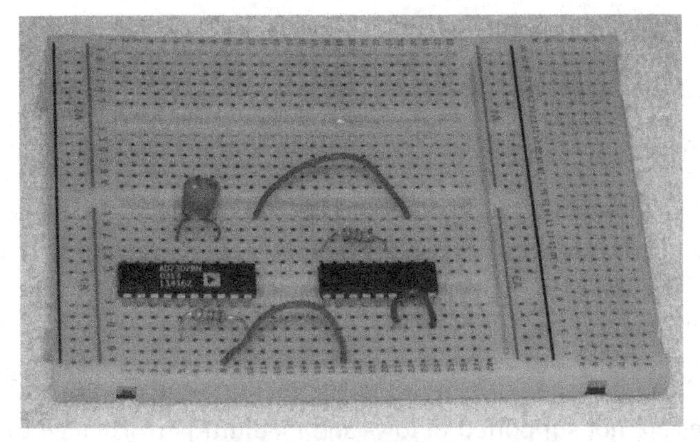

Figure 5.29: A Breadboard With Some Components

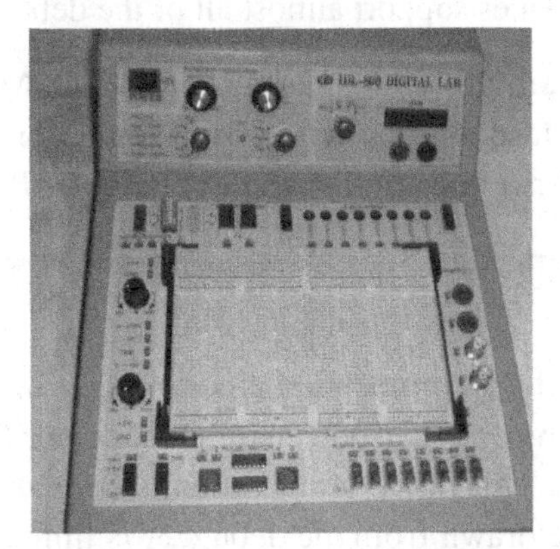

Figure 5.30: IDL-800 Digital Lab

components can easily be removed, and the breadboard can be used for other projects after the circuit has been tested and works satisfactorily.

Some breadboards have built-in power supplies, LEDs, switches, LCD, and so on, making it easier to build and test circuits. Figure 5.30 shows such a complex breadboard, the IDL-800 Digital Lab.

5.3 Using the MPLAB ICD 3 In-Circuit Debugger

The MPLAB ICD 3 in-circuit debugger is a PC-based complex debugger system used for hardware and software development of PIC microcontrollers and dsPIC Digital Signal Controllers based on in-circuit serial programming technology and is manufactured by Microchip Inc. The MPLAB ICD 3 can help the embedded system developer to

- Insert hardware and software breakpoints
- Debug the application in real time

- Set breakpoints based on internal events
- Monitor internal file registers
- Program the device

An ICD is similar to a simulator, but in addition, the ICD connects to the target system hardware and helps the developer to debug the hardware and software in real time. In addition to debugging functions, the MPLAB ICD 3 can be used to program the *program memory* of the target microcontroller device.

The MPLAB ICD 3 can be used with all of the Microchip 8-bit, 16-bit, and 32-bit devices. Some of the features are not supported in low- and medium-performance PIC12F and PIC16F 8-bit devices (see the *MPLAB ICD 3 In-circuit Debugger User's Guide* for full details). 8-bit PIC18F, 16-bit, and 32-bit devices support almost all of the debugger features.

Figure 5.31 is a block diagram of the debugging setup. The MPLAB ICD 3 is connected to the target development board with an RJ-11 type 6-pin connector. The MPLAB ICD 3 is controlled from a PC and is connected to the PC via a USB connector.

The MLAB ICD 3 can be operated in two modes: *target-powered* mode and *debugger powered* mode.

The target-powered mode is the recommended mode, where the source of power for the target hardware is external. In debugger-powered mode, power for the target hardware is derived from the debugger. This mode of operation is not recommended, because the maximum current that can be drawn from the debugger is limited to 100 mA, and this current is drawn from the PC via the USB connection. The debugger-powered mode should be used only during the development of microcontroller-based applications requiring very little current. In addition, the voltage range is limited in the debugger-powered mode. In both modes, analog and digital voltage and ground lines should be connected to the appropriate levels.

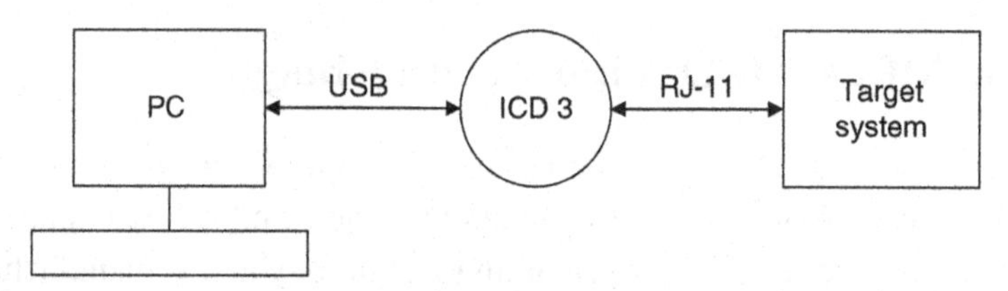

Figure 5.31: The Debugging Setup

Figure 5.32 shows the standard connection of the MPLAB ICD 3 debugger to a PIC microcontroller (in some applications, a header may be required). The RJ-11 pin configuration is as follows:

MPLAB ICD 3 pin	Microcontroller pin
1	Vpp/MCLR
2	Vdd
3	Vss
4	PGD
5	PGC
6	LVP (not used in most applications)

A pull-up resistor (approximately 10 K) is recommended to be connected from the Vpp/MCLR line to the Vdd so that the microcontroller can be reset by the debugger when required.

For programming the microcontroller, no clock is needed on the microcontroller because the debugger sends clock pulses on the PGC line and data on the PGD line, while placing the programming voltage on the Vpp/MCLR line.

For debugging, the target microcontroller must be fully functional, with its power supply connected and its clock running. The requirements for debugging are

- The MPLAB ICD 3 debugger must be connected to a PC via its USB cable, and the PC must be loaded with the MPLAB IDE software (version 8.15 or higher).
- The MPLAB ICD 3 debugger must be connected to the target system as shown in Figure 5.32.
- An external power should be connected to the target system, and both analog and digital power and ground lines of the microcontroller must be connected appropriately.
- The microcontroller must be fully functional with a clock source (e.g., crystal, RC, external oscillator etc.).

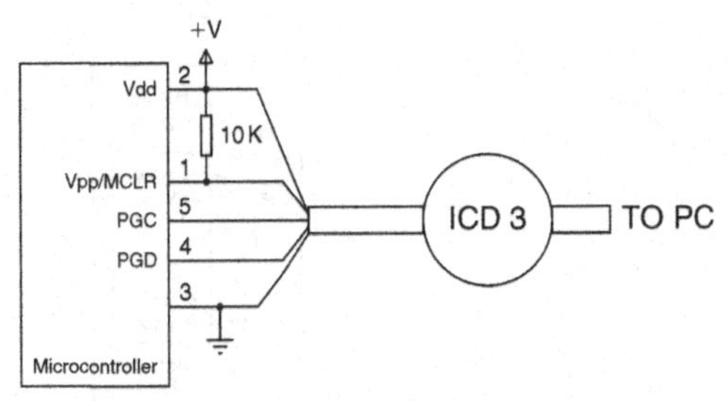

Figure 5.32: Standard Connection of MPLAB ICD 3 to a Microcontroller

- The microcontroller configuration words must be programmed correctly.
    - Watchdog timer must be disabled.
    - Code protection must be disabled.
    - Table read protection must be disabled.
- LVP must be disabled.

5.3.1 The Debugging Process

The debugging process is shown in Figure 5.33. The debugger copies a small program called the *Debug Executive* to the target microcontroller. This executive runs like an application in the program memory and uses some of the file registers and some stack locations of the target microcontroller. The resources used by the debugger depend on the processor type, and more information can be found in the debugger user guide.

To find out whether or not an application program will run correctly, breakpoints are usually set in the target microcontroller device. The debugger sends commands via the PGC and PGD pins to set and store breakpoint addresses in the *internal debug registers* area of the target microcontroller.

The user application program starts to run from reset vector (address 0) of the program memory and will execute until a breakpoint address is encountered. At this point, control is transferred to the debug executive, and the user application program is halted. The MPLAB IDE communicates using the PGC and PGD pins of the MPLAB ICD 3 debugger and gets information about the state of the CPU and value of registers in the register file of the target microcontroller at the halted breakpoint and then displays this information on the PC as requested. The user can examine and also change the value of any register of the register file at the breakpoint if desired.

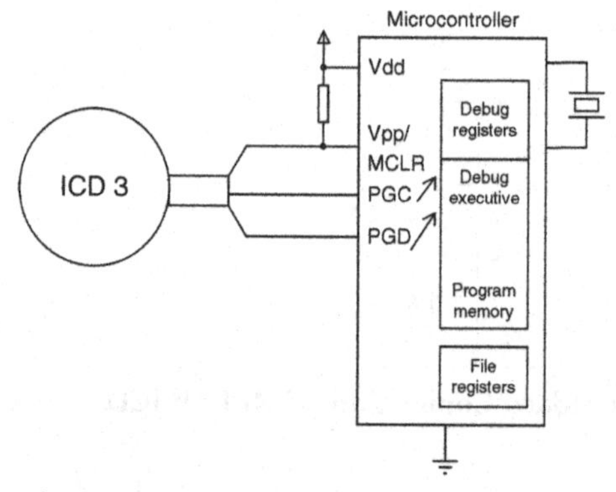

Figure 5.33: The Debugging Process

5.3.2 The MPLAB ICD 3 Test Interface Board

A small PCB called the *ICD 3 Test Interface Board* is supplied with the MPLAB ICD 3 debugger device, which can be used to test that the debugger is functioning correctly. The steps to use this board are as follows:

- Disconnect the debugger from the PC.
- Connect the MPLAB ICD 3 to the test interface board.
- Connect the debugger to the PC.
- Start MPLAB IDE and select ICD 3 as the programmer (or debugger).
- MPLAB IDE runs the self-test and gives a status of pass or fail.

The MPLAB ICD 3 debugger device has three indicator lights:

- Power light: *Green* when debugger is connected to the PC and is receiving power via its USB cable
- Active light: *Blue* when power is first applied or when target is connected
- Status light: *Green* when the debugger is operating normally (in standby)
  - *Red* when failed
  - *Orange* when the debugger is busy

5.3.3 Programming with the MPLAB ICD 3 Debugger

The MPLAB ICD 3 debugger can be used to program a microcontroller. It is important to realize that all the debug features are disabled when the debugger is used as a programmer and the debug executive is not loaded into the microcontroller. The debugger can, however, toggle the MCLR line to restart the microcontroller. Clock is not required when the debugger is used as a programmer.

An example is given below.

Example 5.1

Eight LEDs are connected to PORTD of a PIC18F8722 type microcontroller (any other type of PIC18 series microcontroller can be used in this example, e.g., PIC18F4520) operating with a 10-MHz clock frequency (see Figure 5.34). Write a program to flash the LEDs on and off five times with 1-s delay between each flashing.

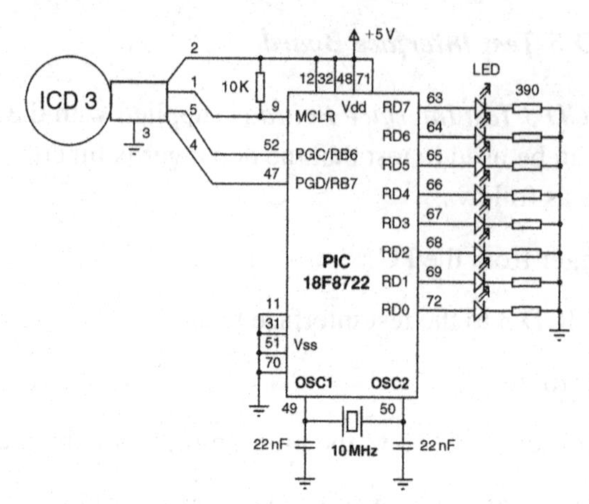

Figure 5.34: Circuit Diagram of Example 5.1

Solution

Although the circuit can be built on a breadboard, in this example, the PICDEM 18 Explorer Development board is used for simplicity. This board is based on the PIC18F8722 80-pin microcontroller and has eight LEDs connected to PORTD of the microcontroller, and the board is compatible with the MPLAB ICD 3 debugger, providing an RJ-11 type socket for the debugger interface.

The required program listing is shown in Figure 5.35. PORTD is configured as output and a **for** loop is used with variable *k* to flash the LEDs five times with a 1-s delay between each output. Note that the delay function **Delay10KTCYx(250)** creates a $250 \times 10{,}000$ cycle time delay. With a 10-MHz clock frequency, the clock period is 0.1 μs, and the cycle time is 0.4 μs. Thus, 2,500,000 cycle delay is equivalent to $2{,}500{,}000 \times 0.4\ \mu s = 1{,}000{,}000\ \mu s$ or 1 s.

The steps to build and compile the program are given in detail in Section 4.3. After the successful compilation, the steps to program the microcontroller with the MPLAB ICD 3 debugger are as follows:

- Connect the MPLAB ICD 3 to the target board (or to the PICDEM 18 Explorer Development board) via the RJ-11 connector.
- If using the PICDEM 18 Explorer Development board, set the following:
  - Switch S4 to PIC MCU.
  - Place jumper JP1 to enable LEDs.

- Connect the MPLAB ICD 3 to the PC via the USB cable.
- Connect +5-V power supply to the target board (or 9–12-V supply to the PICDEM 18 Explorer Development board).
- Select the MPLAB ICD 3 programmer. **Programmer** -> **Select Programmer** -> **MPLAB ICD 3.**
- Select **Project** -> **Build Configuration** -> **Release.**
- Select **Project** -> **Build All** to rebuild the project in Release mode.

```
/*****************************************************************************************************************

                              FLASHING LEDs
                              =============

This program flashes 8 LEDs connected to PortD of a microcontroller 5 times with 1-s intervals. C18 library
function Delay10KTCYx is used to create a 1-s delay between the flashes.

A PIC18F8722 microcontroller is used with a 10-MHz clock.


Programmer:     Dogan Ibrahim
File:           FLASH.C
Version:        1.0
Date:           May, 2009
*****************************************************************************************************************/

#include <p18f8722.h>
#include <delays.h>
#pragma config WDT = OFF, OSC = HS

void main(void)
{
        unsigned char k;

        TRISD = 0;                              // Configure PORTD as output
        for(k=0; k<5; k++)                      // Do 5 times
        {
                PORTD = 0;                      // Turn OFF LED
                Delay10KTCYx(250);              // 1 second delay
                PORTD = 0xFF;                   // Turn ON LED
                Delay10KTCYx(250);              // 1 second delay
        }

        while(1);                               // wait here forever
}
```

Figure 5.35: Program Listing of Example 5.1

- Select **Programmer** -> **Program** to program the target microcontroller. The following message will be displayed after successful programming:
  - Programming...
  - Programming/Verify complete
- Press the **Reset** button to start the program. The eight LEDs connected to PORTD should flash five times with a 1-s delay between each output

■

5.3.4 MPLAB ICD 3 Debugging Example I

An example is given in this section to show how the MPLAB ICD 3 debugger can be used.

■ Example 5.2

Repeat Example 5.1, but compile the program in debug mode, and use the MPLAB ICD 3 debugger to run the program in single-step mode.

Solution

The circuit diagram and the program listing are as in Figures 5.34 and 5.35, respectively. The program should be compiled and run in single-step debug mode. The steps are

- Select the MPLAB ICD 3 debugger. **Debugger** -> **Select Tool** -> **MPLAB ICD 3.**
- Select **Project** -> **Build Configuration** -> **Debug.**
- Select **Project** -> **Build All** to rebuild the project in Debug mode.
- Select **Debugger** -> **Program** to load the code into the target microcontroller.
- Press **F7** key several times until the C code is displayed in the debug window.
- Press **F7** to single-step through the program.
- Set the values of PORTD to be displayed during the debugging. Select **View** -> **Watch**. Select PORTD and click **Add SFR** to add PORTD to the watch window (see Figure 5.36).
- When the debug cursor (green arrow) is on the Delay10KTCYx function, press **F8** to skip to the next instruction without displaying the contents of this function.

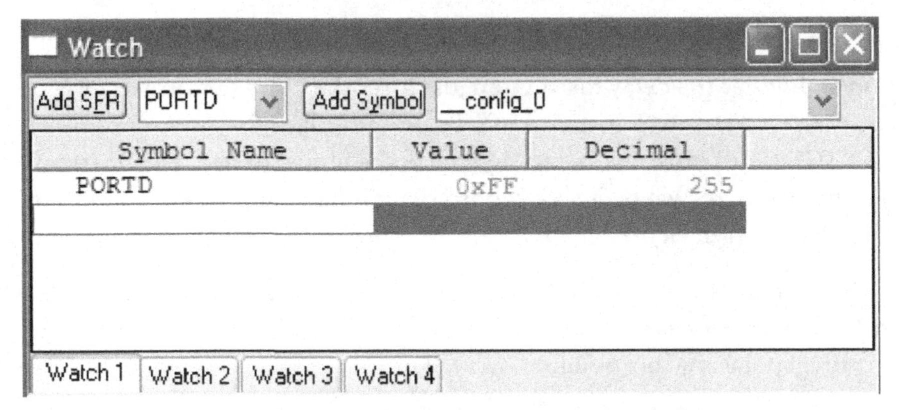

Figure 5.36: The Debug Watch Window

- Press **F7** to execute the PORTD = 0xFF instruction (the LEDs will turn ON).
- When the cursor is again on the Delay10KTCYx function, press **F8** to skip to the next instruction without displaying the contents of this function.
- Repeat the above sequence as required.

To run the program without single-stepping, press **F5** to halt the program, press **F6** to reset, and then press **F9** to run continuously. ■

5.3.5 MPLAB ICD 3 Debugging Example II

This example will show how breakpoints can be used with the debugger.

■ Example 5.3

Repeat Example 5.2, but set a breakpoint after the LEDs are turned ON and run the program up to this breakpoint.

Solution

The circuit diagram and the program listing are as in Figures 5.34 and 5.35, respectively. The required steps are as follows:

- Select the MPLAB ICD 3 debugger. **Debugger** -> **Select Tool** -> **MPLAB ICD 3.**
- Select **Project** -> **Build Configuration** -> **Debug.**
- Select **Project** -> **Build All** to rebuild the project in Debug mode.

- Select **Debugger** -> **Program** to load the code into the target microcontroller.
- Set breakpoint at the delay instruction just after the LEDs are turned ON (after the PORTD = 0xFF instruction). To do this, place the cursor on the second DelayKTCYx(250) instruction and right click the mouse. Select **Set Breakpoint**. A red character "B" will be inserted onto the left-hand side of the code to indicate the breakpoint (see Figure 5.37).
- Run the program by pressing **F9**. The program will run until the second DelayKT-CYx(250) instruction is encountered and then will halt with the green arrow pointing to the breakpoint. At this point, all the LEDs will turn ON.
- Press **F5** to halt the program.

■

5.3.6 MPLAB ICD 3 Debugging Example III

This example will show how more than one breakpoint can be used with the debugger.

■ Example 5.4

Repeat Example 5.3, but set two breakpoints, one before turning the LEDs ON and another after turning the LEDs ON, so that when **F9** is pressed to run the program, the states of the LEDs alternate.

```
C:\MYC\FLASH.C

#include <p18f8722.h>
#include <delays.h>
#pragma config WDT = OFF, OSC = HS

void main(void)
{
    unsigned char k;

    TRISD = 0;                      // Configure PORT B as
    for(k=0; k<5; k++)              // Endless loop
    {
        PORTD = 0;                  // Turn ON LED
        Delay10KTCYx(250);          // 1 second delay
        PORTD = 0xFF;               // Turn OFF LED
B       Delay10KTCYx(250);          // 1 second delay
    }
    while(1);
}
```

Figure 5.37: Setting a Breakpoint

Solution

- The circuit diagram and the program listing are as in Figures 5.34 and 5.35, respectively. The required steps are
- Select the MPLAB ICD 3 debugger. **Debugger -> Select Tool -> MPLAB ICD 3.**
- Select **Project -> Build Configuration -> Debug.**
- Select **Project -> Build All** to rebuild the project in Debug mode.
- Select **Debugger -> Program** to load the code into the target microcontroller.
- Set two breakpoints, one at each delay instruction. To do this, place the cursor on the first DelayKTCYx(250) instruction and right click the mouse. Select **Set Breakpoint**. A red character "B" will be inserted into the left-hand side of the code to indicate the breakpoint. Then, place the cursor on the second DelayKTCYx(250) instruction and right click the mouse. Select **Set Breakpoint**. A red character "B" will be inserted into the left-hand side of the code to indicate the breakpoint (see Figure 5.38).
- Press **F9** to run the program. The LEDs will be OFF. Press **F9** again, the LEDs will be ON, and repeat as necessary.

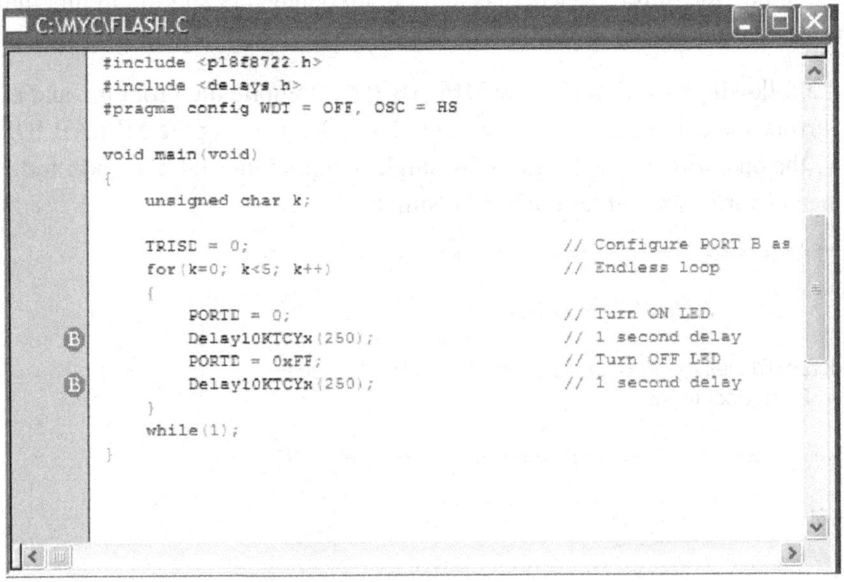

C:\MYC\FLASH.C

```
#include <p18f8722.h>
#include <delays.h>
#pragma config WDT = OFF, OSC = HS

void main(void)
{
    unsigned char k;

    TRISD = 0;                              // Configure PORT B as
    for(k=0; k<5; k++)                      // Endless loop
    {
        PORTD = 0;                          // Turn ON LED
B       Delay10KTCYx(250);                  // 1 second delay
        PORTD = 0xFF;                       // Turn OFF LED
B       Delay10KTCYx(250);                  // 1 second delay
    }
    while(1);
}
```

Figure 5.38: Setting Two Breakpoints

5.4 Summary

This chapter has described the PIC microcontroller software and hardware development tools. It shows that software tools like text editors, assemblers, compilers, and simulators may be required for system development. The required hardware tools include development boards/kits, programming devices, ICDs, or ICEs. In this book, the MPLAB C18 compiler is used in the examples and projects.

Steps in developing and testing MPLAB C18-based C programs are given in the chapter with and without a hardware ICD. In addition, examples of using the PICDEM 18 Explorer development board are shown with the MPLAB ICD 3 ICD.

5.5 Exercises

1. Describe various phases of the microcontroller-based system development cycle.
2. Give a brief description of the microcontroller development tools.
3. Explain the advantages and disadvantages of assemblers and compilers.
4. Explain why a simulator can be a useful tool during the development of a microcontroller-based product.
5. Explain in detail what a device programmer is. Give a few examples of device programmers for the PIC18 series of microcontrollers.
6. Describe briefly the differences between in-circuit debuggers and in-circuit emulators. List the advantages and disadvantages of each type of debugging tool.
7. Enter the following program into the MPLAB IDE, compile the program, and correct syntax errors and any other errors you might have. Then, using the MPLAB IDE, simulate the operation of the program by single-stepping through the code and observe the values of various variables during the simulation.

```
/*============================================

                A SIMPLE LED PROJECT

This program flashes the eight LEDs connected to PORTC of a
PIC18F452 microcontroller.

============================================*/

void main()
{

        TRISC = 0;                      //PORTC is output
```

```
        do
        {
                PORTC = 0xFF;           //Turn ON LEDs on PORTC
                PORTC = 0;              //Turn OFF LEDs on PORTC
        } while(1);                     //Endless loop

}
```

8. Describe the steps necessary to use the MPLAB ICD 3 in-circuit debugger.

9. The following C program contains some deliberately introduced errors. Compile the program to find these errors and correct the errors.

```
void main()
{
        unsigned char i,j,k
        i = 10;
        j = i + 1;

        for(i = 0; i < 10; i++)
        {
                Sum = Sum + i;
                j++
        }
        }
}
```

10. The following C program contains some deliberately introduced errors. Compile the program to find these errors and correct the errors.

```
int add(int a, int b)
{
        result = a + b
}

void main()
{
        int p,q;
        p = 12;
        q = 10;
        z = add(p, q)
        z++;
        for(i = 0; i < z; i++)p++
}
}
```

CHAPTER 6

PIC18 Microcontroller MPLAB C18-Based Simple Projects

In this chapter, we shall be looking at the design of simple PIC18 microcontroller-based projects using the MPLAB C18 language. Here, the idea is to familiarize ourselves with the basic interfacing techniques and also to learn how to use the various microcontroller peripheral registers. We shall be looking at the design of projects using LEDs, push-button switches, keyboards, LED arrays, sound devices, etc., and we shall be developing programs in C language using the MPLAB C18 language. The hardware will be designed on a low-cost breadboard, but development kits such as the PICDEM PIC18 Explorer Development board, BIGPIC5, or others can be used for the projects. We shall be starting with very simple projects and then moving to more complex ones. It is recommended that the reader follow the projects in the order given in the book. The following will be given for each project:

- Circuit diagram
- Description of the hardware
- Algorithm description
- Program listing
- Description of the program
- Suggestions for further development

A program's algorithm can be described using many different graphical or text-based methods. Some of the commonly used methods are flow diagram, structure chart, and program description language (PDL). In this book, we shall be using PDL, which is basically the description of the flow of control in a program using simple English-like commands or keywords.

6.1 Program Description Language

A PDL is a free-format English-like text, which describes the flow of control in a program. PDL is not a programming language, but it is a tool that helps the programmer to think about the logic of the program *before* the program has been developed. Commonly used PDL keywords are described below.

D.O.I.: 10.1016/B978-1-85617-719-1.00010-5

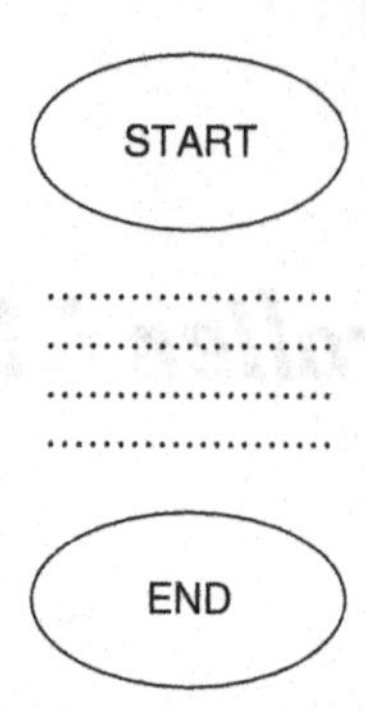

Figure 6.1: START-END in Flow Diagram

6.1.1 START-END

Every PDL program description (or subprogram) should start with a START keyword and terminate with an END keyword. The keywords in a PDL code should be highlighted in bold to make the code more clear. It is also a good practice to indent program statements between PDL keywords in order to enhance the readability of the code.

Example

```
START
    ......
    ......
END
```

The flow-diagram representation of the START-END construct is shown in Figure 6.1.

6.1.2 Sequencing

For normal sequencing in a program, write the statements as short English text as if you are describing the program.

Example

```
Turn ON the LED
Wait 1 second
Turn OFF the LED
```

The flow-diagram representation of the SEQUENCING construct is shown in Figure 6.2.

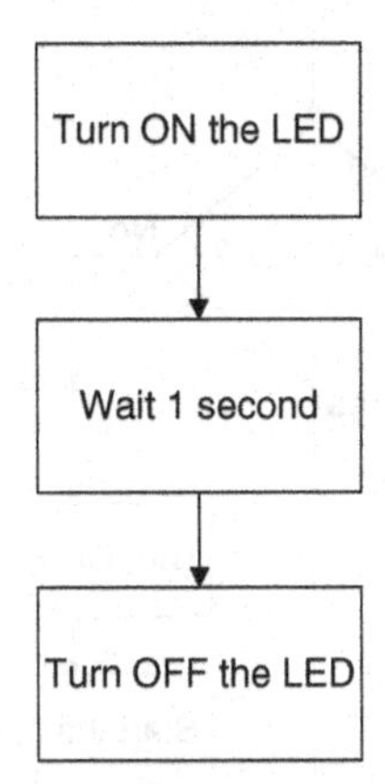

Figure 6.2: SEQUENCING in Flow Diagram

6.1.3 IF-THEN-ELSE-ENDIF

Use IF, THEN, ELSE, and ENDIF keywords to describe the flow of control in your program.

Example

IF switch = 1 **THEN**
  Turn ON LED 1
ELSE
  Turn ON LED 2
  Start the motor
ENDIF

The flow-diagram representation of the IF-THEN-ELSE-ENDIF construct is shown in Figure 6.3.

6.1.4 DO-ENDDO

Use DO and ENDDO keywords to show iteration in your PDL code.

Example

To create an unconditional loop in a program, we can write

Turn ON LED
DO 10 times

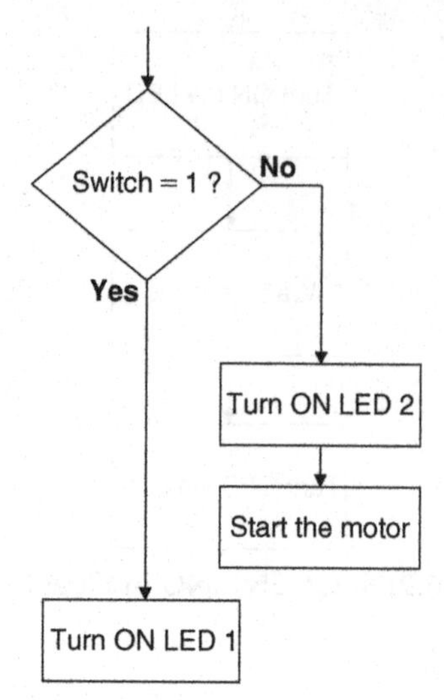

Figure 6.3: IF-THEN-ELSE-ENDIF in Flow Diagram

```
        Set clock to 1
        Wait for 10 ms
        Set clock to 0
ENDDO
```

The flow-diagram representation of the DO-ENDDO construct is shown in Figure 6.4. ■

A variation of the DO-ENDDO construct is to use other keywords such as DO-FOREVER, DO-UNTIL, etc., as shown in the following examples.

■ Example

To create a conditional loop in a program, we can write

```
Turn OFF buzzer
IF switch = 1 THEN
        DO UNTIL Port 1 = 1
             Turn ON LED
             Wait for 10 ms
             Read Port 1
        ENDDO
ENDIF
```

or the following construct can be used when an endless loop is required:

```
DO FOREVER
    Read data from Port 1
    Send data to Port 2
    Wait for 1 s
ENDDO
```

■

6.1.5 REPEAT-UNTIL

This is another control construct, which can be used in PDL codes. An example is given below, where the program waits until a switch value is equal to 1.

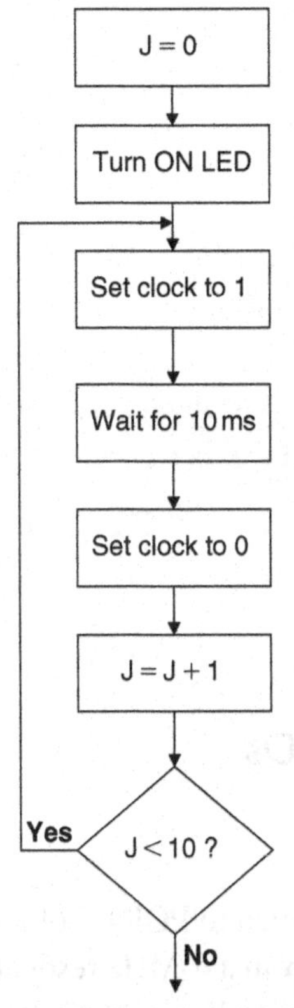

Figure 6.4: DO-ENDDO in Flow Diagram

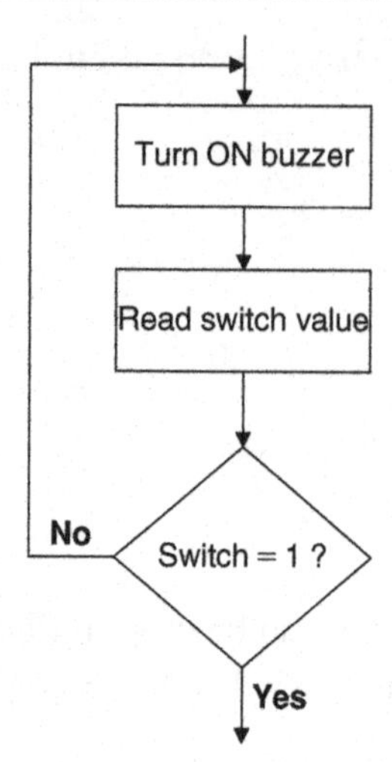

Figure 6.5: REPEAT-UNTIL in Flow Diagram

Example

REPEAT
  Turn ON buzzer
  Read switch value
UNTIL switch = 1

Note that the REPEAT-UNTIL loop is always executed at least once and more than once if the condition at the end of the loop is not met.

The flow-diagram representation of the REPEAT-UNTIL construct is shown in Figure 6.5.

6.2 Project 1 – Chasing LEDs

6.2.1 Project Description

In this project, eight LEDs are connected to PORTC of a PIC18F452-type microcontroller, and the microcontroller is operated from a 4-MHz resonator. When power is applied to the microcontroller (or when the microcontroller is reset), the LEDs turn ON alternately in an anticlockwise manner, where only one LED is ON at any time. A 1-s delay is used between each output so that the LEDs can be seen turning ON and OFF.

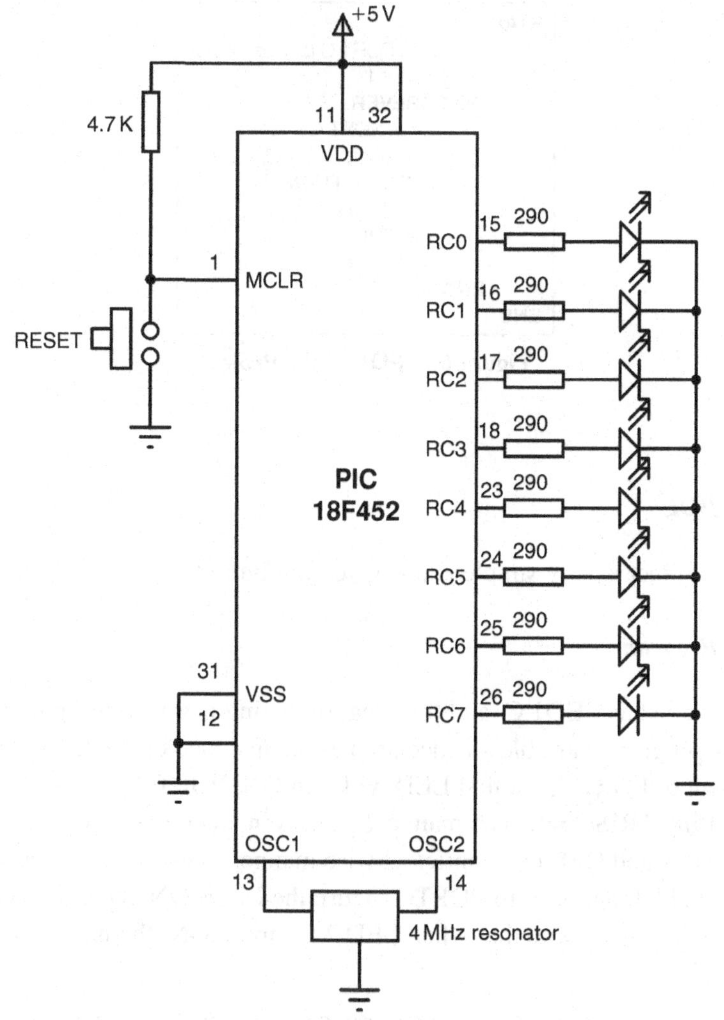

Figure 6.6: Circuit Diagram of the Project

6.2.2 Project Hardware

The circuit diagram of the project is shown in Figure 6.6. LEDs are connected to PORTC in current-sourcing mode with eight 290-Ω resistors. A 4-MHz resonator is connected between the OSC1 and OSC2 pins. In addition, an external **Reset**-push button is connected to the MCLR input to reset the microcontroller when required.

```
START
        Configure PORT C pins as output
        Initialise J = 1
        DO FOREVER
                Set PORT C = J
                Shift left J by 1 digit
                IF J = 0 THEN
                        J = 1
                ENDIF
                Wait 1 second
        ENDDO
END
```

Figure 6.7: PDL of the Project

6.2.3 Project PDL

The PDL of this project is very simple and is given in Figure 6.7.

6.2.4 Project Program

The program is named as LED1.C, and the program listing is given in Figure 6.8. At the beginning of the program, variable *J* is declared as an unsigned character and is set to 1 so that when *J* is sent to PORTC, the first LED will turn ON. PORTC pins are then configured as outputs by setting TRISC = 0. The main program is in an endless **for** loop, where the LEDs are turned ON and OFF in an anticlockwise manner to give the chasing effect. Variable *J* is shifted left and sent to PORTC to turn the LEDs ON and OFF as required. The program checks continuously so that when LED 7 is turned ON, the next LED to be turned ON is LED 0.

The program can be compiled using the MPLAB C18 compiler. The HEX file (LED1.HEX) should be loaded to the PIC18F452 microcontroller using either an in-circuit debugger or a programming device.

Figure 6.9 shows the same program with the 1-s delay configured as a user function and called as **One_Second_Delay**. The modified program is called LED2.C.

6.2.5 Further Development

The project can be modified so that the LEDs chase each other in both directions. For example, if the LEDs are moving in an anticlockwise direction, the direction can be changed so that, when LED RB7 is ON, the next LED to turn ON is RB6; when RB6 is ON, the next LED is RB5; and so on. The LED flashing rate could also be modified to give a different effect to the project.

```
/*******************************************************************************************************************

                                        CHASING LEDS
                                        ============

In this project, 8 LEDs are connected to PORT C of a PIC18F452 microcontroller and the microcontroller is operated
from a 4-MHz resonator. The program turns on the LEDs in an anticlockwise manner with a one-second delay
between each output. The net result is that the LEDs seem to be chasing each other.

Programmer:     Dogan Ibrahim
File:           LED1.C
Version:        1.0
Date:           June, 2009
*******************************************************************************************************************/

#include <p18f452.h>
#include <delays.h>
#pragma config WDT = OFF, OSC = XT

void main(void)
{
        unsigned char J = 1;

        TRISC = 0;                              // Configure PORT C as output

        for(;;)                                 // Endless loop
        {
                PORTC = J;                      // Send J to PORT C
                Delay10KTCYx(100);              // 1 second delay
                J = J << 1;                     // Shift left J
                if(J == 0) J = 1;               // If last LED, move to first LED
        }
}
```

Figure 6.8: Program Listing

```
/*******************************************************************************************************************

                                        CHASING LEDS
                                        ============

In this project, 8 LEDs are connected to PORT C of a PIC18F452 microcontroller and the microcontroller is operated
from a 4-MHz resonator. The program turns on the LEDs in an anticlockwise manner with a one-second delay
between each output. The net result is that the LEDs seem to be chasing each other.

The one second delay is implemented as a user function.

Programmer:     Dogan Ibrahim
File:           LED2.C
Version:        1.0
Date:           June, 2009
*******************************************************************************************************************/

#include <p18f452.h>
#include <delays.h>
```

Figure 6.9: Program Listing Using a User Delay Function

```
#pragma config WDT = OFF, OSC = XT


void One_Second_Delay( )
{
                Delay10KTCYx(100);          // 1 second delay
}



void main(void)
{
        unsigned char J = 1;

        TRISC = 0;                          // Configure PORT C as output

        for(;;)                             // Endless loop
        {
                PORTC = J;                  // Send J to PORT C
                One_Second_Delay( );        // One second delay
                J = J << 1;                 // Shift left J
                if(J == 0) J = 1;           // If last LED, move to first LED
        }
}
```

Figure 6.9: *Cont'd*

6.3 Project 2 – LED Dice

6.3.1 Project Description

This is a simple dice project based on LEDs, a push-button switch, and a PIC18F452 microcontroller operating with a 4-MHz resonator. The block diagram of the project is shown in Figure 6.10.

As shown in Figure 6.11, the LEDs are organized so that when they turn ON, they indicate numbers as in real dice. Operation of the project is as follows: normally, the LEDs are all OFF to indicate that the system is ready to generate a new number. Pressing the switch generates a random dice number between 1 and 6 and displays on the LEDs for 3 s. After 3 s, the LEDs turn OFF again.

6.3.2 Project Hardware

The circuit diagram of the project is shown in Figure 6.12. Seven LEDs representing the faces of dice are connected to PORTC of a PIC18F452 microcontroller in current-sourcing mode using 290-Ω current-limiting resistors. A push-button switch is connected to bit 0 of PORTB (RB0) using a 4.7-K pull-up resistor. Input pin RB0 is normally at logic 1, and pressing the push-button switch forces this pin to logic 0. The MCLR input of the microcontroller is tied

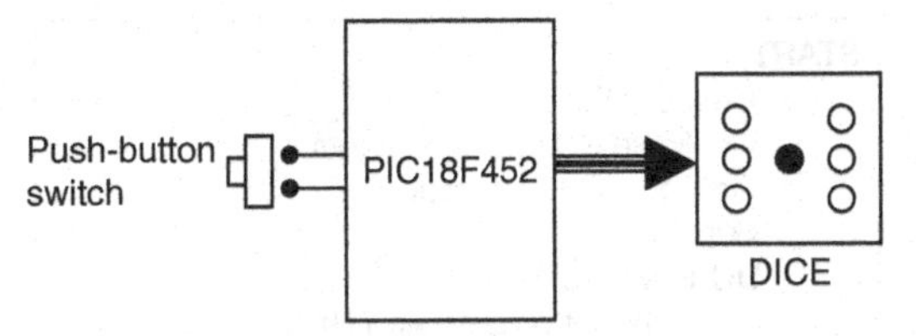

Figure 6.10: Block Diagram of the Project

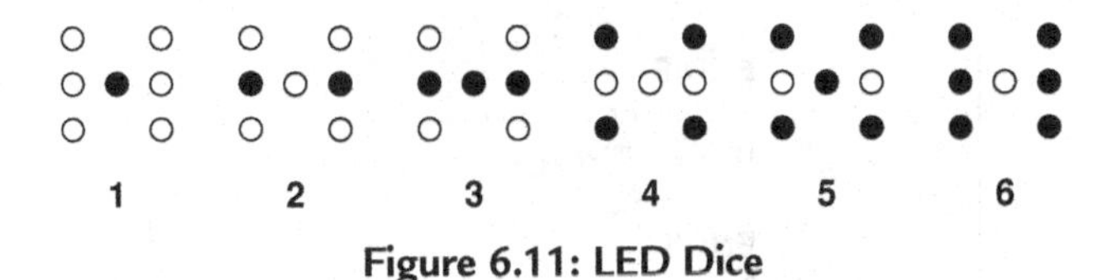

Figure 6.11: LED Dice

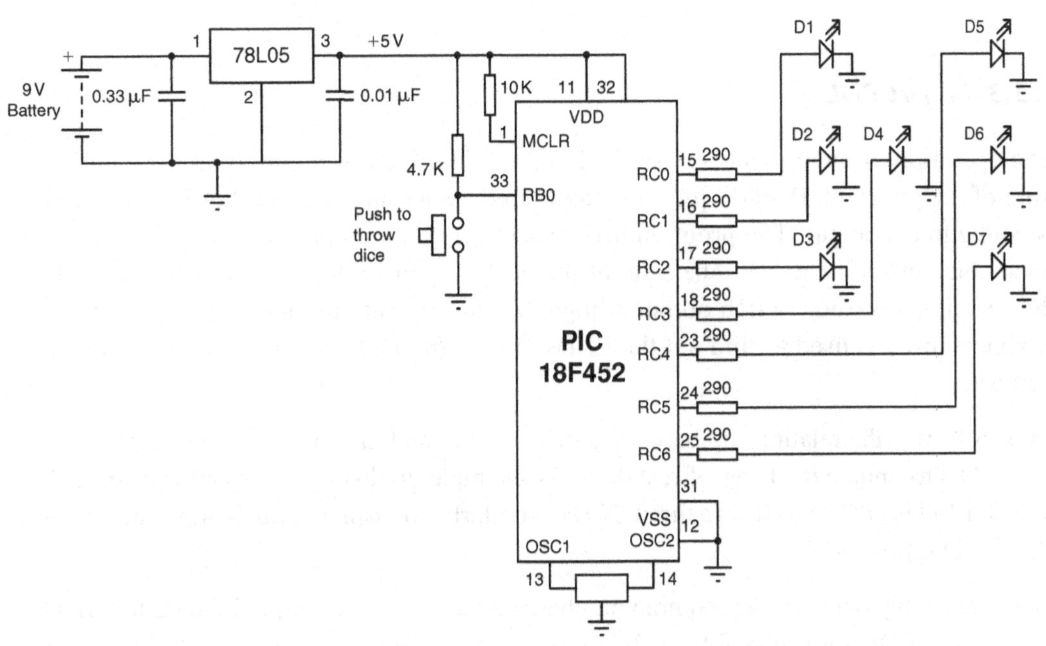

Figure 6.12: Circuit Diagram of the Project

to the +5 V via a 10-K resistor because external reset is not required. The microcontroller is operated from a 4-MHz resonator connected between pins OSC1 and OSC2. The microcontroller is powered from a +9-V battery, and a 78L05-type voltage regulator IC is used to obtain the +5-V supply required for the microcontroller.

```
START
        Create DICE table
        Configure PORT C as outputs
        Configure RB0 as input
        Set J = 1
        DO FOREVER
                IF button pressed THEN
                        Get LED pattern from DICE table
                        Turn ON required LEDs
                        Wait 3 seconds
                        Set J = 0
                        Turn OFF all LEDs
                ENDIF
                Increment J
                IF J = 7 THEN
                        Set J = 1
                ENDIF
        ENDDO
END
```

Figure 6.13: PDL of the Project

6.3.3 Project PDL

The operation of the project is described in the PDL given in Figure 6.13. At the beginning of the program, PORTC pins are configured as outputs, and bit 0 of PORTB (RB0) is configured as input. The program then executes in a loop continuously and increments a variable between 1 and 6. The state of the push-button switch is checked, and when the switch is pressed (switch output at logic 0), the current number is sent to the LEDs. A simple array is used to find out the LEDs to be turned ON corresponding to the dice number.

Table 6.1 gives the relationship between a dice number and the corresponding LEDs to be turned ON to imitate the faces of real dice. For example, to display number 1 (i.e., only the middle LED is ON), we have to turn ON D4. Similarly, to display number 4, we have to turn ON D1, D3, D5, and D7.

The relationship between the required number and the data to be sent to PORTC to turn ON the correct LEDs is given in Table 6.2. For example, to display dice number 2, we have to send hexadecimal 0x22 to PORTC. Similarly, to display number 5, we have to send hexadecimal 0x5D to PORTC, and so on.

6.3.4 Project Program

The program is called DICE.C, and the program listing is given in Figure 6.14. At the beginning of the program, **PBSwitch** is defined as bit 0 of PORTB, and **Pressed** is defined as 0.

Table 6.1: Dice Number and LEDs to be Turned ON

Required Number	LEDs to Be Turned ON
1	D4
2	D2, D6
3	D2, D4, D6
4	D1, D3, D5, D7
5	D1, D3, D4, D5, D7
6	D1, D2, D3, D5, D6, D7

Table 6.2: Required Number and PORTC Data

Required Number	PORTC Data (Hex)
1	0x08
2	0x22
3	0x2A
4	0x55
5	0x5D
6	0x77

The relationships between the dice numbers and the LEDs to be turned ON are stored in an array called **DICE**. Variable *J* is used as the dice number. Variable **Pattern** is the data sent to the LEDs. Program then enters an endless **for** loop, where the value of variable *J* is incremented very fast between 1 and 6. When the push-button switch is pressed, the LED pattern corresponding to the current value of *J* is read from the array and sent to the LEDs. The LEDs remain at this state for 3 s (using a variable-seconds delay function with an argument), and after this time, they all turn OFF to indicate that the system is ready to generate a new dice number.

6.3.5 Using a Pseudorandom Number Generator

In the preceding project, the value of the variable *J* changes very fast between 1 and 6, and when the push-button switch is pressed, the current value of this variable is taken and used as the dice number. Because the values of *J* are changing very fast, we can say that the numbers generated are random (i.e., new numbers do not depend on the previous numbers).

In this section, we shall see how a pseudorandom number generator function can be used to generate the dice numbers. The modified program listing (called DICE2.C) is shown

```
\*****************************************************************************************************************

                                   SIMPLE DICE
                                   ==========

In this project, 7 LEDs are connected to PORT C of a PIC18F452 microcontroller and the microcontroller is
operated from a 4-MHz resonator. The LEDs are organised as the faces of a real dice. When a push-button
switch connected to RB0 is pressed, a dice pattern is displayed on the LEDs. The display remains in this state
for 3 seconds and after this period the LEDs all turn OFF to indicate that the system is ready for the button to be
pressed again.

Author:   Dogan Ibrahim
Date:     June 2009
File:     DICE.C
*****************************************************************************************************************/

#include <p18f452.h>
#include <delays.h>
#pragma config WDT = OFF, OSC = XT

#define PBSwitch PORTBbits.RB0
#define Pressed 0

void N_Seconds_Delay(unsigned char n)
{
     unsigned char k;
     for(k=0; k<n; k++) Delay10KTCYx(100);
}


void main(void)
{
     unsigned char J = 1;
     unsigned char Pattern;
     unsigned char DICE[ ] = {0,0x08,0x22,0x2A,0x55,0x5D,0x77};

     TRISC = 0;                                        // PORT C outputs
     TRISB = 1;                                        // RB0 input
     PORTC = 0;                                        // Turn OFF all LEDs

     for(;;)                                           // Endless loop
     {
          if(PBSwitch == Pressed)                      // Is switch pressed ?
          {
           Pattern = DICE[J];                          // Get LED pattern
           PORTC = Pattern;                            // Turn on LEDs
           N_Seconds_Delay(3);                         // 3 seconds delay
           PORTC = 0;                                  // Turn OFF all LEDs
           J = 0;                                      // Initialise J
          }

          J++;                                         // Increment J
          if(J == 7) J = 1;                            // Back to 1 if > 6
     }
}
```

Figure 6.14: Program Listing

in Figure 6.15. In this program, a function called **Number** generates the dice numbers. The function receives the upper limit of the numbers to be generated (6 in this example). Every time the function is called, a number will be generated between 1 and 6 and returned by the function.

```
/*****************************************************************************************************
                                    SIMPLE DICE
                                    ===========

In this project, 7 LEDs are connected to PORT C of a PIC18F452 microcontroller and the microcontroller is
operated from a 4-MHz resonator. The LEDs are organised as the faces of real dice. When a push-button
switch connected to RB0 is pressed, a dice pattern is displayed on the LEDs. The display remains in this
state for 3 seconds and after this period the LEDs all turn OFF to indicate that the system is ready for the button to
be pressed again.

In this version of the program a pseudorandom number is generated between 1 and 6
for the dice numbers.


Author:   Dogan Ibrahim
Date:     June 2009
File:     DICE2.C
*****************************************************************************************************/
#include <p18f452.h>
#include <delays.h>
#pragma config WDT = OFF, OSC = XT

#define PBSwitch PORTBbits.RB0
#define Pressed 0


//
// This function generates an "n" seconds delay where "n" is an integer
//
void N_Seconds_Delay(unsigned char n)
{
        unsigned char k;
        for(k=0; k<n; k++) Delay10KTCYx(100);
}



//
// This function generates a Pseudo Random integer number between 1 and Lim
//
unsigned char Number(int Lim)
{
        unsigned char Result;
        static unsigned int Y = 1;
```

Figure 6.15: Dice Program Using a Pseudorandom Number Generator

```
            Y = (Y*32719 + 3) % 32749;
            Result = ((Y % Lim) + 1);
            return (Result);
}

//
// Start of MAIN program
//
void main(void)
{
   unsigned char J, Pattern;
   unsigned char DICE[ ] = {0,0x08,0x22,0x2A,0x55,0x5D,0x77};

   TRISC = 0;                              // PORT C outputs
   TRISB = 1;                              // RB0 input
   PORTC = 0;                              // Turn OFF all LEDs

   for(;;)                                 // Endless loop
   {
      if(PBSwitch == Pressed)              // Is switch pressed ?
      {
                  J = Number(6);
        Pattern = DICE[J];                 // Get LED pattern
        PORTC = Pattern;                   // Turn on LEDs
        N_Seconds_Delay(3);                // 3 seconds delay
        PORTC = 0;                         // Turn OFF all LEDs
      }
   }
}
```

Figure 6.15: *Cont'd*

The operation of the program is basically the same as in Figure 6.14. When the push-button switch is pressed, function **Number** is called to generate a new dice number between 1 and 6, and this number is used as an index in array **DICE** in order to find the bit pattern to be sent to the LEDs.

6.4 Project 3 – Two-Dice Project

6.4.1 Project Description

This project is similar to Project 2, but here, a pair of dice are used instead of just one. In many dice games, such as backgammon, a pair of dice are thrown together and then the player takes the action based on the outcome.

The circuit given in Figure 6.12 can be modified by adding another set of seven LEDs for the second dice. For example, the first set of LEDs can be driven from PORTC, the second set from PORTD, and the push-button switch can be connected to RB0 as before. Such a design

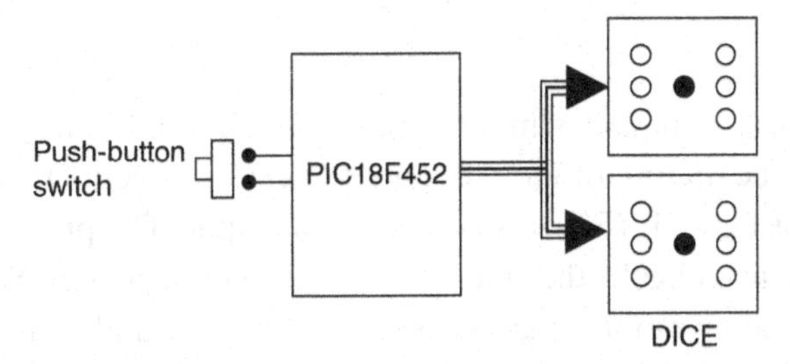

Figure 6.16: Block Diagram of the Project

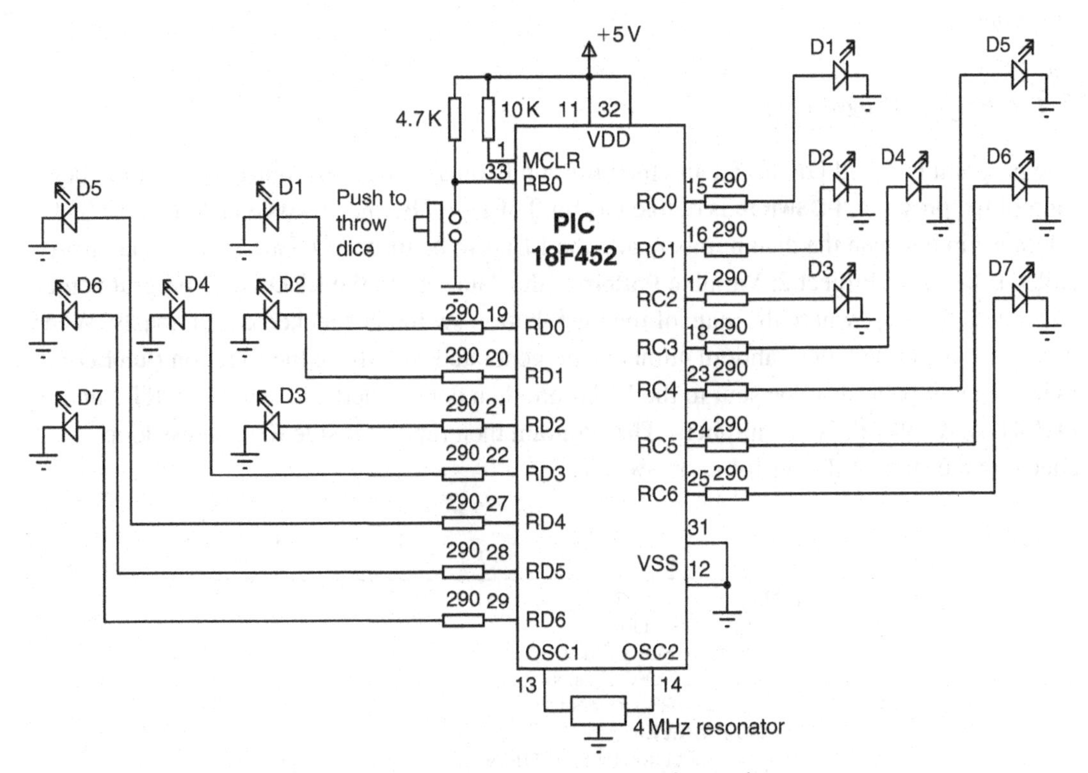

Figure 6.17: Circuit Diagram of the Project

will require the use of 14 output ports just for the LEDs. Later on, we will see how the LEDs can be combined in order to reduce the I/O requirements. Figure 6.16 shows the block diagram of the project.

6.4.2 Project Hardware

The circuit diagram of the project is shown in Figure 6.17. The circuit is basically the same as Figure 6.12, with the addition of another set of LEDs connected to PORTD.

6.4.3 Project PDL

The operation of the project is very similar to that of Project 2. Figure 6.18 shows the PDL for this project. At the beginning of the program, PORTC and PORTD pins are configured as outputs, and bit 0 of PORTB (RB0) is configured as input. The program then executes in a loop continuously and checks the state of the push-button switch. When the switch is pressed, two pseudorandom numbers are generated between 1 and 6, and these numbers are sent to PORTC and PORTD. The LEDs remain at this state for 3 s, and after this time, all the LEDs are turned OFF to indicate that the push button can be pressed again for the next pair of numbers.

6.4.4 Project Program

The program is called DICE3.C, and the program listing is given in Figure 6.19. At the beginning of the program, **PBSwitch** is defined as bit 0 of PORTB, and **Pressed** is defined as 0. The relationship between the dice numbers and the LEDs to be turned ON are stored in an array called **DICE,** as in Project 2. Variable **Pattern** is the data sent to the LEDs. The program enters an endless **for** loop, where the state of the push-button switch is checked continuously. When the switch is pressed, two random numbers are generated by calling the function **Numbers** twice. The bit patterns to be sent to the LEDs are then determined and sent to PORTC and PORTD to display the dice numbers. The program then repeats inside the endless loop, checking the state of the push-button switch.

```
START
        Create DICE table
        Configure PORT C as outputs
        Configure PORT D as outputs
        Configure RB0 as input
        DO FOREVER
                IF button pressed THEN
                        Get a random number between 1 and 6
                        Find bit pattern
                        Turn ON LEDs on PORT C
                        Get second random number between 1 and 6
                        Find bit pattern
                        Turn ON LEDs on PORT D
                        Wait 3 seconds
                        Turn OFF all LEDs
                ENDIF
        ENDDO
END
```

Figure 6.18: PDL of the Project

```
/**********************************************************************************************************
                                    TWO DICE
                                    =========

In this project, 7 LEDs are connected to PORT C of a PIC18F452 microcontroller and the microcontroller is oper-
ated from a 4-MHz resonator. The LEDs are organised as the faces of a real dice. When a push-button switch
connected to RB0 is pressed, a dice pattern is displayed on the LEDs. The display remains in this state for
3 seconds and after this period the LEDs all turn OFF to indicate that the system is ready for the button to be
pressed again.

In this version of the program, a pseudorandom number generator function is used to generate numbers
between 1 and 6 for the dice numbers, and two dice are used connected to PORT C and PORT D of the
microcontroller.


Author:   Dogan Ibrahim
Date:     June 2009
File:     DICE3.C
**********************************************************************************************************/
#include <p18f452.h>
#include <delays.h>
#pragma config WDT = OFF, OSC = XT

#define PBSwitch PORTBbits.RB0
#define Pressed 0



//
// This function generates an "n" seconds delay where "n" is an integer
//
void N_Seconds_Delay(unsigned char n)
{
        unsigned char k;
        for(k=0; k<n; k++) Delay10KTCYx(100);
}



//
// This function generates a Pseudo Random integer number between 1 and Lim
//
unsigned char Number(int Lim)
{
        unsigned char Result;
        static unsigned int Y = 1;

        Y = (Y*32719 + 3) % 32749;
        Result = ((Y % Lim) + 1);
        return (Result);

}
```

Figure 6.19: Program Listing

```
//
// Start of MAIN program
//
void main(void)
{
    unsigned char J, Pattern;
    unsigned char DICE[ ] = {0,0x08,0x22,0x2A,0x55,0x5D,0x77};

    TRISC = 0;                          // PORT C outputs
    TRISD = 0;                          // PORT D outputs
    TRISB = 1;                          // RB0 input
    PORTC = 0;                          // Turn OFF all LEDs
    PORTD = 0;                          // Turn OFF all LEDs

    for(;;)                             // Endless loop
    {
        if(PBSwitch == Pressed)         // Is switch pressed ?
        {
          J = Number(6);                // Generate first number
          Pattern = DICE[J];            // Get LED pattern
          PORTC = Pattern;              // Turn ON first LEDs
          J = Number(6);                // Generate second number
          Pattern = DICE[J];            // Get LED pattern
          PORTD = Pattern;              // Turn ON second LED
          N_Seconds_Delay(3);           // 3 seconds delay
          PORTC = 0;                    // Turn OFF all LEDs
          PORTD = 0;                    // Turn OFF all LEDs
        }
    }
}
```

Figure 6.19: *Cont'd*

6.5 Project 4 – Two Dice Project – Fewer I/O Pins

6.5.1 Project Description

This project is similar to Project 3, but here LEDs are shared and fewer I/O pins are used.

The LEDs in Table 6.1 can be grouped as shown in Table 6.3, and looking at this table, we can say that

- D4 can appear on its own.
- D2 and D6 are always together.
- D1 and D3 are always together.
- D5 and D7 are always together.

Table 6.3: Grouping the LEDs

Required Number	LEDs to Be Turned ON
1	D4
2	D2, D6
3	D2, D6, D4
4	D1, D3, D5, D7
5	D1, D3, D5, D7, D4
6	D2, D6, D1, D3, D5, D7

Thus, we can drive D4 on its own and then drive the D2, D6 pair together in series, the D1, D3 pair together in series, and also the D5, D7 pair together in series. (Actually, we could share D1, D3, D5, D7, but this would require 8 V to drive if the LEDs are connected in series. Connecting these LEDs in parallel will require excessive current, and a driver IC will be required.) All together, four lines will be required to drive seven LEDs. Similarly, four lines will be required to drive the second die. Thus, a pair of dice can be easily driven from an 8-bit output port.

6.5.2 Project Hardware

The circuit diagram of the project is shown in Figure 6.20. PORTC of a PIC18F452 micro-controller is used to drive the LEDs as follows:

- RC0 drives D2, D6 of first die
- RC1 drives D1, D3 of first die
- RC2 drives D5, D7 of first die
- RC3 drives D4 of first die
- RC4 drives D2, D6 of second die
- RC5 drives D1, D3 of second die
- RC6 drives D5, D7 of second die
- RC7 drives D4 of second die

Because we are driving two LEDs on some outputs, we can calculate the required value of the current-limiting resistors. Assuming that the voltage drop across each LED is 2 V, the current through the LED is 10 mA, and the output high voltage of the microcontroller is 4.85 V, the required resistors are

$$R = \frac{4.85 - 2 - 2}{10} = 85\ \Omega.$$

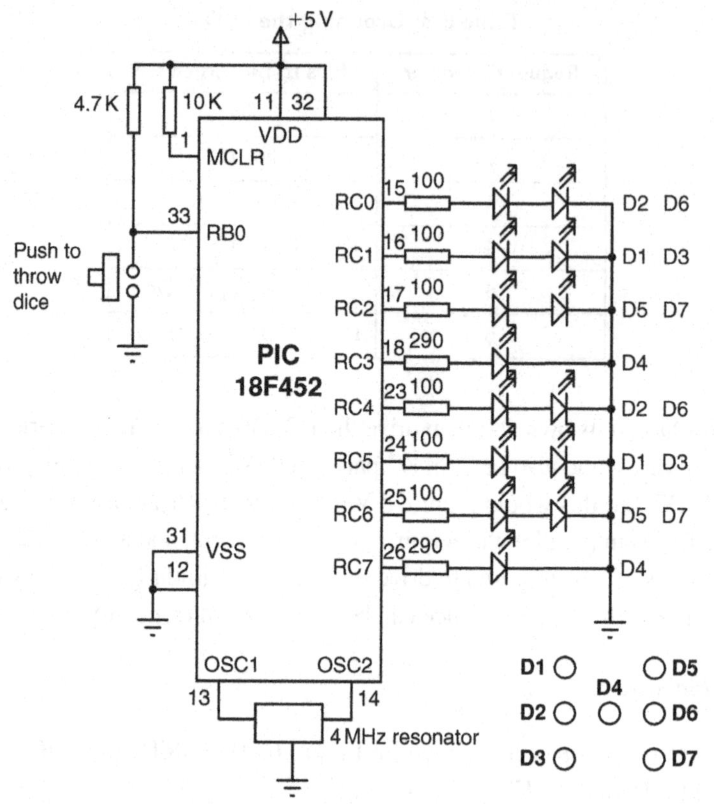

Figure 6.20: Circuit Diagram of the Project

We will choose 100-Ω resistors.

We now need to find the relationship between the dice numbers and the bit pattern to be sent to the LEDs for each die. Table 6.4 shows the relationship between the first dice numbers and the bit pattern to be sent to port pins RC0–RC3. Similarly, Table 6.5 shows the relationship between the second dice numbers and the bit pattern to be sent to port pins RC4–RC7.

We can now find the 8-bit number to be sent to PORTC to display both dice numbers as follows:

- Get the first number from the number generator, call this P.
- Index the DICE table to find the bit pattern for low nibble, i.e., $L = \text{DICE[P]}$.
- Get the second number from the number generator, call this P.
- Index the DICE table to the find bit pattern for high nibble, i.e., $U = \text{DICE[P]}$.
- Multiply high nibble by 16 and add low nibble to find the number to be sent to PORTC, i.e., $R = 16 \times U + L$, where R is the 8-bit number to be sent to PORTC to display both the dice values.

Table 6.4: First Die Bit Patterns

Dice Number	RC3 RC2 RC1 RC0	Hex Value
1	1 0 0 0	8
2	0 0 0 1	1
3	1 0 0 1	9
4	0 1 1 0	6
5	1 1 1 0	E
6	0 1 1 1	7

Table 6.5: Second Die Bit Patterns

Dice Number	RC7 RC6 RC5 RC4	Hex Value
1	1 0 0 0	8
2	0 0 0 1	1
3	1 0 0 1	9
4	0 1 1 0	6
5	1 1 1 0	E
6	0 1 1 1	7

6.5.3 Project PDL

The operation of the project is very similar to that of Project 2. Figure 6.21 shows the PDL for this project. At the beginning of the program, PORTC pins are configured as outputs, and bit 0 of PORTB (RB0) is configured as input. The program then executes in a loop continuously and checks the state of the push-button switch. When the switch is pressed, two pseudorandom numbers are generated between 1 and 6, and the bit pattern to be sent to PORTC is found using the method described above. This bit pattern is then sent to PORTC to display both the dice numbers at the same time. The display shows the dice numbers for 3 s, and then, all the LEDs turn OFF to indicate that the system is waiting for the push button to be pressed again to display the next set of numbers.

6.5.4 Project Program

The program is called DICE4.C, and the program listing is given in Figure 6.22. At the beginning of the program, **PBSwitch** is defined as bit 0 of PORTB, and **Pressed** is defined as 0. The relationships between the dice numbers and the LEDs to be turned ON are stored

```
START
        Create DICE table
        Configure PORT C as outputs
        Configure RB0 as input
        DO FOREVER
              IF button pressed THEN
                        Get a random number between 1 and 6
                        Find low nibble bit pattern
                        Get second random number between 1 and 6
                        High high nibble bit pattern
                        Calculate data to be sent to PORT C
                        Wait 3 seconds
                        Turn OFF all LEDs
              ENDIF
        ENDDO
END
```

Figure 6.21: PDL of the Project

in an array called **DICE** as in Project 2. Variable **Pattern** is the data sent to the LEDs. The program enters an endless **for** loop, where the state of the push-button switch is checked continuously. When the switch is pressed, two random numbers are generated by calling the function **Numbers**. Variables *L* and *U* store the lower and the higher nibbles of the bit pattern to be sent to PORTC. The bit pattern to be sent to PORTC is then determined using the method described in the project hardware section and is stored in variable *R*. This bit pattern is then sent to PORTC to display both the dice numbers at the same time. The dice are displayed for 3 s, and after this period, the LEDs are turned OFF to indicate that the system is ready.

6.5.5 Modifying the Program

The program given in Figure 6.22 can be modified and made more efficient by combining the two dice nibbles into a single table value. The new program is described in this section.

There are 36 possible combinations of two dice values. Referring to Tables 6.4 and 6.5 and Figure 6.20, we can create Table 6.6 to show all the possible two-dice values and the corresponding numbers to be sent to PORTC.

The modified program (program name DICE5.C) is given in Figure 6.23. In this program, the array **DICE** contains the 36 possible dice values. The program enters an endless **for** loop and inside this loop, the state of the push-button switch is checked. In addition, a variable is incremented from 1 to 36 and when the button is pressed, the value of this variable is used as an index to array **DICE** to determine the bit pattern to be sent to PORTC. As before, the program displays the dice numbers for 3 s and then turns OFF all LEDs to indicate that it is ready.

```
/*****************************************************************************************************************************
                                   TWO DICE - USING FEWER I/O PINS
                                   ============================

In this project, LEDs are connected to PORT C of a PIC18F452 microcontroller and the microcontroller is operated
from a 4-MHz resonator. The LEDs are organised as the faces of real dice. When a push-button switch connected to
RB0 is pressed, a dice pattern is displayed on the LEDs. The display remains in this state for 3 seconds and after this
period the LEDs all turn OFF to indicate that the system is ready for the button to be pressed again.

In this version of the program, a pseudorandom number generator function us used to generate numbers between
1 and 6 for the dice numbers.



Author:    Dogan Ibrahim
Date:      June 2009
File:      DICE4.C
*****************************************************************************************************************************/
#include <p18f452.h>
#include <delays.h>
#pragma config WDT = OFF, OSC = XT

#define PBSwitch PORTBbits.RB0
#define Pressed 0


//
// This function generates an "n" seconds delay where "n" is an integer
//
void N_Seconds_Delay(unsigned char n)
{
          unsigned char k;
          for(k=0; k<n; k++) Delay10KTCYx(100);
}



//
// This function generates a Pseudo Random integer number between 1 and Lim
//
unsigned char Number(int Lim)
{
          unsigned char Result;
          static unsigned int Y = 1;

          Y = (Y*32719 + 3) % 32749;
          Result = ((Y % Lim) + 1);
          return (Result);
}


//
// Start of MAIN program
//
```

Figure 6.22: Program Listing

```
void main(void)
{
    unsigned char J, L, U, R;
    unsigned char DICE[ ] = {0,0x08,0x01,0x09,0x06,0x0E,0x07};

    TRISC = 0;                              // PORT C outputs
    TRISB = 1;                              // RB0 input
    PORTC = 0;                              // Turn OFF all LEDs

    for(;;)                                 // Endless loop
    {
        if(PBSwitch == Pressed)             // Is switch pressed ?
        {
                J = Number(6);              // Generate first number
                L = DICE[J];                // Get LED pattern
                J = Number(6);              // Generate second number
                U = DICE[J];                // Get LED pattern
                R = 16*U + L;               // Bit pattern to send to PORTC
                PORTC = R;                  // Turn ON LEDs for both dice
                N_Seconds_Delay(3);         // 3 seconds delay
                PORTC = 0;                  // Turn OFF all LEDs
        }
    }
}
```

Figure 6.22: *Cont'd*

Table 6.6: Two-Dice Combinations and Numbers to be Sent to PORTC

Dice Numbers	PORTC Value	Dice Numbers	PortC Value
1,1	0x88	4,1	0x86
1,2	0x18	4,2	0x16
1,3	0x98	4,3	0x96
1,4	0x68	4,4	0x66
1,5	0xE8	4,5	0xE6
1,6	0x78	4,6	0x76
2,1	0x81	5,1	0x8E
2,2	0x11	5,2	0x1E
2,3	0x91	5,3	0x9E
2,4	0x61	5,4	0x6E
2,5	0xE1	5,5	0xEE
2,6	0x71	5,6	0x7E
3,1	0x89	6,1	0x87
3,2	0x19	6,2	0x17

Table 6.6: Two-Dice Combinations and Numbers to be Sent to PORTC —cont'd

Dice Numbers	PORTC Value	Dice Numbers	PortC Value
3,3	0x99	6,3	0x97
3,4	0x69	6,4	0x67
3,5	0xE9	6,5	0xE7
3,6	0x79	6,6	0x77

```
/*************************************************************************************************

                          TWO DICE - USING FEWER I/O PINS
                          ============================

In this project, LEDs are connected to PORT C of a PIC18F452 microcontroller and the microcontroller is operated
from a 4-MHz resonator. The LEDs are organised as the faces of real dice. When a push-button switch connected to
RB0 is pressed, a dice pattern is displayed on the LEDs. The display remains in this state for 3 seconds and after this
period the LEDs all turn OFF to indicate that the system is ready for the button to be pressed again.

In this version of the program, a pseudorandom number generator function is used to generate numbers between
1 and 6 for the dice numbers.


Author:   Dogan Ibrahim
Date:     June 2009
File:     DICE5.C
*************************************************************************************************/
#include <p18f452.h>
#include <delays.h>
#pragma config WDT = OFF, OSC = XT

#define PBSwitch PORTBbits.RB0
#define Pressed 0


//
// This function generates an "n" seconds delay where "n" is an integer
//
void N_Seconds_Delay(unsigned char n)
{
          unsigned char k;
          for(k=0; k<n; k++) Delay10KTCYx(100);
}



//
// This function generates a Pseudo Random integer number between 1 and Lim
//
unsigned char Number(int Lim)
{
```

Figure 6.23: Modified Program

```
        unsigned char Result;
        static unsigned int Y = 1;

        Y = (Y*32719 + 3) % 32749;
        Result = ((Y % Lim) + 1);
        return (Result);
}


//
// Start of MAIN program
//
void main(void)
{
    unsigned char Pattern, J = 1;
    unsigned char DICE[ ] = {0, 0x88, 0x18, 0x98, 0x68, 0xE8, 0x78,
                             0x81, 0x11, 0x91, 0x61, 0xE1, 0x71,
                             0x89, 0x19, 0x99, 0x69, 0xE9, 0x79,
                             0x86, 0x16, 0x96, 0x66, 0xE6, 0x76,
                             0x8E, 0x1E, 0x9E, 0x6E, 0xEE, 0x7E,
                             0x87, 0x17, 0x97, 0x67, 0xE7, 0x77};

    TRISC = 0;                              // PORT C outputs
    TRISB = 1;                              // RB0 input
    PORTC = 0;                              // Turn OFF all LEDs

    for(;;)                                 // Endless loop
    {
        if(PBSwitch == Pressed)             // Is switch pressed ?
        {
                Pattern = DICE[J];          // Number to send to PORTC
                PORTC = Pattern;            // Send to PORTC
                N_Seconds_Delay(3);         // 3 seconds delay
                PORTC = 0;                  // Turn OFF all LEDs
        }

                J++;                        // Increment J
                if(J == 37) J = 1;          // If J = 37, reset to 1
    }
}
```

Figure 6.23: *Cont'd*

6.6 Project 5 – Seven-Segment LED Counter

6.6.1 Project Description

This project describes the design of a seven-segment LED-based counter, which counts from 0 to 9 continuously with a 1-s delay between each count. The project shows how a seven-segment LED can be interfaced and used in a PIC microcontroller project.

Seven-segment displays are used frequently in electronic circuits to show numeric or alphanumeric values. As shown in Figure 6.24, a seven-segment display basically consists of

Figure 6.24: Some Seven-Segment Displays

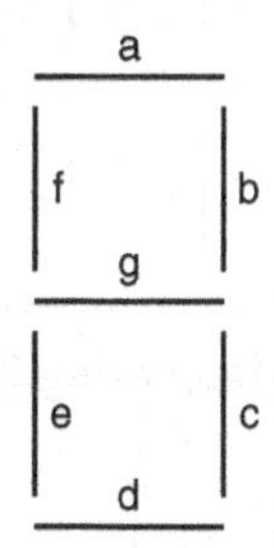

Figure 6.25: Segment Names of a Seven-Segment Display

seven LEDs connected so that numbers 0 to 9 and some letters can be displayed. Segments are identified by letters from **a** to **g,** and Figure 6.25 shows the segment names of a typical seven-segment display.

Figure 6.26 shows how numbers from 0 to 9 can be obtained by turning ON the different segments of the display.

Seven-segment displays are available in two different configurations: **common cathode** and **common anode**. As shown in Figure 6.27, in common cathode configuration, all the cathodes of all segment LEDs are connected together to ground. The segments are turned ON by applying the logic 1 to the required segment LED via current-limiting resistors. In common cathode configuration, the seven-segment LED is connected to the microcontroller in current-sourcing mode.

In a common anode configuration, the anode terminals of all the LEDs are connected together, as shown in Figure 6.28. This common point is then normally connected to the supply voltage. A segment is turned ON by connecting its cathode terminal to logic 0 via a current-limiting resistor. In common anode configuration, the seven-segment LED is connected to the microcontroller in current-sinking mode.

In this project, a *Kingbright SA52-11* model red common anode seven-segment display is used. This is a 13-mm (0.52-inch) display with 10 pins, and it also has a segment LED for the decimal point. Table 6.7 shows the pin configuration of this display.

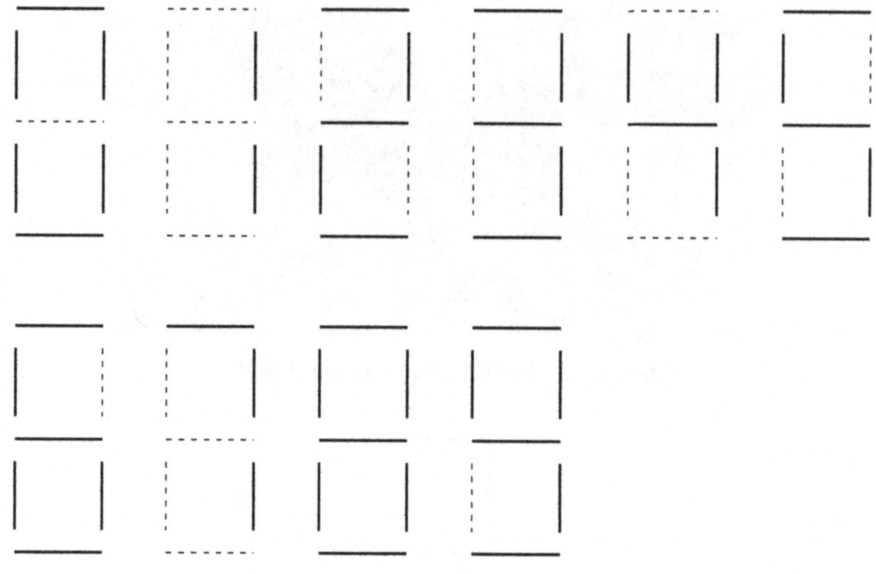

Figure 6.26: Displaying Numbers 0–9

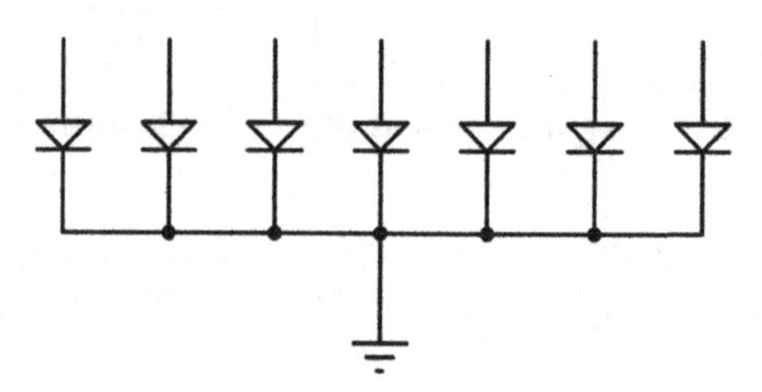

Figure 6.27: Common Cathode Seven-Segment Display

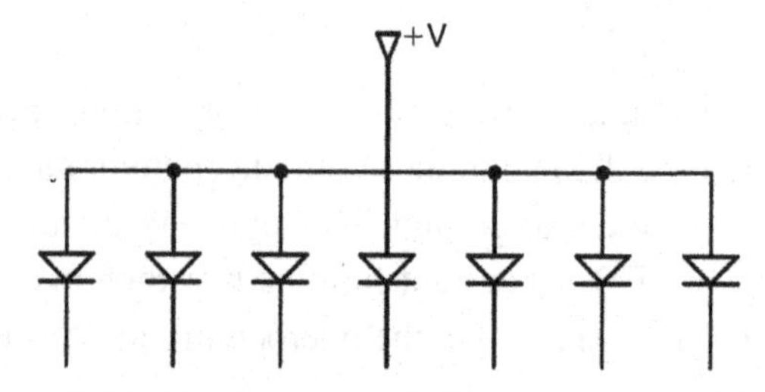

Figure 6.28: Common Anode Seven-Segment Display

6.6.2 Project Hardware

The circuit diagram of the project is shown in Figure 6.29. A PIC18F452-type microcontroller is used with a 4-MHz resonator. Segments **a** to **g** of the display are connected to PORTC of the microcontroller through 290-Ω current-limiting resistors. Before driving the display, we have to know the relationship between the numbers to be displayed and the corresponding segments to be turned ON, and this is shown in Table 6.8. For example, to display number 3,

Table 6.7: SA52-11 Pin Configuration

Pin Numbers	Segments
1	E
2	D
3	Common anode
4	C
5	Decimal point
6	B
7	A
8	Common anode
9	F
10	G

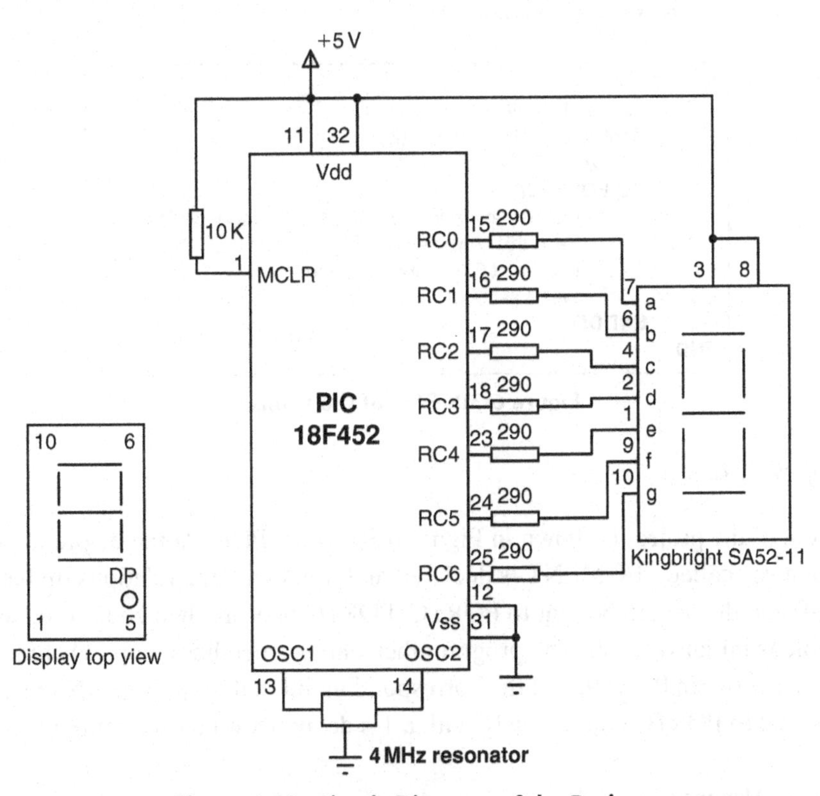

Figure 6.29: Circuit Diagram of the Project

we have to send the hexadecimal number 0x4F to PORTC, which turns ON segments **a**, **b**, **c**, **d**, and **g**. Similarly, to display number 9, we have to send the hexadecimal number 0x6F to PORTC, which turns ON segments **a**, **b**, **c**, **d**, **f**, and **g**.

Table 6.8: Displayed Number and Data Sent to PORTC

Number	x g f e d c b a	PORTC Data
0	0 0 1 1 1 1 1 1	0x3F
1	0 0 0 0 0 1 1 0	0x06
2	0 1 0 1 1 0 1 1	0x5B
3	0 1 0 0 1 1 1 1	0x4F
4	0 1 1 0 0 1 1 0	0x66
5	0 1 1 0 1 1 0 1	0x6D
6	0 1 1 1 1 1 0 1	0x7D
7	0 0 0 0 0 1 1 1	0x07
8	0 1 1 1 1 1 1 1	0x7F
9	0 1 1 0 1 1 1 1	0x6F

If x is not used, it is considered to be 0.

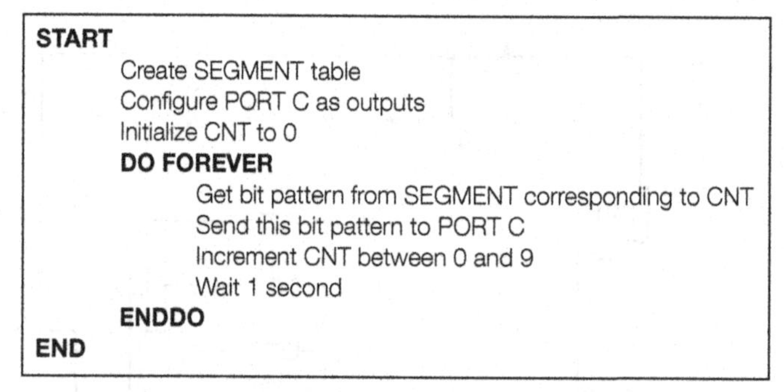

START
Create SEGMENT table
Configure PORT C as outputs
Initialize CNT to 0
DO FOREVER
Get bit pattern from SEGMENT corresponding to CNT
Send this bit pattern to PORT C
Increment CNT between 0 and 9
Wait 1 second
ENDDO
END

Figure 6.30: PDL of the Project

6.6.3 Project PDL

The operation of the project is shown in Figure 6.30 with a PDL. At the beginning of the program, an array called SEGMENT is declared and filled with the relationship between the numbers 0–9 and the data to be sent to PORTC. PORTC pins are then configured as outputs and a variable is initialized to 0. The program then enters an endless loop, where the variable is incremented between 0 and 9, and the corresponding bit pattern to turn ON the appropriate segments is sent to PORTC continuously with a 1-s delay between each output.

6.6.4 Project Program

The program is called SEVEN1.C, and the listing is given in Figure 6.31. At the beginning of the program, character variables **Pattern** and **Cnt** are declared, and **Cnt** is cleared to 0. Then, Table 6.8 is implemented using array SEGMENT. After configuring PORTC pins as outputs,

```
/*****************************************************************************************************************
                                    7-SEGMENT DISPLAY
                                    =================

In this project, a common anode 7-segment LED display is connected to PORT C of a PIC18F452 microcontroller and
the microcontroller is operated from a 4-MHz resonator. The program displays numbers 0 to 9 on the display with a
one-second delay between each output.

Author:   Dogan Ibrahim
Date:     June 2009
File:     SEVEN1.C
*****************************************************************************************************************/
#include <p18f452.h>
#include <delays.h>
#pragma config WDT = OFF, OSC = XT

//
// This function generates 1 second delay
//
void One_Second_Delay( )
{
          Delay10KTCYx(100);
}


//
// Start of MAIN program
//
void main(void)
{
   unsigned char Pattern, Cnt = 0;
   unsigned char SEGMENT[ ] = {0x3F,0x06,0x5B,0x4F,0x66,0x6D,0x7D, 0x07,0x7F,0x6F};

   TRISC = 0;                                   // PORT C are outputs

   for(;;)                                      // Endless loop
   {
     Pattern = SEGMENT[Cnt];                    // Number to send to PORT C
     Pattern = ~Pattern;                        // Invert bit pattern
     PORTC = Pattern;                           // Send to PORT C
     Cnt++;                                     // Increment Cnt
     if(Cnt == 10) Cnt = 0;                     // Cnt is between 0 and 9
     One_Second_Delay( );                       // 1 second delay
   }
}
```

Figure 6.31: Program Listing

the program enters an endless loop using the **for** statement. Inside the loop, the bit pattern corresponding to the contents of **Cnt** is found and stored in variable **Pattern**. Because we are using a common anode display, a segment is turned ON when it is at logic 0, and thus, the bit pattern is inverted before sending to PORTC. The value of **Cnt** is then incremented between 0 and 9, after which the program waits for a second before repeating the above sequence.

6.6.5 Modified Program

Note that the program can be made more readable if we create a function to display the required number and then call this function from the main program. Figure 6.32 shows the modified program (called SEVEN2.C). A function called **Display** is created with an argument called **no**. The function gets the bit pattern from the local array SEGMENT indexed by **no**, inverts it, and then returns the resulting bit pattern to the calling program.

```
/*******************************************************************************************************************
                                    7-SEGMENT DISPLAY
                                    =================

In this project, a common anode 7-segment LED display is connected to PORT C of a PIC18F452 microcontroller and
the microcontroller is operated from a 4-MHz resonator. The program displays numbers 0 to 9 on the display with a
one-second delay between each output.

In this version of the program, a function called "Display" is used to display the number.

Author:    Dogan Ibrahim
Date:      June 2009
File:      SEVEN2.C
*******************************************************************************************************************/
#include <p18f452.h>
#include <delays.h>
#pragma config WDT = OFF, OSC = XT

//
// This function generates 1 second delay
//
void One_Second_Delay( )
{
        Delay10KTCYx(100);
}


//
// This function displays a number on the 7-segment LED.
// The number is passed in the argument list of the function.
//
unsigned char Display(unsigned char no)
{
        unsigned char Pattern;
        unsigned char SEGMENT[ ] = {0x3F, 0x06, 0x5B, 0x4F, 0x66, 0x6D, 0x7D, 0x07, 0x7F, 0x6F};

        Pattern = SEGMENT[no];
        Pattern = ~Pattern;
        return (Pattern);
}
```

Figure 6.32: Modified Program Listing

```
//
// Start of MAIN program
//
void main(void)
{
    unsigned char Cnt = 0;

    TRISC = 0;                              // PORT C are outputs

    for(;;)                                 // Endless loop
    {
        PORTC = Display(Cnt);               // Send to PORT C
        Cnt++;                              // Increment Cnt
        if(Cnt == 10) Cnt = 0;              // Cnt is between 0 and 9
        One_Second_Delay( );                // 1 second delay
    }
}
```

Figure 6.32: *Cont'd*

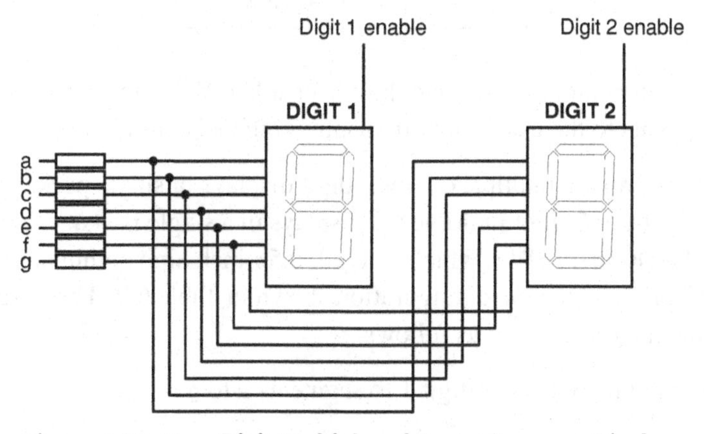

Figure 6.33: Two-Digit Multiplexed Seven-Segment Displays

6.7 Project 6 – Two-Digit Multiplexed Seven-Segment LED

6.7.1 Project Description

This project is similar to Project 5, but here, a two-digit multiplexed 7-segment LED is used instead of just one digit and a fixed number, and in this project, the number 25 is displayed. In multiplexed LED applications (see Figure 6.33), the LED segments of all the digits are tied together, and the common pins of each digit are turned ON separately by the microcontroller. When each digit is displayed for several milliseconds, the eye cannot differentiate that the digits are not ON all the time. In this manner, we can multiplex any number of seven-segment displays together. For example, to display the number 53, we have to send number 5 to the first digit and enable its common pin. After a few milliseconds, number 3 is sent to the second

Table 6.9: Pin Configuration of DC56-11EWA Dual Display

Pin Nos.	Segments
1,5	E
2,6	D
3,8	C
14	Digit 1 Enable
17,7	G
15,10	B
16,11	A
18,12	F
13	Digit 2 Enable
4	Decimal Point1
9	Decimal Point 2

digit, and the common point of the second digit is enabled. When this process is repeated continuously, it appears to the user that both displays are ON continuously.

Some manufacturers provide multiplexed multidigit displays in single packages. For example, we can purchase 2-, 4-, or 8-digit multiplexed displays in a single package. The display used in this project is the DC56-11EWA, which is a red, 0.56-inch high common-cathode two-digit display, having 18 pins and the pin configuration shown in Table 6.9. This display can be controlled from the microcontroller as follows:

- Send the segment bit pattern for digit 1 to segments **a** to **g**.
- Enable digit 1.
- Wait for a few milliseconds.
- Disable digit 1.
- Send the segment bit pattern for digit 2 to segments **a** to **g**.
- Enable digit 2.
- Wait for a few milliseconds.
- Disable digit 2.
- Repeat the above process continuously.

The segment configuration of DC56-11EWA display is shown in Figure 6.34. In a multiplexed display application, the segment pins of the corresponding segments are connected

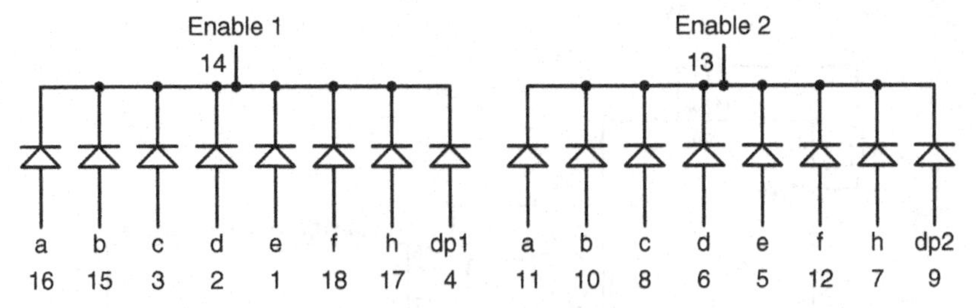

Figure 6.34: DC56-11EWA Display Segment Configuration

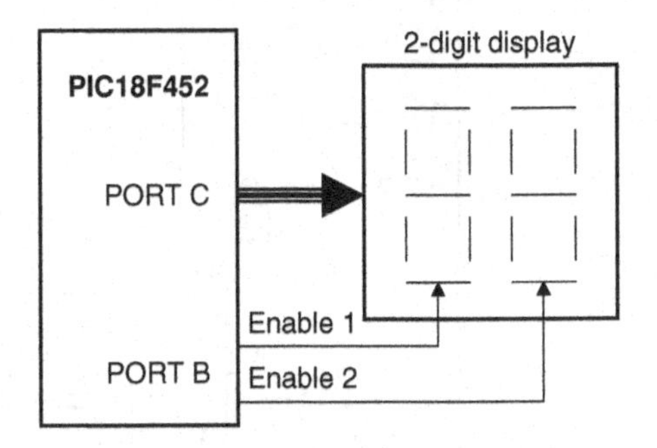

Figure 6.35: Block Diagram of the Project

together. For example, pins 11 and 16 are connected as a common **a** segment. Similarly, pins 15 and 10 are connected as a common **b** segment, and so on.

6.7.2 Project Hardware

The block diagram of this project is shown in Figure 6.35. The circuit diagram is given in Figure 6.36. The segments of the display are connected to PORTC of a PIC18F452-type microcontroller, operated with a 4-MHz resonator. Current-limiting resistors are used on each segment of the display. Each digit is enabled using a BC108-type transistor switch connected to the RB0 and RB1 port pins of the microcontroller. A segment is turned ON when a logic 1 is applied to the base of the corresponding segment transistor.

6.7.3 Project PDL

At the beginning of the program, PORTB and PORTC pins are configured as outputs. The program then enters an endless loop, where first the most significant digit (MSD) of the number is calculated, next function **Display** is called to find the bit pattern and then sent to the display, and digit 1 is enabled. Then, after a small delay, digit 1 is disabled, the least significant digit (LSD) of the number is calculated, function **Display** is called to find the bit pattern and then sent to the display, and digit 2 is enabled. Then, again after a small delay, digit 2 is disabled, and the above process repeats indefinitely. Figure 6.37 shows the PDL of the project.

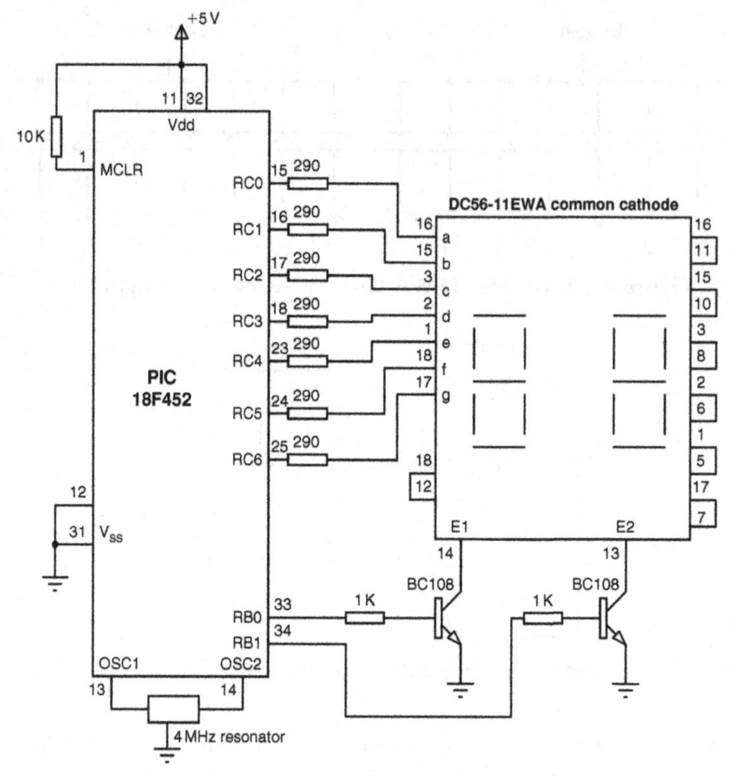

Figure 6.36: Circuit Diagram of the Project

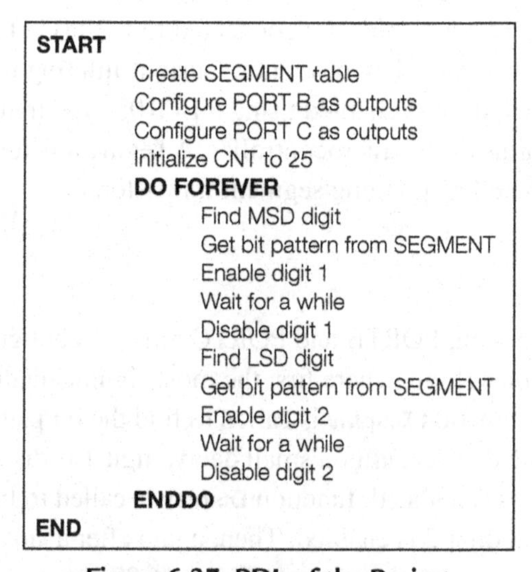

```
START
        Create SEGMENT table
        Configure PORT B as outputs
        Configure PORT C as outputs
        Initialize CNT to 25
        DO FOREVER
              Find MSD digit
              Get bit pattern from SEGMENT
              Enable digit 1
              Wait for a while
              Disable digit 1
              Find LSD digit
              Get bit pattern from SEGMENT
              Enable digit 2
              Wait for a while
              Disable digit 2
        ENDDO
END
```

Figure 6.37: PDL of the Project

6.7.4 Project Program

The program is named SEVEN3.C, and the listing is shown in Figure 6.38. **DIGIT1** and **DIGIT2** are defined to be equal to bit 0 and bit 1 of PORTB, respectively. The value to be displayed (number 25) is stored in variable **Cnt**. An endless loop is formed using a **for** statement. Inside the loop, the MSD of the number is calculated by dividing the number by 10. Function **Display** is then called to find the bit pattern to send to PORTC. Then, digit 1 is enabled by setting DIGIT1 = 1, and the program waits for 10 ms. After this, digit 1 is disabled, and the LSD of the number is calculated using the mod operator ("%") and sent to PORTC. At the same time, digit 2 is enabled by setting DIGIT2 = 1, and the program waits for 10 ms. After this waiting time, digit 2 is disabled and the program repeats forever.

```
/*****************************************************************************************************
                              Dual 7-SEGMENT DISPLAY
                              ======================

In this project, two common cathode 7-segment LED displays are connected to PORT C of a PIC18F452
microcontroller and the microcontroller is operated from a 4-MHz resonator. Digit 1 (left digit) enable pin is connected
to port pin RB0 and digit 2 (right digit) enable pin is connected to port pin RB1 of the microcontroller. The program
displays the number 25 on the displays.

Author:   Dogan Ibrahim
Date:     June 2009
File:     SEVEN3.C
*****************************************************************************************************/
#include <p18f452.h>
#include <delays.h>
#pragma config WDT = OFF, OSC = XT

#define DIGIT1 PORTBbits.RB0
#define DIGIT2 PORTBbits.RB1

//
// This function finds the bit pattern to be sent to the port to display a number
// on the 7-segment LED. The number is passed in the argument list of the function.
//
unsigned char Display(unsigned char no)
{
     unsigned char Pattern;
     unsigned char SEGMENT[ ] = {0x3F,0x06,0x5B,0x4F,0x66,0x6D,0x7D,0x07, 0x7F,0x6F};

     Pattern = SEGMENT[no];                              // Pattern to return
     return (Pattern);
}

//
// This function generates 10 ms delay
//
void Ten_Ms_Delay( )
```

Figure 6.38: Program Listing

```
{
        Delay10KTCYx(1);
}


//
// Start of MAIN Program
//
void main(void)
{
      unsigned char Msd, Lsd, Cnt = 25;

      TRISC = 0;                              // PORT C are outputs
      TRISB = 0;                              // RB0, RB1 are outputs

      DIGIT1 = 0;                             // Disable digit 1
      DIGIT2 = 0;                             // Disable digit 2

      for(;;)                                 // Endless loop
      {
      Msd = Cnt / 10;                         // MSD digit
      PORTC = Display(Msd);                   // send to PORT C
      DIGIT1 = 1;                             // Enable digit 1
      Ten_Ms_Delay( );                        // Wait a while

      DIGIT1 = 0;                             // Disable digit 1
      Lsd = Cnt % 10;                         // LSD digit
      PORTC = Display(Lsd);                   // Send to PORT C
      DIGIT2 = 1;                             // Enable digit 2
      Ten_Ms_Delay( );                        // Wait a while
      DIGIT2 = 0;                             // Disable digit 2
      }
}
```

Figure 6.38: *Cont'd*

6.8 Project 7 – Two-Digit Multiplexed Seven-Segment LED Counter With Timer Interrupt

6.8.1 Project Description

This project is similar to Project 6, but here, the timer interrupt of the microcontroller is used to refresh the displays. In Project 6, the microcontroller was busy updating the displays continuously every 10 ms, and therefore it could not perform any other tasks. For example, if we wish to make a counter with 1-s delay between each count, the program given in Project 6 cannot be used as the displays cannot be updated while the program waits for 1 s.

In this project, a counter will be designed to count from 0 to 99, and the display will be refreshed every 5 ms inside the timer interrupt service routine (ISR); the main program can perform other tasks, in this example, incrementing the count and waiting for a second between each count.

In this project, Timer 0 is operated in an 8-bit mode. The time for an interrupt is given by

$$\text{Time} = (4 \times \text{clock period}) \times \text{Prescaler} \times (256 - \text{TMR0L}),$$

where Prescaler is the selected prescaler value, and TMR0L is the value loaded into timer register TMR0L to generate timer interrupts every time period. In our application, the clock frequency is 4 MHz, i.e., clock period = 0.25 μs, and Time = 5 ms. By selecting a prescaler value of 32, the number to be loaded into TMR0L can be calculated as follows:

$$\text{TMR0L} = 256 - \frac{\text{Time}}{4 \times \text{clockperiod} \times \text{prescaler}}$$

or

$$\text{TMR0L} = 256 - \frac{5000}{4 \times 0.25 \times 32} = 100$$

Thus, TMR0L should be loaded with 100. The value to be loaded into the TMR0 control register T0CON can then be found as:

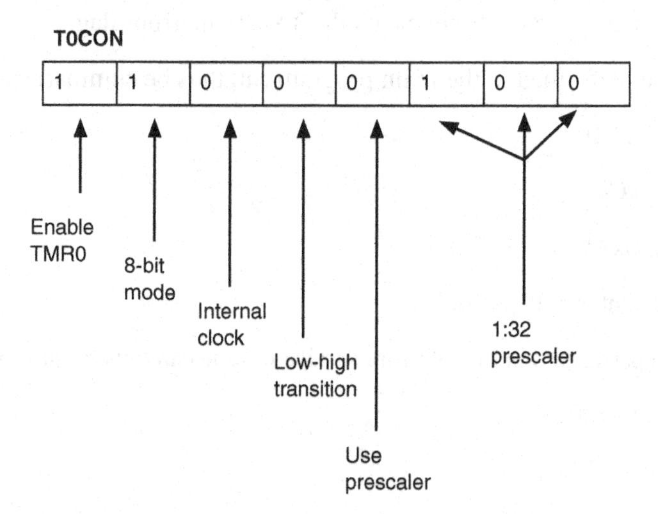

Thus, the T0CON register should be loaded with hexadecimal 0xC4. The next register to be configured is the interrupt control register INTCON, where we will disable priority-based interrupts and enable the global interrupts and TMR0 interrupts:

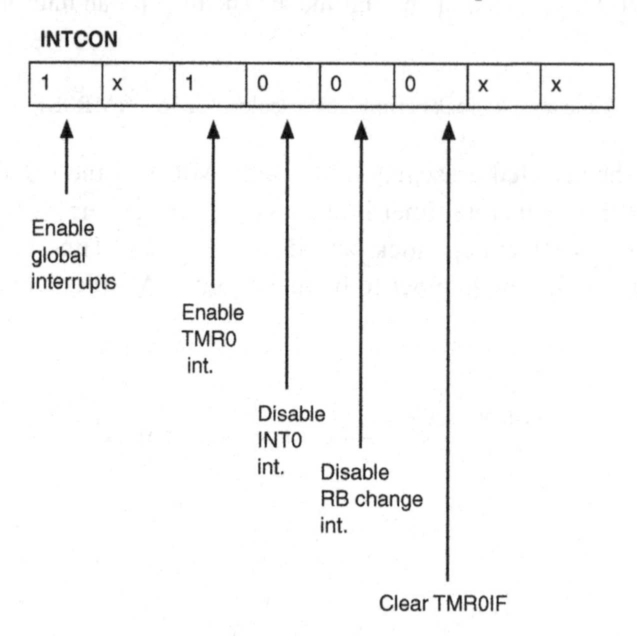

Considering the don't care entries (X) as 0, the hexadecimal value to be loaded into register INTCON is thus 0xA0.

When an interrupt occurs, the program automatically jumps to the ISR. Inside this routine, we have to reload register TMR0L, we also have to re-enable the TMR0 interrupts, and we have to clear the TMR0 interrupt flag bit. Setting the INTCON register to 0x20 re-enables the TMR0 interrupts and, at the same time, clears the TMR0 interrupt flag.

The operations to be performed in the main program can thus be summarized as follows:

- Load TMR0L with 100
- Set T0CON to 0xC4
- Set INTCON to 0xA0
- Increment the counter with 1-s delays

The operations to be performed in the interrupt service routine can thus be summarized as follows:

- Reload TMR0L to 100
- Refresh displays
- Set INTCON to 0x20 (re-enable TMR0 interrupts and clear the timer interrupt flag)

6.8.2 Project Hardware

The circuit diagram of this project is same as in Figure 6.36, where a dual seven-segment display is connected to PORTB and PORTC of a PIC18F452 microcontroller.

6.8.3 Project PDL

The PDL of the project is shown in Figure 6.39. The program is in two sections: the main program and the ISR. Inside the main program, TMR0 is configured to generate interrupts every 5 ms. In addition, the counter is incremented with 1-s delay. Inside the ISR, the timer interrupt is re-enabled, and the display digits are refreshed alternately at every 5 ms.

6.8.4 Project Program

Before looking at the program code, it is worthwhile to describe how the interrupts are handled in the MPLAB C18 programs. The C18 compiler does not automatically start the ISR at the high or low interrupt vectors in program memory. The C18 compiler uses several pragmas, first to locate the start of the ISR at the reset vector and then to distinguish the ISR from a standard user function.

```
MAIN PROGRAM:

    START
        Configure PORT B as outputs
        Configure PORT C as outputs
        Clear variable Cnt to 0
        Configure TMR0 to generate interrupts every 5ms
        DO FOREVER
            Increment Cnt between 0 and 99
            Delay 1 second
        ENDO
    END


INTERRUPT SERVICE ROUTINE:
    START
        Re-configure TMR0
        IF Digit 1 updated THEN
            Update digit 2
        ELSE
            Update digit 1
        END
    END
```

Figure 6.39: PDL of the Project

The interrupt vector is specified in the program using the **#pragma code** statement. The high-priority vector is at address 0x08, and the low-priority vector is at address 0x18. To set the high-priority vector, the following statement is used:

```
#pragma code high_vector = 0x08
```

where the name high_vector is not a reserved name and any other name can be chosen by the programmer.

The above statement specifies that the code section that follows the statement starts at memory location 0x08. After the above statement, we have to write the following single line of assembly code to force a jump to the ISR:

```
_asm GOTO timer_ISR _endasm
```

where timer_ISR is the name of the ISR function. To return to the normal program code, we have to use the statement

```
#pragma code
```

Now, we have to use the following pragma statement to specify a high-priority ISR function:

```
#pragma interrupt timer_ISR
```

where interrupt is a reserved name, and timer_ISR is the name of the ISR function declared by the user. This function must follow the above statement.

If it is required to use a low-priority interrupt, then the above pragma statements should be changed to (assuming the name of the ISR function is timer_ISR)

```
#pragma code low_vector = 0x18
```

and

```
#pragma interruptlow timer_ISR
```

The program (called SEVEN4.C) is given in Figure 6.40. In this program function, prototypes are used at the beginning of the program to declare the functions called by the main program. At the beginning of the main program, PORTB and PORTC are configured as outputs. Then, register T0CON is loaded with 0xC4 to enable the TMR0 and set the prescaler to 32. TMR0L register is loaded with 100 so that an interrupt can be generated after 5 ms. The program then enters an endless loop, where the value of **Cnt** is incremented every second using a delay function.

Inside the ISR register, TMR0L is reloaded, TMR0 interrupts are re-enabled, and the timer interrupt flag is cleared so that further timer interrupts can be generated. The display digits are then updated alternately based on the value of variable **Cnt**. A variable called **Flag** is used to determine which digit to update. The function **Display** is called as in Project 6 to find the bit pattern to be sent to PORTC.

```
/*****************************************************************************************************************
                              Dual 7-SEGMENT DISPLAY COUNTER
                              ==============================

In this project, two common cathode 7-segment LED displays are connected to PORT C of a PIC18F452
microcontroller and the microcontroller is operated from a 4-MHz resonator. Digit 1 (left digit) enable pin is connected
to port pin RB0 and digit 2 (right digit) enable pin is connected to port pin RB1 of the microcontroller. The program
counts up from 0 to 99 with a one-second delay between each count.

The display is updated in a timer interrupt service routine every 5 ms.

Author:   Dogan Ibrahim
Date:     June 2009
File:     SEVEN4.C
*****************************************************************************************************************/
#include <p18f452.h>
#include <delays.h>
#pragma config WDT = OFF, OSC = XT

#define DIGIT1 PORTBbits.RB0
#define DIGIT2 PORTBbits.RB1

unsigned char Cnt = 0;
unsigned char Flag = 0;

void One_Second_Delay(void);
unsigned char Display(unsigned char);
void timer_ISR(void);


//
// Define the high interrupt vector to be at 0x08
//
#pragma code high_vector=0x08                               //Following code at address 0x08
void interrupt(void)
{
        _asm GOTO timer_ISR _endasm                         // Jump to ISR
}
#pragma code                                                // Return to default code section

//
// timer_ISR is an interrupt service routine (jumps here every 5ms)
//
#pragma interrupt timer_ISR
void timer_ISR( )
{
        unsigned char Msd, Lsd;
        TMR0L = 100;                                        // Re-load TMR0
        INTCON = 0x20;                                      // Set T0IE and clear T0IF
        Flag = ~ Flag;                                      // Toggle Flag
        if(Flag == 0)                                       // Do digit 1
        {
          DIGIT2 = 0;
          Msd = Cnt / 10;                                   // MSD digit
          PORTC = Display(Msd);                             // Send to PORT C
          DIGIT1 = 1;                                       // Enable digit 1
        }
        else
        {                                                   // Do digit 2
          DIGIT1 = 0;                                       // Disable digit 1
          Lsd = Cnt % 10;                                   // LSD digit
```

Figure 6.40: Program of the Project

```
            PORTC = Display(Lsd);                       // Send to PORT C
            DIGIT2 = 1;                                 // Enable digit 2
        }
}


//
// This function generates 1 second delay
//
void One_Second_Delay( )
{
        Delay10KTCYx(100);
}


//
// This function finds the bit pattern to be sent to the port to display a number
// on the 7-segment LED. The number is passed in the argument list of the function.
//
unsigned char Display(unsigned char no)
{
      unsigned char Pattern;
      unsigned char SEGMENT[ ] = {0x3F,0x06,0x5B,0x4F,0x66,0x6D,0x7D,0x07, 0x7F,0x6F};

      Pattern = SEGMENT[no];                            // Pattern to return
      return (Pattern);
}



//
// Start of MAIN Program. configure PORT B and PORT C as outputs. In addition,
// configure TMR0 to interrupt at every 5ms
//
void main(void)
{
        TRISC = 0;                                      // PORT C are outputs
        TRISB = 0;                                      // RB0, RB1 are outputs

        DIGIT1 = 0;                                     // Disable digit 1
        DIGIT2 = 0;                                     // Disable digit 2
//
// Configure TMR0 timer interrupt
//
        T0CON = 0xC4;                                   // Prescaler = 32
        TMR0L = 100;                                    // Load TMR0L with 100
        INTCON = 0xA0;                                  // Enable TMR0 interrupt
        One_Second_Delay( );

        for(;;)                                         // Endless loop
        {
            Cnt++;                                      // Increment Cnt
            if(Cnt == 100) Cnt = 0;                     // Count between 0 and 99
            One_Second_Delay( );                        // Wait 1 second
        }
}
```

Figure 6.40: *Cont'd*

6.8.5 Modifying the Program

In Figure 6.40, the display counts as 00 01…09 10 11…99 00 01…i.e., the first digit is shown as 0 for numbers less than 10. The program could be modified so that the first digit is blanked if the number to be displayed is less than 10. The modified program (called SEVEN5.C) is shown in Figure 6.41. Here, the first digit (MSD) is not enabled if the number to be displayed is 0.

```
/*****************************************************************************************************************
                              Dual 7-SEGMENT DISPLAY COUNTER
                              ================================

In this project, two common cathode 7-segment LED displays are connected to PORT C of a PIC18F452
microcontroller and the microcontroller is operated from a 4-MHz resonator. Digit 1 (left digit) enable pin is connected
to port pin RB0 and digit 2 (right digit) enable pin is connected to port pin RB1 of the microcontroller. The program
counts up from 0 to 99 with a one-second delay between each count.

The display is updated in a timer interrupt service routine every 5 ms.

In this version of the program the first digit is blanked if the number is 0.


Author:   Dogan Ibrahim
Date:     June 2009
File:     SEVEN5.C
*****************************************************************************************************************/
#include <p18f452.h>
#include <delays.h>
#pragma config WDT = OFF, OSC = XT

#define DIGIT1 PORTBbits.RB0
#define DIGIT2 PORTBbits.RB1

unsigned char Cnt = 0;
unsigned char Flag = 0;

void One_Second_Delay(void);
unsigned char Display(unsigned char);
void timer_ISR(void);


//
// Define the high interrupt vector to be at 0x08
//
#pragma code high_vector=0x08                       //Following code at address 0x08
void interrupt(void)
{
        _asm GOTO timer_ISR _endasm                 // Jump to ISR
}
```

Figure 6.41: Modified Program

```
#pragma code                                            // Return to default code section


//
// timer_ISR is an interrupt service routine (jumps here every 5ms)
//
#pragma interrupt timer_ISR
void timer_ISR( )
{
        unsigned char Msd, Lsd;
        TMR0L = 100;                                    // Re-load TMR0
        INTCON = 0x20;                                  // Set T0IE and clear T0IF
        Flag = ~ Flag;                                  // Toggle Flag
        if(Flag == 0)                                   // Do digit 1
        {
          DIGIT2 = 0;
          Msd = Cnt / 10;                               // MSD digit
          if(Msd != 0)                                  // blank MSD
          {
          PORTC = Display(Msd);                         // Send to PORT C
          DIGIT1 = 1;                                   // Enable digit 1
          }
        }
        else
        {                                               // Do digit 2
          DIGIT1 = 0;                                   // Disable digit 1
          Lsd = Cnt % 10;                               // LSD digit
          PORTC = Display(Lsd);                         // Send to PORT C
          DIGIT2 = 1;                                   // Enable digit 2
        }
}
//
// This function generates 1 second delay
//
void One_Second_Delay( )
{
        Delay10KTCYx(100);
}



//
// This function finds the bit pattern to be sent to the port to display a number
// on the 7-segment LED. The number is passed in the argument list of the function.
//
unsigned char Display(unsigned char no)
{
  unsigned char Pattern;
  unsigned char SEGMENT[ ] = {0x3F,0x06,0x5B,0x4F,0x66,0x6D,0x7D,0x07, 0x7F,0x6F};

  Pattern = SEGMENT[no];                                // Pattern to return
  return (Pattern);
}
```

Figure 6.41: *Cont'd*

```
//
// Start of MAIN Program. configure PORT B and PORT C as outputs. In addition,
// configure TMR0 to interrupt at every 5ms
//
void main(void)
{
    TRISC = 0;                              // PORT C are outputs
    TRISB = 0;                              // RB0, RB1 are outputs

    DIGIT1 = 0;                             // Disable digit 1
    DIGIT2 = 0;                             // Disable digit 2
//
// Configure TMR0 timer interrupt
//
    T0CON = 0xC4;                           // Prescaler = 32
    TMR0L = 100;                            // Load TMR0L with 100
    INTCON = 0xA0;                          // Enable TMR0 interrupt
    One_Second_Delay( );

    for(;;)                                 // Endless loop
    {
      Cnt++;                                // Increment Cnt
      if(Cnt == 100) Cnt = 0;               // Count between 0 and 99
      One_Second_Delay( );                  // Wait 1 second
    }

}
```

Figure 6.41: *Cont'd*

6.9 Project 8 – Four-Digit Multiplexed Seven-Segment LED Counter With Timer Interrupt

6.9.1 Project Description

This project is similar to Project 7, but a 4-digit display is used instead of a 2-digit display. Thus, any integer in the range from 0 to 9999 can be displayed. As in Project 7, the display is updated in the ISR; consequently, the processor can perform other tasks while the display is updated in the background. In this project, the value of a counter is displayed as it counts up every second. The operation of this project is very similar to that of Project 7; therefore, the project is not described again here in detail.

6.9.2 Project Hardware

The circuit diagram of this project is shown in Figure 6.42, where four 7-segment displays are connected to PORTB and PORTC of a PIC18F452 microcontroller. The segments are connected in parallel to PORTC, and they are enabled from PORTB.

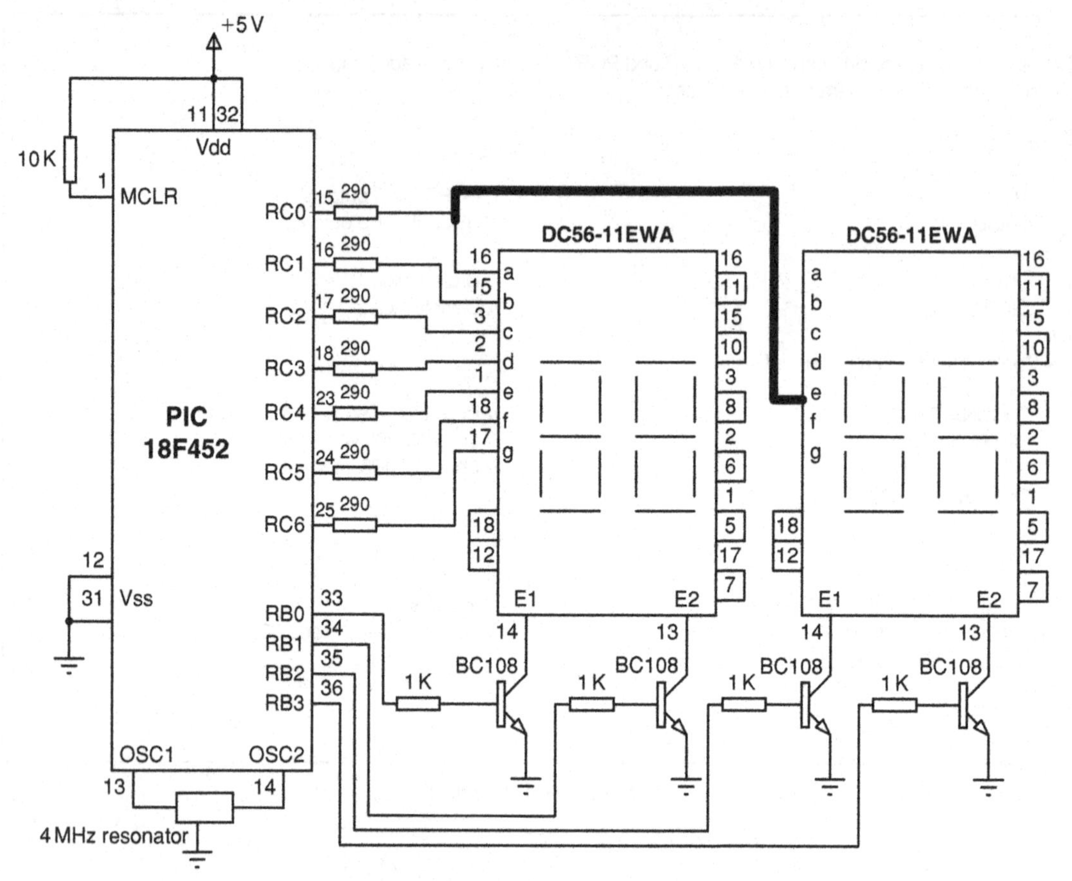

Figure 6.42: Circuit Diagram of the Project

6.9.3 Project PDL

The PDL of the project is shown in Figure 6.43. The program is in two sections: the main program and the ISR. Inside the main program, TMR0 is configured to generate interrupts every 5 ms. In addition, the counter is incremented with a 1-s delay. Inside the ISR, the timer interrupt is re-enabled, and the display digits are refreshed alternately every 5 ms.

6.9.4 Project Program

The program (called SEVEN6.C) is given in Figure 6.44. In this program, function prototypes are used at the beginning of the program to declare the functions called by the main program. At the beginning of the main program, PORTB and PORTC are configured as outputs. Then, register T0CON is loaded with 0xC4 to enable the TMR0 and set the prescaler to 32. The TMR0L register is loaded with 100 so that an interrupt can be generated after 5 ms. The program then enters an endless loop, where the value of **Cnt** is incremented every second using a delay function.

```
MAIN PROGRAM:

        START
                Configure PORT B as outputs
                Configure PORT C as outputs
                Clear variable Cnt to 0
                Configure TMR0 to generate interrupts every 5ms
                DO FOREVER
                        Increment Cnt between 0 and 99
                        Delay 1 second
                ENDO
        END


INTERRUPT SERVICE ROUTINE:
        START
                Re-configure TMR0
                IF Digit 1 updated THEN
                        Update digit 2
                ELSE IF Digit 2 is updated THEN
                        Update digit 3
                ELSE IF Digit 3 is updated THEN
                        Update digit 4
                ELSE IF Digit 4 is updated THEN
                        Update digit 1
                END
        END
```

Figure 6.43: PDL of the Project

```
/******************************************************************************************************
                              Four 7-SEGMENT DISPLAY COUNTER
                              ==============================

In this project, four common cathode 7-segment LED displays are connected to PORT C of a PIC18F452
microcontroller and the microcontroller is operated from a 4-MHz resonator. Digit 1 (left digit) enable pin is connected
to  port pin RB0, digit 2 enable pin is connected to port pin RB1. Digit 3 enable pin is connected to port pin RB2 and
digit 4 (right digit) enable pin is connected to port pin RB3.

The program counts up from 0 to 9999 with a one-second delay between each count.

The display is updated in a timer interrupt service routine every 5 ms.



Author:   Dogan Ibrahim
Date:     June 2009
File:     SEVEN6.C
******************************************************************************************************/
#include <p18f452.h>
#include <delays.h>
#pragma config WDT = OFF, OSC = XT
#define DIGIT1 PORTBbits.RB0
```

Figure 6.44: Program of the Project

```
#define DIGIT2 PORTBbits.RB1
#define DIGIT3 PORTBbits.RB2
#define DIGIT4 PORTBbits.RB3

unsigned int Cnt = 0;
unsigned char Flag = 0;

void One_Second_Delay(void);
unsigned char Display(unsigned char);
void timer_ISR(void);


//
// Define the high interrupt vector to be at 0x08
//
#pragma code high_vector=0x08                    //Following code at address 0x08
void interrupt(void)
{
        _asm GOTO timer_ISR _endasm              // Jump to ISR
}
#pragma code                                     // Return to default code section

//
// timer_ISR is an interrupt service routine (jumps here every 5ms)
//
#pragma interrupt timer_ISR
void timer_ISR( )
{
    unsigned int Dig;
    TMR0L = 100;                                 // Re-load TMR0
    INTCON = 0x20;                               // Set T0IE and clear T0IF

          DIGIT1 = 0;
          DIGIT2 = 0;
          DIGIT3 = 0;
          DIGIT4 = 0;

          switch(Flag)
          {
          case 0:
                Dig = Cnt / 1000;                // MSD digit
                PORTC = Display(Dig);            // Send to PORT C
                DIGIT1 = 1;                      // Enable digit 1
                break;
          case 1:
                Dig = Cnt / 100;                 // 2nd digit
                Dig = Dig % 10;
                PORTC = Display(Dig);            // Send to PORT C
                DIGIT2 = 1;                      // Enable Digit 2
                break;
          case 2:
                Dig = Cnt / 10;                  // 3rd digit
                Dig = Dig % 10;
                PORTC = Display(Dig);            // Send to PORT C
```

Figure 6.44: *Cont'd*

```
                    DIGIT3 = 1;                                  // Enable Digit 3
                    break;
          case 3:
                    Dig = Cnt % 10;                              // 4th digit
                    PORTC = Display(Dig);                        // Send to PORT C
                    DIGIT4 = 1;                                  // Enable Digit 4
                    break;
          }

          Flag++;
          if(Flag == 4)Flag = 0;
}



//
// This function generates 1 second delay
//
void One_Second_Delay( )
{
          Delay10KTCYx(100);
}



//
// This function finds the bit pattern to be sent to the port to display a number
// on the 7-segment LED. The number is passed in the argument list of the function.
//
unsigned char Display(unsigned char no)
{
          unsigned char Pattern;
          unsigned char SEGMENT[ ] = {0x3F,0x06,0x5B,0x4F,0x66,0x6D,0x7D,0x07, 0x7F,0x6F};

          Pattern = SEGMENT[no];                                 // Pattern to return
          return (Pattern);
}



//
// Start of MAIN Program. configure PORT B and PORT C as outputs. In addition,
// configure TMR0 to interrupt at every 5ms
//
void main(void)
{
     TRISC = 0;                          // PORT C are outputs
     TRISB = 0;                          // RB0, RB1 are outputs

     DIGIT1 = 0;                         // Disable digit 1
     DIGIT2 = 0;                         // Disable digit 2
     DIGIT3 = 0;                         // Disable digit 3
     DIGIT4 = 0;                         // Disable digit 4
//
```

Figure 6.44: *Cont'd*

```
// Configure TMR0 timer interrupt
//
      T0CON = 0xC4;                          // Prescaler = 32
      TMR0L = 100;                           // Load TMR0L with 100
      INTCON = 0xA0;                         // Enable TMR0 interrupt
      One_Second_Delay( );

      for(;;)                                // Endless loop
   {
      Cnt++;                                 // Increment Cnt
      if(Cnt == 10000) Cnt = 0;              // Count between 0 and 9999
      One_Second_Delay( );                   // Wait 1 second
   }
}
```

Figure 6.44: *Cont'd*

Inside the ISR register, TMR0L is reloaded, TMR0 interrupts are re-enabled, and the timer interrupt flag is cleared so that further timer interrupts can be generated. The display digits are then updated based on the value of variable **Cnt**. A variable called **Flag** is used to determine which digit to update. *Flag* varies between 0 and 3, and the display digits are updated in a **switch** statement as follows:

Flag	Digit Updated
0	1 (MSD)
1	2
2	3
3	4 (LSD)

Function **Display** is called as in Project 7 to find the bit pattern to be sent to PORTC.

6.9.5 Modifying the Program

In Figure 6.44, the display counts as 0000 0001…0009 0010 0011…9999 0000 0001…i.e., the leading digits are shown even when they are zero. The program could be modified so that the leading digits are blanked if the corresponding digits are zero. The modified program (called SEVEN7.C) is shown in Figure 6.45. Note that the first digit is enabled if the number is greater than or equal to 1000, the second digit is enabled if the number is greater than or equal to 100, and the third digit is enabled if the number is greater than or equal to 10.

6.9.6 Using MPLAB C18 Compiler Timer Library Routines

In the previous examples, the timer TMR0 was configured by writing into the timer registers. MPLAB C18 compiler provides timer library routines that simplify the configuration of timers. The program given in Figure 6.45 has been modified to use these timer library

```
/*****************************************************************************************************************************
                    Four 7-SEGMENT DISPLAY COUNTER
                    ===============================

In this project, four common cathode 7-segment LED displays are connected to PORT C of a PIC18F452
microcontroller and the microcontroller is operated from a 4-MHz resonator. Digit 1 (left digit) enable pin is connected
to port pin RB0, digit 2 enable pin is connected to port pin RB1. Digit 3 enable pin is connected to port pin RB2 and
digit 4 (right digit) enable pin is connected to port pin RB3.

The program counts up from 0 to 9999 with a 1-second delay between each count.

The display is updated in a timer interrupt service routine at every 5 ms.
In this version of the program, leading zeroes are blanked.



Author:   Dogan Ibrahim
Date:     June 2009
File:     SEVEN7.C
*****************************************************************************************************************************/
#include <p18f452.h>
#include <delays.h>
#pragma config WDT = OFF, OSC = XT

#define DIGIT1 PORTBbits.RB0
#define DIGIT2 PORTBbits.RB1
#define DIGIT3 PORTBbits.RB2
#define DIGIT4 PORTBbits.RB3

unsigned int Cnt = 0;
unsigned char Flag = 0;

void One_Second_Delay(void);
unsigned char Display(unsigned char);
void timer_ISR(void);


//
// Define the high interrupt vector to be at 0x08
//
#pragma code high_vector=0x08                          //Following code at address 0x08
void interrupt(void)
{
          _asm GOTO timer_ISR _endasm                  // Jump to ISR
}
#pragma code                                           // Return to default code section

//
// timer_ISR is an interrupt service routine (jumps here every 5ms)
//
#pragma interrupt timer_ISR
void timer_ISR( )
{
      unsigned int Dig;
      TMR0L = 100;                                     // Re-load TMR0
      INTCON = 0x20;                                   // Set T0IE and clear T0IF
```

Figure 6.45: Modified Program

```
            DIGIT1 = 0;
            DIGIT2 = 0;
            DIGIT3 = 0;
            DIGIT4 = 0;

            switch(Flag)
            {
            case 0:
                if(Cnt >= 1000)
                {
                    Dig = Cnt / 1000;                   // MSD digit
                    PORTC = Display(Dig);               // Send to PORT C
                    DIGIT1 = 1;                         // Enable digit 1
                }
                break;
            case 1:
                if(Cnt >= 100)
                {
                    Dig = Cnt / 100;                    // 2nd digit
                    Dig = Dig % 10;
                    PORTC = Display(Dig);               // Send to PORT C
                    DIGIT2 = 1;                         // Enable Digit 2
                }
                break;
            case 2:
                if(Cnt >= 10)
                {
                    Dig = Cnt / 10;                     // 3rd digit
                    Dig = Dig % 10;
                    PORTC = Display(Dig);               // Send to PORT C
                    DIGIT3 = 1;                         // Enable Digit 3
                }
                break;
            case 3:
                    Dig = Cnt % 10;                     // 4th digit
                    PORTC = Display(Dig);               // Send to PORT C
                    DIGIT4 = 1;                         // Enable Digit 4
                    break;
                }
                Flag++;
                if(Flag == 4)Flag = 0;
}


//
// This function generates 1 second delay
//
void One_Second_Delay( )
{
        Delay10KTCYx(100);
}
```

Figure 6.45: *Cont'd*

```
//
// This function finds the bit pattern to be sent to the port to display a number
// on the 7-segment LED. The number is passed in the argument list of the function.
//
unsigned char Display(unsigned char no)
{
   unsigned char Pattern;
   unsigned char SEGMENT[ ] = {0x3F,0x06,0x5B,0x4F,0x66,0x6D,0x7D,0x07,0x7F,0x6F};

   Pattern = SEGMENT[no];                              // Pattern to return
   return (Pattern);
}


//
// Start of MAIN Program. configure PORT B and PORT C as outputs. In addition,
// configure TMR0 to interrupt at every 5ms
//
void main(void)
{
    TRISC = 0;                                         // PORT C are outputs
    TRISB = 0;                                         // RB0, RB1 are outputs

    DIGIT1 = 0;                                        // Disable digit 1
    DIGIT2 = 0;                                        // Disable digit 2
    DIGIT3 = 0;                                        // Disable digit 3
    DIGIT4 = 0;                                        // Disable digit 4
//
// Configure TMR0 timer interrupt
//
    T0CON = 0xC4;                                      // Prescaler = 32
    TMR0L = 100;                                       // Load TMR0L with 100
    INTCON = 0xA0;                                     // Enable TMR0 interrupt
    One_Second_Delay( );

    for(;;)                                            // Endless loop
    {
     Cnt++;                                            // Increment Cnt
     if(Cnt == 10000) Cnt = 0;                         // Count between 0 and 9999
     One_Second_Delay( );                              // Wait 1 second
    }

}
```

Figure 6.45: *Cont'd*

routines. The modified program (called SEVEN8.C) is given in Figure 6.46. As can be seen in this program, the header file <timers.h> should be declared as the beginning of the program. In the main program, the timer TMR0 is configured by calling the library function OpenTimer0

```
OpenTimer0(TIMER_INT_ON & T0_8BIT & T0_SOURCE_INT & T0_PS_1_32);
```

```
/*****************************************************************************************************************
                              Four 7-SEGMENT DISPLAY COUNTER
                              ==============================

In this project, four common cathode 7-segment LED displays are connected to PORT C of a PIC18F452
microcontroller and the microcontroller is operated from a 4-MHz resonator. Digit 1 (left digit) enable pin is connected
to port pin RB0, digit 2 enable pin is connected to port pin RB1. Digit 3 enable pin is connected to port pin RB2 and
digit 4 (right digit) enable pin is connected to port pin RB3.

The program counts up from 0 to 9999 with a 1-second delay between each count.

The display is updated in a timer interrupt service routine every 5 ms.

In this version of the program, leading zeroes are blanked and MPLAB C18
compiler TIMER library routines are used.



Author:   Dogan Ibrahim
Date:     June 2009
File:     SEVEN8.C
*****************************************************************************************************************/
#include <p18f452.h>
#include <delays.h>
#include <timers.h>

#pragma config WDT = OFF, OSC = XT

#define DIGIT1 PORTBbits.RB0
#define DIGIT2 PORTBbits.RB1
#define DIGIT3 PORTBbits.RB2
#define DIGIT4 PORTBbits.RB3

unsigned int Cnt = 0;
unsigned char Flag = 0;

void One_Second_Delay(void);
unsigned char Display(unsigned char);
void timer_ISR(void);


//
// Define the high interrupt vector to be at 0x08
//
#pragma code high_vector=0x08                          //Following code at address 0x08
void interrupt(void)
{
          _asm GOTO timer_ISR _endasm                  // Jump to ISR
}
#pragma code                                           // Return to default code section

//
// timer_ISR is an interrupt service routine (jumps here every 5ms)
//
```

Figure 6.46: Using the Timer Library Functions

```
#pragma interrupt timer_ISR
void timer_ISR( )
{
    unsigned int Dig;
    WriteTimer0(100);                                   // Re-load TMR0
    INTCON = 0x20;                                      // Set T0IE and clear T0IF

        DIGIT1 = 0;
        DIGIT2 = 0;
        DIGIT3 = 0;
        DIGIT4 = 0;

        switch(Flag)
        {
            case 0:
                if(Cnt >= 1000)
                {
                    Dig = Cnt / 1000;                   // MSD digit
                    PORTC = Display(Dig);               // Send to PORT C
                    DIGIT1 = 1;                         // Enable Digit 1
                }
                break;
            case 1:
                if(Cnt >= 100)
                {
                    Dig = Cnt / 100;                    // 2nd digit
                    Dig = Dig % 10;
                    PORTC = Display(Dig);               // Send to PORT C
                    DIGIT2 = 1;                         // Enable Digit 2
                }
                break;
            case 2:
                if(Cnt >= 10)
                {
                    Dig = Cnt / 10;                     // 3rd digit
                    Dig = Dig % 10;
                    PORTC = Display(Dig);               // Send to PORT C
                    DIGIT3 = 1;                         // Enable Digit 3
                }
                break;
            case 3:
                Dig = Cnt % 10;                         // 4th digit
                PORTC = Display(Dig);                   // Send to PORT C
                DIGIT4 = 1;                             // Enable Digit 4
                break;
        }
            Flag++;
            if(Flag == 4)Flag = 0;
}


//
```

Figure 6.46: *Cont'd*

```
// This function generates 1 second delay
//
void One_Second_Delay( )
{
         Delay10KTCYx(100);
}



//
// This function finds the bit pattern to be sent to the port to display a number
// on the 7-segment LED. The number is passed in the argument list of the function.
//
unsigned char Display(unsigned char no)
{
   unsigned char Pattern;
   unsigned char SEGMENT[ ] = {0x3F,0x06,0x5B,0x4F,0x66,0x6D,0x7D, 0x07,0x7F,0x6F};

   Pattern = SEGMENT[no];                                   // Pattern to return
   return (Pattern);
}



//
// Start of MAIN Program. configure PORT B and PORT C as outputs. In addition,
// configure TMR0 to interrupt at every 5ms
//
void main(void)
{
   TRISC = 0;                                               // PORT C are outputs
   TRISB = 0;                                               // RB0, RB1 are outputs

        DIGIT1 = 0;                                         // Disable digit 1
        DIGIT2 = 0;                                         // Disable digit 2
        DIGIT3 = 0;                                         // Disable digit 3
        DIGIT4 = 0;                                         // Disable digit 4
//
// Configure timer TMR0
//
   OpenTimer0(TIMER_INT_ON & T0_8BIT & T0_SOURCE_INT & T0_PS_1_32);
   WriteTimer0(100);
   INTCON = 0xA0;                                           // Enable interrupts
   One_Second_Delay( );

   for(;;)                                                  // Endless loop
   {
     Cnt++;                                                 // Increment Cnt
     if(Cnt == 10000) Cnt = 0;                              // Count between 0 and 9999
     One_Second_Delay( );                                   // Wait 1 second
   }

}
```

Figure 6.46: *Cont'd*

This function has been called with the following arguments:

Enable timer interrupts.

Use timer in 8-bit mode.

Use internal clock source for the timer.

Use 1:32 prescaler value.

The timer is loaded by calling function WriteTimer0 and specifying the timer value as the argument.

```
WriteTimer0(100);
```

6.10 Summary

In this chapter, working examples of microcontroller C programs developed using the MPLAB C18 compiler are given. These examples should be very useful to readers who are new to programming the PIC microcontrollers using the C language.

6.11 Exercises

1. It is required to connect eight LEDs to PORTB of a PIC18F452-type microcontroller. Draw the circuit diagram of the project assuming that the microcontroller is to be operated from a 4-MHz crystal. Write a C program to turn ON the odd-numbered LEDs (i.e., LEDs connected to RB1, RB3, RB5, and RB7).
2. Eight LEDs are connected to a microcontroller as in Exercise 1. Write a C program to turn the LEDs ON and OFF every minute. Use C18 delay routines in your program.
3. Eight LEDs are connected to a microcontroller as in Exercise 1. Write a C program using TMR0 interrupts to turn the LEDs ON and OFF every 500 ms.
4. It is required to test a two-input NAND gate using a microcontroller. Assume that the inputs of the NAND gate are connected to ports RB0 and RB1, and the output is connected to port RB2. In addition, an LED is connected to port RB7. Draw the circuit diagram of the project and also write a C program to test the gate. If the gate is faulty, turn ON the LED, otherwise the LED should remain OFF.
5. Repeat Exercise 4, but this time, write a C program to test a NOR gate.
6. Write a C program to use TMR0 in a 16-bit mode and generate timer interrupts at 1-s intervals. Assume that an LED is connected to port RB7 of the microcontroller and turn this LED ON and OFF every 5 s.

CHAPTER 7

Serial Peripheral Interface Bus Operation

As described in Chapter 3, the secure digital (SD) card will be used in serial peripheral interface (SPI) bus mode. Before looking at the MPLAB C18 SD card functions and subprograms, it is worthwhile to see how the PIC18 microcontrollers can be configured and used in SPI mode. With a good understanding of the operation of the SPI bus, the SD card functions and subprograms can easily be modified if required.

In this chapter, the configuration and operation of the PIC18 microcontrollers in SPI mode will be described and then an MPLAB C18-based example will be given to show how the microcontrollers can be programmed and used for SPI operation using a real external device.

7.1 The Master Synchronous Serial Port Module

The master synchronous serial port (MSSP) module is a serial interface module on the PIC18 series of microcontrollers used for communicating with other serial devices, such as EEPROMs, display drivers, A/D converters, D/A converters, and so on.

The MSSP module can operate in one of two modes:

- SPI
- Inter-integrated circuit (I^2C)

Both SPI and I^2C are serial bus protocol standards. The SPI protocol was initially developed and proposed by Motorola for use in microprocessor- and microcontroller-based interface applications. In a system that uses the SPI bus, one device acts as the master and other devices act as slaves. The master initiates a communication and also provides clock pulses to the slave devices.

The I^2C is a two-wire bus and was initially developed by Philips for use in low-speed microprocessor-based communication applications.

In this chapter, we shall be looking at the operation of the MSSP module in SPI mode.

7.2 MSSP in SPI Mode

The SPI mode allows 8 bits of data to be transmitted synchronously and received simultaneously. In master mode, the device uses three signals, and in slave mode, a fourth signal is used. In this chapter, we shall be looking at how to use the MSSP module in master

D.O.I.: 10.1016/B978-1-85617-719-1.00011-7

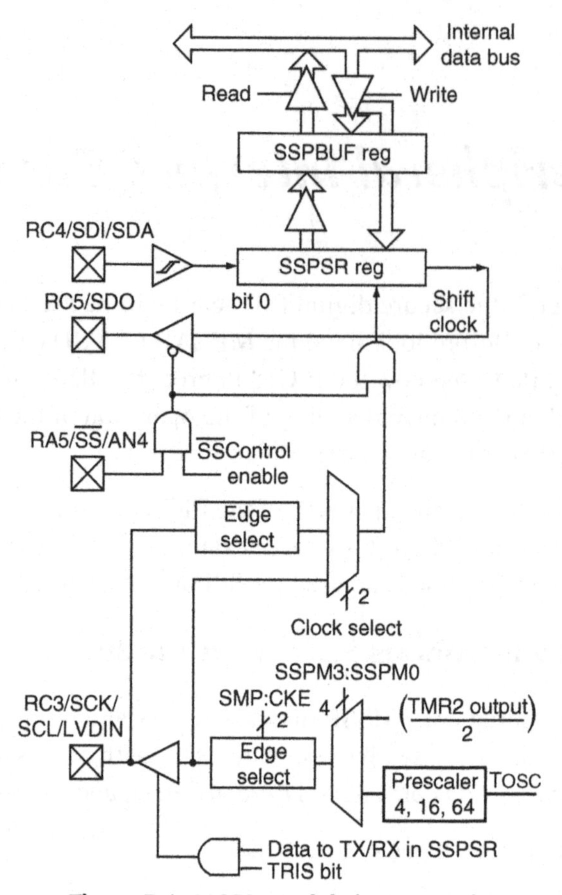

Figure 7.1: MSSP Module in SPI Mode

mode only because in a microcontroller and SD card communication, the microcontroller operates as an SPI master device and the SD card operates as an SPI slave device.

In master mode, the following microcontroller pins are used:

- Serial data out (SDO) – Pin RC5
- Serial data in (SDI) – Pin RC4
- Serial clock (SCK) – Pin RC3

Figure 7.1 shows the block diagram of the MSSP module when operating in SPI mode.

7.3 SPI Mode Registers

The MSSP module has four registers when operating in SPI mode:

- MSSP control register (SSPCON1)
- MSSP status register (SSPSTAT)

- Serial receive/transmit buffer register (SSPBUF)
- MSSP shift register (not accessible by the programmer)

7.3.1 SSPSTAT

SSPSTAT is the status register, with the lower 6 bits read only and the upper 2 bits read/write. Figure 7.2 shows the bit definitions of this register. Only bits 0, 6, and 7 are related to operation in SPI mode. Bit 7 (SMP) allows the user to select the input data sample time. When SMP = 0, input data is sampled at the middle of data output time, and when SMP = 1, the sampling is done at the end. Bit 6 (CKE) allows the user to select the transmit clock

R/W-0	R/W-0	R-0	R-0	R-0	R-0	R-0	R-0
SMP	CKE	D/$\overline{A}$	P	S	R/$\overline{W}$	UA	BF
bit 7							bit 0

bit 7 **SMP:** Sample bit
<u>SPI Master mode:</u>
1 = Input data sampled at end of data output time
0 = Input data sampled at middle of data output time
<u>SPI Slave mode:</u>
SMP must be cleared when SPI is used in slave mode

bit 6 **CKE:** SPI clock edge select
<u>When CKP = 0:</u>
1 = Data transmitted on the rising edge of SCK
0 = Data transmitted on the falling edge of SCK
<u>When CKP = 1:</u>
1 = Data transmitted on the falling edge of SCK
0 = Data transmitted on the rising edge of SCK

bit 5 **D/$\overline{A}$:** Data $\overline{\text{address}}$ bit
Used in I²C mode only

bit 4 **P:** STOP bit
Used in I²C mode only. This bit is cleared when the MSSP module is disabled, SSPEN is cleared.

bit 3 **S:** START bit
Used in I²C mode only

bit 2 **R/$\overline{W}$:** Read write bit information
Used in I²C mode only

bit 1 **UA:** Update address
Used in I²C mode only

bit 0 **BF:** Buffer full status bit (receive mode only)
1 = Recevie complete, SSPBUF is full
0 = Receive not complete, SSPBUF is empty

Legend:			
R = Readable bit	W = Writable bit	U = Unimplemented bit, read as '0'	
-n = Value at POR	'1' = Bit is set	'0' = Bit is cleared	x = Bit is unknown

Figure 7.2: SSPSTAT Register Bit Configuration

edge. When CKE = 0, transmit occurs on transition from idle to active clock state, and when CKE = 1, transmit occurs on transition from active to idle clock state. Bit 0 (BF) is the buffer full status bit. When BF = 1, receive is complete (i.e., SSPBUF is full), and when BF = 0, receive is not complete (i.e., SSPBUF is empty).

7.3.2 SSPCON1

SSPCON1 is the control register (see Figure 7.3) used to enable the SPI mode and to set the clock polarity (CP) and the clock frequency. In addition, the transmit collision detection (bit 7) and the receive overflow detection (bit 6) are indicated by this register.

R/W-0	R/W-0	R-0	R-0	R-0	R-0	R-0	R-0
SMP	CKE	D/$\overline{A}$	P	S	R/$\overline{W}$	UA	BF
bit 7							bit 0

bit 7 **SMP:** Sample bit
SPI Master mode:
1 = Input data sampled at end of data output time
0 = Input data sampled at middle of data output time
SPI Slave mode:
SMP must be cleared when SPI is used in slave mode

bit 6 **CKE:** SPI clock edge select
When CKP = 0:
1 = Data transmitted on the rising edge of SCK
0 = Data transmitted on the falling edge of SCK
When CKP = 1:
1 = Data transmitted on the falling edge of SCK
0 = Data transmitted on the rising edge of SCK

bit 5 **D/$\overline{A}$:** Data $\overline{\text{address}}$ bit
Used in I²C mode only

bit 4 **P:** STOP bit
Used in I²C mode only. This bit is cleared when the MSSP module is disabled, SSPEN is cleared.

bit 3 **S:** START bit
Used in I²C mode only

bit 2 **R/$\overline{W}$:** Read write bit information
Used in I²C mode only

bit 1 **UA:** Update address
Used in I²C mode only

bit 0 **BF:** Buffer full status bit (receive mode only)
1 = Receive complete, SSPBUF is full
0 = Receive not complete, SSPBUF is empty

Legend:			
R = Readable bit	W = Writable bit	U = Unimplemented bit, read as '0'	
-n = Value at POR	'1' = Bit is set	'0' = Bit is cleared	x = Bit is unknown

Figure 7.3: SSPCON1 Register Bit Configuration

7.4 Operation in SPI Mode

Figure 7.4 shows a simplified block diagram with a master and a slave device communicating over the SPI bus. The SDO (pin RC5) output of the master device is connected to the SDI (pin RC4) input of the slave device, and the SDI input of the master device is connected to the SDO output of the slave device. The clock SCK (pin RC3) is derived by the master device. The data communication is as follows:

Sending Data to the Slave: To send data from the master to the slave, the master writes the data byte into its SSPBUF register. This byte is also written automatically into the SSPSR register of the master. As soon as a byte is written into the SSPBUF register, eight clock pulses are sent out from the master SCK pin and at the same time, the data bits are sent out from the master SSPSR into the slave SSPSR, i.e., the contents of master and slave SSPSR registers are swapped. At the end of this data transmission, the SSPIF flag (PIR1 register) and the BF flag (SSPSTAT) will be set to show that the transmission is complete. Care should be taken not to write a new byte into SSPBUF before the current byte is shifted out, otherwise an overflow error will occur (indicated by bit 7 of SSPCON1).

Receiving Data From the Slave: To receive data from the slave device, the master has to write a "dummy" byte into its SSPBUF register to start the clock pulses that have to be sent out from the master. The received data is then clocked into SSPSR of the master, bit-by-bit. When the complete eight bits are received, the byte is transferred to the SSPBUF register and flags SSPIF and BF are set. It is interesting to note that the received data is double-buffered.

7.4.1 Configuration of MSSP for SPI Master Mode

The following MSSP parameters for the master device must be set up before the SPI communication can take place successfully:

- Set data clock rate
- Set clock edge mode

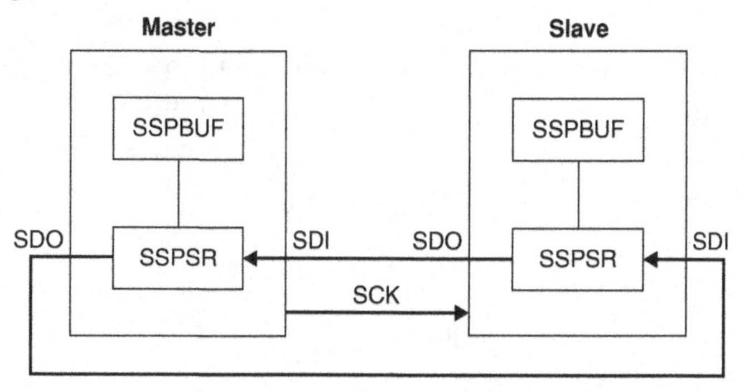

Figure 7.4: A Master and a Slave Device on the SPI Bus

- Clear bit 5 of TRISC (i.e., SDO = RC5 is output)
- Clear bit 3 of TRISC (i.e., SCK = RC3 is output)
- Enable the SPI mode

Note that the SDI pin (pin RC4) direction is automatically controlled by the SPI module.

Data Clock Rate

The clock is derived by the master, and the clock rate is user programmable to one of the following values via the SSPCON1 register bits 0–3 (see Figure 7.3):

- Fosc/4
- Fosc/16
- Fosc/64
- Timer 2 output/2

Clock Edge Mode

The clock edge is user programmable via register SSPCON1 (bit 4), and the user can either set the clock edge as *idle high* or *idle low*. In the idle high mode, the clock is high when the device is not transmitting, and in the idle low mode, the clock is low when the device is not transmitting. Data can be transmitted either at the rising or at the falling edge of the clock. The CKE bit of SSPSTAT (bit 6) is used to select the clock edge.

Enabling the SPI Mode

Bit 5 of SSPCON1 must be set to enable the SPI mode. To reconfigure the SPI parameters, this bit must be cleared, SPI mode configured, and then the SSPEN bit set back to 1.

The example given below demonstrates how to set the SPI parameters.

Example 7.1

It is required to operate the MSSP device of a PIC microcontroller in SPI mode. The data should be shifted on the rising edge of the clock, and the SCK signal must be idle low. The required data rate is at least 1 Mbps. Assume that the microcontroller clock rate is 16 MHz and the input data is to be sampled at the middle of data output time. What should be the settings of the MSSP registers?

Solution

Register SSPCON1 should be set as follows:

- Clear bits 6 and 7 of SSPCON1 (i.e., no collision detect and no overflow).

- Clear bits 4 to 0 to select idle low for the clock.
- Set bits 0 through 3 to 0000 or 0001 to select the clock rate to Fosc/4 (i.e., 16/4 = 4 Mbps data rate) or Fosc/16 (i.e., 16/16 = 1 Mbps).
- Set bit 5 to enable the SPI mode.

Thus, register SSPCON1 should be set to the following bit pattern:

$$00100000 \text{ i.e.,} 0 \times 20$$

Register SSPSTAT should be set as follows:

- Clear bit 7 to sample the input data at the middle of data output time.
- Clear bit 6 to 0 to transmit the data on the rising edge (low-to-high) of the SCK clock.
- Bits 5 through 0 are not used in the SPI mode.

Thus, register SSPSTAT should be set to the following bit pattern:

$$0\,0\,0\,00000 \text{ i.e.,} 0 \times 00$$

■

7.5 SPI Bus MPLAB C18 Library Functions

MPLAB C18 compiler provides useful library functions for programming the MSSP module in SPI mode. These library functions are shown in Table 7.1 (see the Microchip document *MPLAB C18 C Compiler Libraries* for more details).

The header file **spi.h** must be included at the beginning of the C18 programs to use these functions. The available functions are briefly described below.

Table 7.1: MPLAB C18 SPI Library Functions

Functions	Descriptions
CloseSPI	Disable the SPI module
DataRdySPI	Determine whether a data byte is available on the SPI bus
getcSPI	Read a byte from the SPI bus
getsSPI	Read a string from the SPI bus
OpenSPI	Initialize the SPI module
putcSPI	Write a byte to the SPI bus
putsSPI	Write a string to the SPI bus
ReadSPI	Read a byte from the SPI bus
WriteSPI	Write a byte to the SPI bus

7.5.1 CloseSPI

Description:	This function disables the SPI module.
Example code:	closeSPI();

7.5.2 DataRdySPI

Description:	This function determines whether the SPI buffer contains data. 0 is returned if there is no data in the SSPBUF register, and 1 is returned if new data is available in the SSPBUF register.
Example code:	While (DataRdySPI());

7.5.3 getcSPI

Description:	This function reads a byte from the SPI bus.
Example code:	d = getcSPI();

7.5.4 getsSPI

Description:	This function reads a string from the SPI bus. The number of characters to read must be specified in the function argument.
Example code:	getsSPI(dat, 10);

7.5.5 OpenSPI

Description:	This function initializes the SPI module for SPIbus communications.
Example code:	OpenSPI(SPI_FOSC_4, MODE_00, SMPEND);

The first argument is the SCK clock rate, and it can have the following values:

SPI_FOSC_4	Master clock rate is Fosc/4
SPI_FOSC_16	Master clock rate is Fosc/16
SPI_FOSC_64	Master clock rate is Fosc/64
SPI_FOSC_TMR2	Master clock rate is TMR2 output/2
SLV_SSON	(used in slave mode only)
SLV_SSOFF	(used in slave mode only)

The second parameter specifies the clock edge, and it can take one of the following values:

MODE_00	Clock is idle low, transmit on rising edge
MODE_01	Clock is idle low, transmit on falling edge

MODE_10	Clock is idle high, transmit on falling edge
MODE_11	Clock is idle high, transmit on rising edge

The last parameter is the input data sampling time, and it can take one of the following values:

SMPEND	Input data sample at the end of data out
SMPMID	Input data sample at middle of data out

7.5.6 putcSPI

Description: This function writes a byte to the SPI bus. The function returns 0 if there is no collision and 1 if a collision has occurred.

Example code: stat = putcSPI(data);

7.5.7 putsSPI

Description: This function writes a string to the SPI bus.

Example code: unsigned char test[] = "SPI bus"; putsSPI(test);

7.5.8 ReadSPI

Description: This function reads a byte from the SPI bus. The function is same as getcSPI.

Example code: x = readSPI();

7.5.9 WriteSPI

Description: This function writes a byte to the SPI bus. The function is same as the putcSPI.

Example code: stat = writeSPI('c');

7.6 Example of an SPI Bus Project

In this section, an example of an SPI bus project is given to show how the SPI bus can be used to communicate with an SPI bus device. In this example, a TC72-type integrated circuit digital temperature sensor chip is used to read the ambient temperature and then display on an LCD every second.

Figure 7.5 shows the block diagram of the project. The temperature sensor TC72 is connected to the SPI bus pins of a PIC18F452-type microcontroller. In addition, the microcontroller is connected to a standard LCD device to display the temperature.

The specifications and operation of the TC72 temperature sensor are described below in detail.

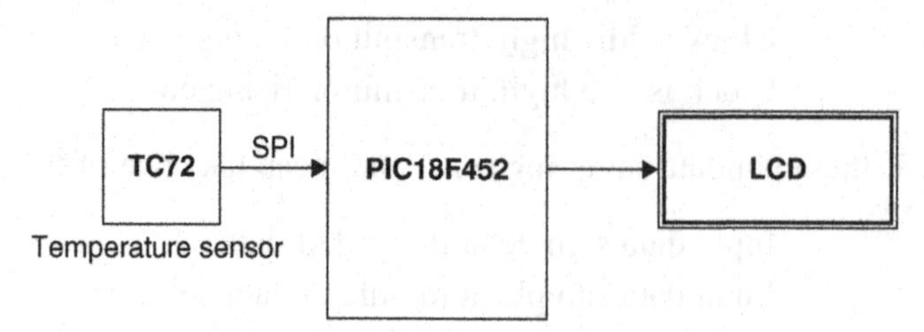

Figure 7.5: Block Diagram of the Project

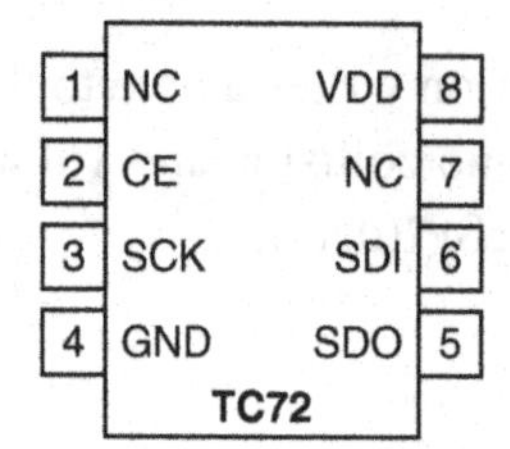

Figure 7.6: TC72 Pin Configuration

7.6.1 TC72 Temperature Sensor

The TC72 is an SPI bus–compatible digital temperature sensor IC that is capable of reading temperatures from –55 to +125°C.

The device has the following features:

- SPI bus compatible
- 10-bit resolution with 0.25°C/bit
- ±2°C accuracy from –40 to +85°C
- 2.65–5.5 V operating voltage
- 250-μA typical operating current
- 1-μA shutdown operating current
- Continuous and one-shot (OS) operating modes

The pin configuration of TC72 is shown in Figure 7.6. The device is connected to an SPI bus via standard SPI bus pins SDI, SDO, and SCK. Pin CE is the chip-enable pin and is used to select a particular device in multiple TC72 applications. CE must be logic 1 for the device to be enabled. The device is disabled (output in tri-state mode) when CE is logic 0.

The TC72 can operate either in *one-shot* (OS) mode or in *continuous* mode. In OS mode, the temperature is read after a request is sent to read the temperature. In the continuous mode, the device measures the temperature approximately every 150 ms.

Temperature data is represented in 10-bit two's complement format with a resolution of 0.25°C/bit. The converted data is available in two 8-bit registers. The MSB register stores the decimal part of the temperature, whereas the LSB register stores the fractional part. Only bits 6 and 7 of this register are used. The format of these registers is shown below:

MSB: S 2^6 2^5 2^4 2^3 2^2 2^1 2^0
LSB: 2^{-1} 2^{-2} 0 0 0 0 0 0

where S is the sign bit. An example is given below.

Example 7.2

The MSB and LSB settings of a TC72 are as follows:

MSB: 00101011
LSB: 10000000

Find the reading of the temperature.

Solution

The temperature is found to be

$$MSB = 2^5 + 2^3 + 2^1 + 2^0 = 43$$
$$LSB = 2^{-1} = 0.5$$

Thus, the temperature is 43.5°C.

Table 7.2 shows the sample temperature output data of the TC72 sensor.

Table 7.2: TC72 Temperature Output Data

Temperature (°C)	Binary (MSB/LSB)	Hex
+125	0111 1101/0000 0000	7D00
+74.5	0100 1010/1000 0000	4A80
+25	0001 1001/0000 0000	1900
+1.5	0000 0001/1000 0000	0180
+0.5	0000 0000/1000 0000	0080
+0.25	0000 0000/0100 0000	0040
0	0000 0000/0000 0000	0000
−0.25	1111 1111/1100 0000	FFC0
−0.5	1111 1111/1000 0000	FF80
−13.25	1111 0010/1100 0000	F2C0
−25	1110 0111/0000 0000	E700
−55	1100 1001/0000 0000	C900

TC72 Read/Write Operations

The SDI input writes data into TC72's control register, whereas SDO outputs the temperature data from the device. The TC72 can operate using either the rising or the falling edge of the clock (SCK). The clock idle state is detected when the CE signal goes high. As shown in Figure 7.7, the CP determines whether the data is transmitted on the rising or on the falling clock edge.

The maximum clock frequency (SCK) of TC72 is specified as 7.5 MHz. Data transfer consists of an address byte, followed by one or more data bytes. The most significant bit (A7) of the address byte determines whether a read or a write operation will occur. If A7 = 0, one or more read cycles will occur, otherwise, if A7 = 1, one or more write cycles will occur. The multibyte read operation will start by writing onto the highest desired register and then by reading from high to low addresses. For example, the temperature high-byte address can be sent with A7 = 0 and then the resulting high-byte, low-byte, and control register can be read as long as the CE pin is held active (CE = 1).

The procedure to read a temperature from the device is as follows:

- Configure the microcontroller SPI bus for the required clock rate and clock edge.
- Enable TC72 by setting CE = 1.
- Send temperature result high-byte read address (0 × 02) to the TC72 (see Table 7.3).
- Write a "dummy" byte into the SSPBUF register to start eight pulses to be sent out from the SCK pin and then read the temperature result high byte.
- Write a "dummy" byte into SSPBUF register to start eight pulses to be sent out from the SCK pin and then read the temperature low byte.
- Set CE = 0 to disable the TC72 so that a new data transfer can begin.

Internal Registers of the TC72

As shown in Table 7.3, the TC72 has four internal registers: Control register, LSB temperature register, MSB temperature register, and Manufacturer's ID register.

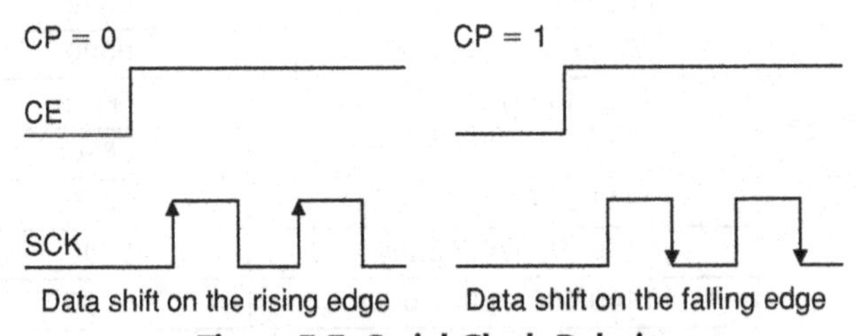

Figure 7.7: Serial Clock Polarity

Table 7.3: TC72 Internal Registers

Register	Read Address	Write Address	Bit 7	Bit 6	Bit 5	Bit 4	Bit 3	Bit 2	Bit 1	Bit 0
Control	0 × 00	0 × 80	0	0	0	OS	0	1	0	SHDN
LSB temperature	0 × 01	N/A	T1	T0	0	0	0	0	0	0
MSB temperature	0 × 02	N/A	T9	T8	T7	T6	T5	T4	T3	T2
Manufacturer's ID	0 × 03	N/A	0	1	0	1	0	1	0	0

Table 7.4: Selecting the Mode of Operation

Operating Mode	One Shot	Shutdown (SHDN)
Continuous	X	0
Shutdown	0	1
One shot	1	1

Control Register

The Control register is a read and write register used to select the mode of operation as shutdown, continuous, or OS. The address of this register is 0 × 00 when reading and 0 × 80 when writing onto the device. Table 7.4 shows how different modes are selected. At power-up, the shutdown bit (SHDN) is set to 1 so that the device is in shutdown mode at start-up, and the device is used in this mode to minimize power consumption.

A temperature conversion is initiated by a write operation to the Control register to select either the continuous mode or the OS mode. The temperature data will be available in the MSB and LSB registers after approximately 150 ms of the write operation. The OS mode performs a single temperature measurement, after which time the device returns to the shutdown mode. In continuous mode, new temperature data is available at 159-ms intervals.

LSB and MSB Registers

The LSB and MSB registers are read-only registers that contain the 10-bit measured temperature data. The address of the MSB register is 0 × 02 and that of the LSB register is 0 × 01.

Manufacturer's ID

The Manufacturer's ID is a read-only register with address 0 × 03. This register identifies the device as a temperature sensor, returning 0 × 054.

7.6.2 The Circuit Diagram

The circuit diagram of the project is shown in Figure 7.8. The TC72 temperature sensor is connected to the SPI bus pins of a PIC18F452 microcontroller, which is operated from a 4-MHz crystal. The CE pin of the TC72 is controlled from pin RC0 of the microcontroller. An LCD is connected to PORTB of the microcontroller in the default configuration, i.e., the connection between the TC72, LCD, and microcontroller is as follows:

Microcontroller	LCD
RB0	D4
RB1	D5
RB2	D6
RB3	D7
RB4	E
RB5	R/S
RB6	RW
Microcontroller	**TC72**
RC0	CE
RC3	SCK
RC4	SDO
RC5	SDI

The operation of the project is very simple. The microcontroller sends control commands to the TC72 sensor to initiate the temperature conversions every second. The temperature data is then read and displayed on the LCD.

7.6.3 The Program

The program listing of the project is shown in Figure 7.9. The program reads the temperature from the TC72 sensor and displays it on the LCD every second. In this version of the program, only the positive temperatures and only the integer part are displayed.

The program consists of a number of functions. At the beginning of the program, some definitions are made to make the program more readable. The LCD delay routines **DelayFor18TCY, DelayPORXLCD**, and **DelayXLCD** are then given for a 4-MHz clock rate. Two further delay routines are also used in the program: **Delay200 ms** creates a delay of 200 ms and is used during the TC72 temperature conversion routine. **One_Second_Delay** creates a 1-s delay and is used to read and display the temperature every second.

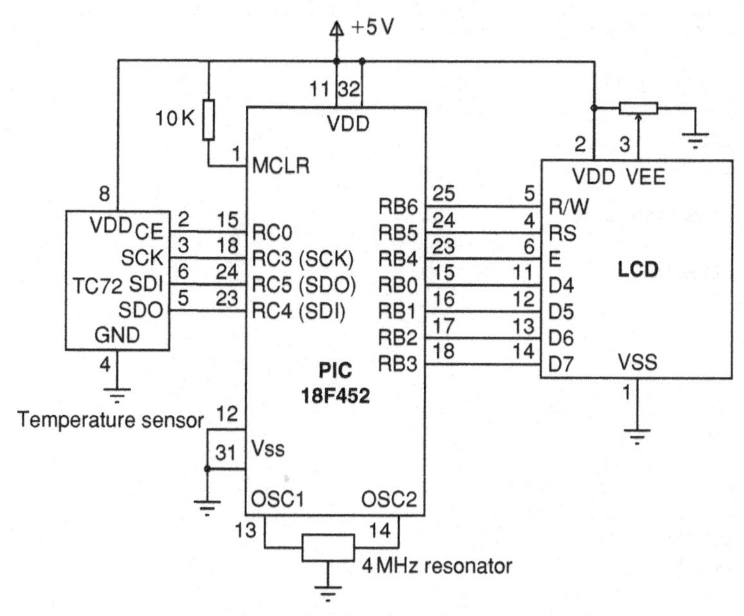

Figure 7.8: Circuit Diagram of the Project

```
/*****************************************************************************************************************
                    SPI BUS-BASED DIGITAL THERMOMETER
                    =================================

In this project, a TC72-type SPI bus-based temperature sensor IC is used.
The IC is connected to the SPI bus pins of a PIC18F452 type microcontroller
(i.e., to pins RC3 = SCK, RC4 = SDI, and RC5 = SDO) and the microcontroller is
operated from a 4-MHz resonator.

In addition, PORT B pins of the microcontroller are connected to a standard
LCD.

The microcontroller reads the temperature every second and displays on the
LCD as a positive number (fractional part of the temperature and negative
temperatures are not displayed in this version of the program).

An example display is:

          23


Author:   Dogan Ibrahim
Date:     July 2009
File:     TC72-1.C
*****************************************************************************************************************/
#include <p18f452.h>
#include <spi.h>
#include <stdlib.h>
#include <xlcd.h>
#include <delays.h>
```

Figure 7.9: Program Listing of the Project

```
#pragma config WDT = OFF, OSC = XT, LVP = OFF

#define CE PORTCbits.RC0

#define CLR_LCD 1
#define HOME_LCD 2

#define Ready SSPSTATbits.BF

unsigned char LSB, MSB;
int result;

//
// LCD Delays
//
void DelayFor18TCY(void)                                    // 18 cycle delay
{
        Nop( ); Nop( ); Nop( ); Nop( );
        Nop( ); Nop( ); Nop( ); Nop( );
        Nop( ); Nop( ); Nop( ); Nop( );
        Nop( ); Nop( );
        return;
}


void DelayPORXLCD(void)                                     // 15ms Delay
{
        Delay1KTCYx(15);
}


void DelayXLCD(void)                                        // 5ms Delay
{
        Delay1KTCYx(5);
}



//
// This function generates 200ms delay for the TC72 conversion
//
void Delay200ms(void)
{
        Delay1KTCYx(200);
}



//
// This function generates 1 second delay
//
void One_Second_Delay(void)
{
        Delay10KTCYx(100);
}
```

Figure 7.9: *Cont'd*

```
//
// Clear LCD and home cursor
//
void Clr_LCD(void)
{
        while(BusyXLCD( ));
        WriteCmdXLCD(CLR_LCD);                    // Clear LCD
        while(BusyXLCD( ));                       // Wait until ready
        WriteCmdXLCD(HOME_LCD);                   // Home the cursor
}


//
// Initialize the LCD, clear and home the cursor
//
void Init_LCD(void)
{
        OpenXLCD(FOUR_BIT & LINE_5X7);     // 8 bit, 5x7 character
        Clr_LCD( );                        // Clear LCD and home cursor
}


//
// Initialize the SPI bus
//
void Init_SPI(void)
{
        OpenSPI(SPI_FOSC_4, MODE_01, SMPEND);              // SPI clk = 1MHz
}


//
// This function sends a control byte to the TC72 and waits until the
// transfer is complete
//
void Send_To_TC72(unsigned char cmd)
{
        SSPBUF = cmd;                      // Send control to TC72
        while(!Ready);                     // Wait until data is shifted out
}



//
// This function reads the temperature from the TC72 sensor
//
// Temperature data is read as follows:
//
// 1. Enable TC72 (CE=1, for single byte write)
// 2. Send Address 0x80 (A7=1)
// 3. Clear BF flag
// 4. Send One-Shot command (Control = 0001 0001)
// 5. Disable TC72 (CE=0, end of single byte write)
// 6. Clear BF flag
// 7. Wait at least 150ms for temperature to be available
```

Figure 7.9: *Cont'd*

```
//  8. Enable TC72 (CE=1, for multiple data transfer)
//  9. Send Read MSB command (Read address=0x02)
// 10. Clear BF flag
// 11. Send dummy output to start clock and read data (Send 0x00)
// 12. Read high temperature into variable MSB
// 13. Send dummy output to start clock and read data (Send 0x00)
// 14. Read low temperature into variable LSB
// 15. Disable TC72 data transfer (CE=0)
// 16. Copy high result into variable "result"
void Read_Temperature(void)
{
        char dummy;

        CE = 1;                              // Enable TC72
        Send_To_TC72(0x80);                  // Send control write with A7=1
        dummy = SSPBUF;                      // Clear BF flag
        Send_To_TC72(0x11);                  // Set for one-shot operation
        CE = 0;                              // Disable TC72
        dummy = SSPBUF;                      // Clear BF flag
        Delay200ms( );                       // Wait 200ms for conversion
        CE = 1;                              // Enable TC72
        Send_To_TC72(0x02);                  // Read MSB temperature address
        dummy = SSPBUF;                      // Clear BF flag
        Send_To_TC72(0x00);                  // Read temperature high byte
        MSB = SSPBUF;                        // save temperature and clear BF
        Send_To_TC72(0x00);                  // Read temperature low byte
        LSB = SSPBUF;                        // Save temperature and clear BF
        CE = 0;                              // Disable TC72
        result = MSB;
}


//
// This function formats the temperature for displaying on the LCD.
// The temperature is read as a byte. We have to convert it to a string
// to display on the LCD.
//
// Only the positive MSB is displayed in this version of the program
//
void Format_Temperature(char *tmp)
{
        itoa(result,tmp);                    // Convert integer to ASCII
}


//
// This function clears the LCD, homes the cursor and then displays the
// temperature on the LCD
//
void Display_Temperature(char *d)
{
        Clr_LCD( );                          // Clear LCD and home cursor
        putsXLCD(d);
}
```

Figure 7.9: *Cont'd*

```
/* ============ START OF MAIN PROGRAM =================== */
//
// Start of MAIN Program. Display a message on the LCD and then
// display the temperature every second
//
void main(void)
{
          char msg[ ] = "Temperature...";
          char tmp[3];

          TRISC = 0;                          // Configure RC0 (CE) as output
          TRISB = 0;                          // PORT B are outputs


          One_Second_Delay( );
//
// Initialize the LCD
//
          Init_LCD( );
//
// Display a message on the LCD
//
          putsXLCD(msg);
//
// Wait 2 seconds before starting
//
          One_Second_Delay( );
          One_Second_Delay( );
//
// Clear the LCD and home cursor
//
          Clr_LCD( );
//
// Initialize the SPI bus
//
          Init_SPI( );


/* =============== ENDLESS PROGRAM LOOP ================= */
//
// Endless loop. Inside this loop read the TC72 temperature, display on the LCD,
// wait for 1 second and repeat the process
//
   for(;;)                                    // Endless loop
   {
            Read_Temperature( );              // Read the TC72 temperature
            Format_Temperature(tmp);          // Format the data for display
            Display_Temperature(tmp);         // Display the temperature
            One_Second_Delay( );              // Wait 1 second
          }
}
```

Figure 7.9: *Cont'd*

Some other functions used in the program are as follows:

Clr_LCD: This function clears the LCD screen and homes the cursor.

Init_LCD: This function initializes the LCD to 4-bit operation with 5×7 characters. The function also calls Clr_LCD to clear the LCD screen and home the cursor.

Init_SPI: This function initializes the microcontroller SPI bus to

Clock rate: Fosc/4 (i.e., 1 MHz)

Clock idle low, shift data on clock falling edge

Input data sample at end of data out.

Send_To_TC72: This function loads a byte to SPI register SSPBUF and then waits until the data is shifted out.

Read_Temperature: This function communicates with the TC72 sensor to read the temperature. The following operations are performed by this function:

1. Enables TC72 (CE = 1, for single-byte write).
2. Sends Address 0×80 (A7 = 1).
3. Clears BF flag.
4. Sends OS command (Control = 0001 0001).
5. Disables TC72 (CE = 0, end of single-byte write).
6. Clears BF flag.
7. Waits at least for 150 ms for the temperature to be available.
8. Enables TC72 (CE = 1 for multiple data transfer).
9. Sends Read MSB command (Read address = 0×02).
10. Clears BF flag.
11. Sends dummy output to start clock and read data (Send 0×00).
12. Reads high temperature into variable MSB.
13. Sends dummy output to start clock and read data (Send 0×00).
14. Reads low temperature into variable LSB.
15. Disables TC72 data transfer (CE = 0).
16. Copies high result into variable "result."

Format_Temperature: This function converts the integer temperature into ASCII string so that it can be displayed on the LCD.

Display_Temperature: This function calls to Clr_LCD to clear the LCD screen and homes the cursor. The temperature is then displayed calling function putsXLCD.

Main Program: At the beginning of the main program, the port directions are configured, LCD is initialized, the message "Temperature..." is sent to the LCD, and the microcontroller SPI bus is initialized. The program then enters an endless loop, where the following functions are called inside this loop:

Read_Temperature();

Format_Temperature(tmp);

Display_Temperature(tmp);

One_Second_Delay();

7.6.4 Displaying Negative Temperatures

The program given in Figure 7.9 displays only the positive temperatures. Negative temperatures are stored in TC72 in two's complement format. If bit 8 of the MSB byte is set, the temperature is negative and two's complement should be taken to find the correct temperature. For example, if the MSB and LSB bytes are "1110 0111/1000 0000," the correct temperature is

1110 0111/1000 0000 -> the complement is 0001 1000/0111 1111

Adding "1" to find the two's complement gives 0001 1000/1000 0000,

i.e., the temperature is "–24.5°C."

Similarly, if the MSB and LSB bytes are "1110 0111/0000 0000," the correct temperature is

1110 0111/0000 0000 -> the complement is 0001 1000/1111 1111

Adding "1" to find the two's complement gives 0001 1001/0000 0000,

i.e., the temperature is "–25°C."

The modified program is shown in Figure 7.10. In this program, both negative and positive temperatures are displayed, and the sign "–" is inserted before the negative temperatures. The temperature is displayed in the integer format with no fractional part in this version of the program.

In this version of the program, the **Format_Temperature** function is modified such that if the temperature is negative, the two's complement is taken, the sign bit is inserted, and then the value is shifted right by eight digits and converted into an ASCII string for the display.

7.6.5 Displaying the Fractional Part

The program in Figure 7.10 displays both the positive and negative temperatures, but it does not display the fractional part of the temperature. The modified program given in Figure 7.11 displays the fractional part as well. The main program is basically the same as that in Figure 7.10, but function **Format_Temperature** is modified. In the new program, the LSB byte of the converted data is taken into consideration and the fractional part is displayed as "0.00," "0.25," "0.50," or "0.75." The two most significant bits of the LSB byte are shifted right by 6 bits. The fractional part then takes one of the following values:

Two-Shifted LSB Bits	Fractional Part
00	0.00
01	0.25
10	0.50
11	0.75

```
/****************************************************************************************************************
                    SPI BUS-BASED DIGITAL THERMOMETER
                    ==================================

In this project, a TC72 type SPI bus-based temperature sensor IC is used.
The IC is connected to the SPI bus pins of a PIC18F452-type microcontroller
(i.e., to pins RC3=SCK, RC4=SDI, and RC5=SDO) and the microcontroller is
operated from a 4-MHz resonator.

In addition, PORT B pins of the microcontroller are connected to a standard
LCD.

The microcontroller reads the temperature every second and displays it on the
LCD as a positive or negative number (fractional part of the temperature is
not displayed in this version of the program).

An example display is:

          -25

Author:   Dogan Ibrahim
Date:     July 2009
File:     TC72-2.C
****************************************************************************************************************/
```

Figure 7.10: Modified Program to Display Negative Temperatures

```
#include <p18f452.h>
#include <spi.h>
#include <stdlib.h>
#include <xlcd.h>
#include <delays.h>

#pragma config WDT = OFF, OSC = XT, LVP = OFF

#define CE PORTCbits.RC0

#define CLR_LCD 1
#define HOME_LCD 2

#define Ready SSPSTATbits.BF

unsigned char LSB, MSB;
unsigned int result;

//
// LCD Delays
//
void DelayFor18TCY(void)                                    // 18 cycle delay
{
            Nop( ); Nop( ); Nop( ); Nop( );
            Nop( ); Nop( ); Nop( ); Nop( );
            Nop( ); Nop( ); Nop( ); Nop( );
            Nop( ); Nop( );
   return;
}


void DelayPORXLCD(void)                                     // 15ms Delay
{
            Delay1KTCYx(15);
}


void DelayXLCD(void)                                        // 5ms Delay
{
            Delay1KTCYx(5);
}



//
// This function generates 200ms delay for the TC72 conversion
//
void Delay200ms(void)
{
            Delay1KTCYx(200);
}
```

Figure 7.10: *Cont'd*

```
//
// This function generates 1 second delay
//
void One_Second_Delay(void)
{
        Delay10KTCYx(100);
}



//
// Clear LCD and home cursor
//
void Clr_LCD(void)
{
        while(BusyXLCD( ));
        WriteCmdXLCD(CLR_LCD);                    // Clear LCD
        while(BusyXLCD( ));                       // Wait until ready
        WriteCmdXLCD(HOME_LCD);                   // Home the cursor
}


//
// Initialize the LCD, clear and home the cursor
//
void Init_LCD(void)
{
        OpenXLCD(FOUR_BIT & LINE_5X7);            // 8 bit, 5x7 character
        Clr_LCD( );                               // Clear LCD, home cursor
}


//
// Initialize the SPI bus
//
void Init_SPI(void)
{
        OpenSPI(SPI_FOSC_4, MODE_01, SMPEND);     // SPI clk = 1MHz
}


//
// This function sends a control byte to the TC72 and waits until the
// transfer is complete
//
void Send_To_TC72(unsigned char cmd)
{
        SSPBUF = cmd;                             // Send control to TC72
        while(!Ready);                            // Wait until shifted out
}



//
// This function reads the temperature from the TC72 sensor
```

Figure 7.10: *Cont'd*

```
//
// Temperature data is read as follows:
//
// 1.     Enable TC72 (CE=1, for single byte write)
// 2.     Send Address 0x80 (A7=1)
// 3.     Clear BF flag
// 4.     Send One-Shot command (Control = 0001 0001)
// 5.     Disable TC72 (CE=0, end of single byte write)
// 6.     Clear BF flag
// 7.     Wait at leat 150ms for temperature to be available
// 8.     Enable TC72 (CE=1, for multiple data transfer)
// 9.     Send Read MSB command (Read address=0x02)
// 10.    Clear BF flag
// 11.    Send dummy output to start clock and read data (Send 0x00)
// 12.    Read high temperature into variable MSB
// 13.    Send dummy output to start clock and read data (Send 0x00)
// 14.    Read low temperature into variable LSB
// 15.    Disable TC72 data transfer (CE=0)
// 16.    Copy high result into variable "result"
void Read_Temperature(void)
{
        char dummy;

        CE = 1;                              // Enable TC72
        Send_To_TC72(0x80);                  // Send control with A7=1
        dummy = SSPBUF;                      // Clear BF flag
        Send_To_TC72(0x11);                  // Set for one-shot operation
        CE = 0;                              // Disable TC72
        dummy = SSPBUF;                      // Clear BF flag
        Delay200ms( );                       // Wait 200ms for conversion
        CE = 1;                              // Enable TC72
        Send_To_TC72(0x02);                  // Read MSB temperature address
        dummy = SSPBUF;                      // Clear BF flag
        Send_To_TC72(0x00);                  // Read temperature high byte
        MSB = SSPBUF;                        // Save temperature and clear BF
        Send_To_TC72(0x00);                  // Read temperature low byte
        LSB = SSPBUF;                        // Save temperature and clear BF
        CE = 0;                              // Disable TC72
        result = MSB*256 + LSB;              // The complete temperature
}



//
// This function formats the temperature for displaying on the LCD.
// The temperature is read as a byte. We have to convert it to a string
// to display on the LCD.
//
// Positive and negative temperatures are displayed in this version of the program
//
void Format_Temperature(char *tmp)
{
        if(result & 0x8000)                  // If negative
        {
                result = ~result;            // Take complement
```

Figure 7.10: *Cont'd*

```
                result++;                                   // Take 2's complement
                result >>= 8;                               // Get integer part
                *tmp++ = '-';                               // Insert "-" sign
        }
        else
        {
                result >>= 8;                               // Get integer part
        }

        itoa(result,tmp);                                   // Convert integer to ASCII
}



//
// This function clears the LCD, homes the cursor and then displays the
// temperature on the LCD
//
void Display_Temperature(char *d)
{
        Clr_LCD( );                                         // Clear LCD and home cursor
        putsXLCD(d);
}



/* ========================== START OF MAIN PROGRAM ========================== */
//
// Start of MAIN Program. Display a message on the LCD and then
// display the temperature every second
//
void main(void)
{
        char msg[ ] = "Temperature...";
        char tmp[4];

        TRISC = 0;                          // Configure RC0 (CE) as output
        TRISB = 0;                          // PORT B are outputs

        One_Second_Delay( );
//
// Initialize the LCD
//
        Init_LCD( );
//
// Display a message on the LCD
//
        putsXLCD(msg);
//
// Wait 2 seconds before starting
//
        One_Second_Delay( );
        One_Second_Delay( );
//
// Clear the LCD and home cursor
```

Figure 7.10: *Cont'd*

```
//
            Clr_LCD( );
//
// Initialize the SPI bus
//
            Init_SPI( );

/* ========================== ENDLESS PROGRAM LOOP ========================== */
//
// Endless loop. Inside this loop, read the TC72 temperature, display on the LCD,
// wait for 1 second and repeat the process
//
    for(;;)                                         // Endless loop
    {
            Read_Temperature( );                    // Read the TC72 temperature
            Format_Temperature(tmp);                // Format the data for display
            Display_Temperature(tmp);               // Display the temperature
            One_Second_Delay( );                    // Wait 1 second
            }
}
```

Figure 7.10: *Cont'd*

```
/*****************************************************************************************************
                    SPI BUS–BASED DIGITAL THERMOMETER
                    =================================

In this project, a TC72-type SPI bus–based temperature sensor IC is used.
The IC is connected to the SPI bus pins of a PIC18F452-type microcontroller
(i.e., to pins RC3=SCK, RC4=SDI, and RC5=SDO) and the microcontroller is
operated from a 4-MHz resonator.

In addition, PORT B pins of the microcontroller are connected to a standard
LCD.

The microcontroller reads the temperature every second and displays it on the
LCD as a positive or negative number. The fractional part of the temperature is
displayed as 2 digits in this version of the program.

An example display is:

        -25.75


Author:  Dogan Ibrahim
Date:    July 2009
File:    TC72-3.C
*****************************************************************************************************/
#include <p18f452.h>
#include <spi.h>
#include <stdlib.h>
#include <xlcd.h>
#include <delays.h>
```

Figure 7.11: Modified Program to Display Fractional Part as Well

```
#pragma config WDT = OFF, OSC = XT, LVP = OFF

#define CE PORTCbits.RC0

#define CLR_LCD 1
#define HOME_LCD 2

#define Ready SSPSTATbits.BF

unsigned char LSB, MSB;
unsigned int result, int_part, fract_part;

//
// LCD Delays
//
void DelayFor18TCY(void)                                    // 18 cycle delay
{
            Nop( ); Nop( ); Nop( ); Nop( );
            Nop( ); Nop( ); Nop( ); Nop( );
            Nop( ); Nop( ); Nop( ); Nop( );
            Nop( ); Nop( );
   return;
}


void DelayPORXLCD(void)                                     // 15ms Delay
{
            Delay1KTCYx(15);
}


void DelayXLCD(void)                                        // 5ms Delay
{
            Delay1KTCYx(5);
}



//
// This function generates 200ms delay for the TC72 conversion
//
void Delay200ms(void)
{
            Delay1KTCYx(200);
}


//
// This function generates 1 second delay
//
void One_Second_Delay(void)
{
            Delay10KTCYx(100);
}
```

Figure 7.11: *Cont'd*

```
//
// Clear LCD and home cursor
//
void Clr_LCD(void)
{
        while(BusyXLCD( ));
        WriteCmdXLCD(CLR_LCD);              // Clear LCD
        while(BusyXLCD( ));                 // Wait until ready
        WriteCmdXLCD(HOME_LCD);             // Home the cursor
}


//
// Initialize the LCD, clear and home the cursor
//
void Init_LCD(void)
{
        OpenXLCD(FOUR_BIT & LINE_5X7);      // 8 bit, 5x7 character
        Clr_LCD( );                         // Clear LCD, home cursor
}


//
// Initialize the SPI bus
//
void Init_SPI(void)
{
        OpenSPI(SPI_FOSC_4, MODE_01, SMPEND);    // SPI clk = 1MHz
}


//
// This function sends a control byte to the TC72 and waits until the
// transfer is complete
//
void Send_To_TC72(unsigned char cmd)
{
        SSPBUF = cmd;                       // Send control to TC72
        while(!Ready);                      // Wait until data is shifted out
}


//
// This function reads the temperature from the TC72 sensor
//
// Temperature data is read as follows:
//
// 1. Enable TC72 (CE=1, for single byte write)
// 2. Send Address 0x80 (A7=1)
// 3. Clear BF flag
// 4. Send One-Shot command (Control = 0001 0001)
// 5. Disable TC72 (CE=0, end of single byte write)
```

Figure 7.11: *Cont'd*

```
// 6. Clear BF flag
// 7. Wait at leat 150ms for temperature to be available
// 8.  Enable TC72 (CE=1, for multiple data transfer)
// 9.  Send Read MSB command (Read address=0x02)
// 10. Clear BF flag
// 11. Send dummy output to start clock and read data (Send 0x00)
// 12. Read high temperature into variable MSB
// 13. Send dummy output to start clock and read data (Send 0x00)
// 14. Read low temperature into variable LSB
// 15. Disable TC72 data transfer (CE=0)
// 16. Copy high result into variable "result"
void Read_Temperature(void)
{
        char dummy;

        CE = 1;                         // Enable TC72
        Send_To_TC72(0x80);             // Send control write with A7=1
        dummy = SSPBUF;                 // Clear BF flag
        Send_To_TC72(0x11);             // Set for one-shot operation
        CE = 0;                         // Disable TC72
        dummy = SSPBUF;                 // Clear BF flag
        Delay200ms( );                  // Wait 200ms for conversion
        CE = 1;                         // Enable TC72
        Send_To_TC72(0x02);             // Read MSB temperature address
        dummy = SSPBUF;                 // Clear BF flag
        Send_To_TC72(0x00);             // Read temperature high byte
        MSB = SSPBUF;                   // Save temperature and clear BF
        Send_To_TC72(0x00);             // Read temperature low byte
        LSB = SSPBUF;                   // Save temperature and clear BF
        CE = 0;                         // Disable TC72
        result = MSB*256 + LSB;         // The complete temperature
}


//
// This function formats the temperature for displaying on the LCD.
// The temperature is read as a byte. We have to convert it to a string
// to display on the LCD.
//
// Positive and negative temperatures are displayed in this version of the program
//
void Format_Temperature(char *tmp)
{
        if(result & 0x8000)             // If negative
        {
                result = ~result;       // Take complement
                result++;               // Take 2's complement
                int_part = result >> 8; // Get integer part
                *tmp++ = '-';           // Insert "-" sign
        }
        else
        {
                int_part = result >> 8; // Get integer part
        }
```

Figure 7.11: *Cont'd*

```
//
// Convert integer part to ASCII string
//
        itoa(int_part,tmp);                         // Convert integer to ASCII


//
// Now find the fractional part. First we must find the end of the string "tmp"
// and then append the fractional part to it
//
                while(*tmp !='\0')tmp++;            // find end of string "tmp"
//
// Now add the fractional part as ".00", ".25", ".50", or ".75"
//
                fract_part = result &0x00C0;                    // fractional part
                fract_part = fract_part >> 6;                   // fract is between 0-3
                switch(fract_part)
                {
                        case 1:                                 // Fractional part = 0.25
                                *tmp++ = '.';                   // decimal point
                                *tmp++ = '2';                   // "2"
                                *tmp++ = '5';                   // "5"
                                break;
                        case 2:                                 // Fractional part = 0.50
                                *tmp++ = '.';                   // decimal point
                                *tmp++ = '5';                   // "5"
                                *tmp++ = '0';                   // "0"
                                break;
                        case 3:                                 // Fractional part = 0.75
                                *tmp++ = '.';                   // decimal point
                                *tmp++ = '7';                   // "7"
                                *tmp++ = '5';                   // "5"
                                break;
                        case 0:                                 // Fractional part = 0.00
                                *tmp++ = '.';                   // decimal point
                                *tmp++ = '0';                   // "0"
                                *tmp++ = '0';                   // "0"
                                break;
                }
                *tmp++ = '\0';                                  // Null terminator
}




//
// This function clears the LCD, homes the cursor and then displays the
// temperature on the LCD
//
void Display_Temperature(char *d)
{
        Clr_LCD( );                                             // Clear LCD and home cursor
        putsXLCD(d);
}
```

Figure 7.11: *Cont'd*

```
/* ================ START OF MAIN PROGRAM ================ */
//
// Start of MAIN Program. Display a message on the LCD and then display the
// temperature every second
//
void main(void)
{
            char msg[ ] = "Temperature...";
            char tmp[8];

            TRISC = 0;                                  // Configure RC0 (CE) as output
            TRISB = 0;                                  // PORT B are outputs

            One_Second_Delay( );
//
// Initialize the LCD
//
            Init_LCD( );
//
// Display a message on the LCD
//
            putsXLCD(msg);
//
// Wait 2 seconds before starting
//
            One_Second_Delay( );
            One_Second_Delay( );
//
// Clear the LCD and home cursor
//
            Clr_LCD( );
//
// Initialize the SPI bus
//
            Init_SPI( );

/* ============================ ENDLESS PROGRAM LOOP============================ */
//
// Endless loop. Inside this loop, read the TC72 temperature, display on the LCD,
// wait for 1 second and repeat the process
//
    for(;;)                                             // Endless loop
    {
             Read_Temperature( );
            // Read the TC72 temperature
             Format_Temperature(tmp);                   // Format the data for display
             Display_Temperature(tmp);                  // Display the temperature
                    One_Second_Delay( );                // Wait 1 second
            }
}
```

Figure 7.11: *Cont'd*

7.7 Summary

In this chapter, the properties of the SPI bus and how it can be used in the PIC microcontroller circuits are described. An example is given to show how the SPI bus can be used to read the temperature from an SPI bus–compatible temperature sensor device.

7.8 Exercises

1. What are the operating modes of the MSSP module?
2. Which pins are used in the SPI mode? Describe the function of each pin.
3. Describe the SPI mode registers in detail. How can the MSSP be configured to operate in SPI mode?
4. Explain in detail, by drawing a diagram, how data can be sent and received by a master SPI device.
5. It is required to operate the MSSP device of a PIC microcontroller in SPI mode. The data should be shifted on rising edge of the clock and the SCK signal must be idle low. The required data rate is at least 2 Mbps. Assume that the microcontroller clock rate is 32 MHz and the input data is to be sampled at the middle of data output time. What should be the settings of the MSSP registers?
6. Describe which MPLAB C18 library functions can be used to read and write to an SPI device.
7. The MSB and LSB registers of a TC72 sensor contain the following bits. What is the temperature reading?

   MSB: 00111101
   LSB: 11000000
8. Modify the program given in Figure 7.9 to display the temperature as a 2-digit integer number between 0 and 99 on a pair of seven-segment displays.

CHAPTER 8

MPLAB C18 SD Card Functions and Procedures

Reading and writing onto secure digital (SD) cards is a complex process and requires the development of a number of rather complex functions and procedures to handle the card I/O operations correctly.

Fortunately, Microchip Inc. provides a library of file I/O functions for implementing the card file operations. This library is named "Microchip MDD File system," where MDD stands for "memory disk drive." The library can be downloaded free of charge from the Microchip Web site. The library is based on the ISO/IEC 9293 specifications and supports the MPLAB C18 and MPLAB C30 compilers. The library can be used for

- FAT16 and FAT32 file systems. FAT16 is an earlier file system usually found in MSDOS-based systems and early Windows systems. The current Windows operating systems (e.g., Windows XP) support both FAT16 and FAT32 file systems. SD cards (and multimedia cards) up to 2 GB use the FAT16 standard filing system. The FAT32 filing system is used for higher-capacity SD cards, usually between 2 GB and 2 TB.
- The MDD library supports SD cards, compact flash cards, and USB thumb drives.

In this book, we shall be looking at how to install the MDD library and how to use the library functions and procedures in the PIC18 series of microcontrollers, using the MPLAB C18 compiler (further information about the MDD library can be obtained from the Microchip application note AN1045, Document no: DS01045B, entitled "Implementing File I/O Functions Using Microchip's Memory Disk Drive File System Library.")

8.1 Installation of the MDD Library

The current version (at the time of writing this book) of the Microchip MDD library has the filename "Microchip MDD File System 1.2.1 Installer.exe." The library can be installed by the following steps:

- Download the library from Microchip Web site
- Double-click the file and follow the instructions to load

D.O.I.: 10.1016/B978-1-85617-719-1.00012-9

- The MDD library creates a directory called "Microchip Solutions" under the **"C:\" root directory**. The following directories are created and files are copied to the "Microchip Solutions" directory:

  Directory: "MDD File System-SD Card" contains demonstration programs for the PIC18, PIC24, and PIC32 microcontrollers.

  Directory: "MDD File System-SD Data Logger" contains an example SD card data logging application.

  Directory: "Microchip" contains various common files, help files, include files, and documentation files.

8.2 MDD Library Functions

8.2.1 File and Disk Manipulation Functions

The MDD library provides a large number of "File and Disk Manipulation" functions that can be called and used from MPLAB C18 programs. The functions can be placed under the following groups:

- Initialize a card
- Open/create/close/delete/locate/rename a file on the card
- Read/write to an opened file
- Create/delete/change/rename a directory on the card
- Format a card
- Set file creation and modification date and time

The summary of each function is given briefly in Tables 8.1–8.6.

8.2.2 Library Options

A number of options are available in the MDD library. These options are enabled or disabled by uncommenting or commenting them, respectively, in include file **FSconfig.h**. The available options are given in Table 8.7.

Table 8.1: Initialize a Card Function

Functions	Descriptions
FSInit	Initialize the card

Table 8.2: Open/Create/Close/Delete/Locate/Rename Functions

Functions	Descriptions
FSfopen/FSfopenpgm	Opens an existing file for reading or for appending at the end of the file or creates a new file for writing.
FSfclose	Updates and closes a file. The file time-stamping information is also updated
FSRemove/FSremovepgm	Deletes a file
FSrename	Changes the name of a file
FindFirst/FindFirstpgm	Locates a file in the current directory that matches the specified name and attributes
FindNext	Locates the next file in the current directory that matches the name and attributes specified earlier

...pgm versions are to be used with the PIC18 microcontrollers where the arguments are specified in ROM.

Table 8.3: Read/Write Functions

Functions	Descriptions
FSfread	Reads data from an open file to a buffer
FSfwrite	Writes data from a buffer onto an open file
FSftell	Returns the current position in a file
FSfprintf	Writes a formatted string onto a file

Table 8.4: Create/Delete/Change/Rename Directory

Functions	Descriptions
FSmkdir	Creates a new subdirectory in the current woking directory
FSrmdir	Deletes the specified directory
FSchdir	Changes the current working directory
FSrename	Changes the name of a directory
FSgetcwd	Returns name of the current working directory

Table 8.5: Format a Card

Function	Description
FSformat	Formats a card

Table 8.6: File Time-Stamping Function

Function	Description
SetClockVars	Sets the date and time that will be applied to files when they are created or modified

Table 8.7: MDD Library Options (in File FSconfig.h)

Library Options	Descriptions
ALLOW_WRITES	Enables write functions to write onto the card
ALLOWS_DIRS	Enables directory functions (writes must be enabled)
ALLOW_FORMATS	Enables card formatting function (writes must be enabled)
ALLOW_FILESEARCH	Enables file and directory search
ALLOW_PGMFUNCTIONS	Enables pgm functions for getting parameters from the ROM
ALLOW_FSFPRINTF	Enables Fsfprintf function (writes must be enabled)
SUPPORT_FAT32	Enables FAT32 functionality

Table 8.8: MPLAB C18 Memory Usage with MDD Library

Functions Included	Program Memory (Bytes)	Data Memory (Bytes)
Read-only mode (basic)	11099	2121
File search enabled	+2098	+0
Write enabled	+7488	+0
Format enabled	+2314	+0
Directories enabled	+8380	+90
pgm functions enabled	+288	+0
FSfprintf enabled	+2758	+0
FAT32 support enabled	+407	+4

8.2.3 Memory Usage

The MPLAB C18 program memory and the data memory usage with the MDD library functions when the MPLAB C18 compiler is used are shown in Table 8.8. Note that 512 bytes of data are used for the data buffer, and an additional 512 bytes are used for the file allocation table buffer. The amount of required memory also depends on the number of files opened at a time. In Table 8.8, it is assumed that two files are opened. The first row shows the minimum memory requirements, and additional memory will be required when any of the subsequent row functionality is enabled.

8.2.4 Library Setup

There are a number of header files that should be customized before compiling a project (see Chapter 9 for details). These files and the type of customization that can be done are given below (it is assumed that we will be using dynamic memory allocation in the filing system):

- File **FSConfig.h** can be modified to change
  1. The maximum number of files open at any time (the default is 2)
  2. FAT sector size (the default is 512 bytes)
  3. Library options. The defaults are (comment the ones not required to save code space):

     ALLOW_FILESEARCH

     ALLOW_WRITES

     ALLOW_DIRS

     ALLOW_PGMFUNCTIONS

     SUPPORT_FAT32

     USERDEFINEDCLOCK

- File **HardwareProfile.h** can be modified
  1. to change system clock (the default is 4 MHz)
  2. to enable SD-SPI interface. The default is

     USE_SD_INTERFACE_WITH_SPI
  3. to define SD card interface pins. The defaults are

SD chip select (SD_CS)	-> RB3
SD card detect (SD_CD)	-> RB4
SD write enable (SD_WE)	-> RA4
SPI clock (SPICLOCKPORT)	-> RC3
SPI input (SPIINPORT)	-> RC4
SPI output (SPIOUTPORT)	-> RC5

(If any of the above is changed, then the corresponding TRIS registers must also be changed accordingly.)

4. Configure main SPI control registers. The defaults are

SPICON1	-> SSP1CON1
SPISTAT	-> SSP1STAT
SPIBUF	-> SSP1BUF
SPISTAT_RBF	-> SSP1STATbits.BF
SPICON1bits	-> SSP1CON1bits
SPISTATbits	-> SSP1STATbits
SPI_INTERRUPT_FLAG	-> PIR1bits.SSPIF
SPIENABLE	-> SPICON1bits.SSPEN

- Make sure that all the I/O pins used in the SD card interface are configured as digital I/O (and not as analog I/O)
- Modify the linker file to include a 512-byte section of RAM to act as a buffer for file read and write operations. In addition, create a section in the linker called **dataBuffer** that maps to this RAM.
- Modify the linker file to include a 512-byte section of RAM to act as a buffer for read and write of FAT. In addition, create a section in the linker called **FATBuffer** that maps to this RAM.
- Select the appropriate microcontroller definition file at the beginning of your program.

8.3 Sequence of Function Calls

The sequence of function calls to read or write data onto a file or to delete an existing file is given in this section.

8.3.1 Reading from an Existing File

The steps to open an existing file and read from it are

Call **FSInit** to initialize the card and SPI bus

Call **FSfopen** or **FSfopenpgm** to open the existing file in **read** mode

Call **FSfread** to read data from the file

Call **FSfclose** to close the file

The **FSread** function can be called as many times as required.

8.3.2 Writing Onto an Existing File

The steps to open an existing file and append data to it are

Call **FSInit** to initialize the card and SPI bus

Call **FSfopen** or **FSfopenpgm** to open the existing file in **append** mode

Call **FSwrite** to write data onto the file

Call **FSfclose** to close the file

The **FSwrite** function can be called as many times as required.

8.3.3 Deleting an Existing File

The steps to delete an existing file are

Call **FSInit** to initialize the card and SPI bus

Call **FSfopen** or **FSfopenpgm** to open the existing file in **write** mode

Call **FSremove** or **FSremovepgm** to delete the file

Call **FSfclose** to close the file

8.4 Detailed Function Calls

This section gives a detailed description of the Microchip file and disk manipulation functions (further information can be obtained from the Microchip Application Note: AN1045, Document no: DS01045B). The functions return and its typical use are shown with a simple call to each function. The functions are given in the order of typical use, i.e., initializing, opening, reading, writing, deleting, renaming, directories, and so on.

8.4.1 FSInit

This function initializes the SPI bus and mounts the SD card, and it must be called before any other MDD functions are called. The function returns an integer status. If the card is detected and the card is formatted with FAT12/FAT16 or FAT32, then TRUE is returned, otherwise a FALSE is returned.

The following example shows how this function can be called:

```
if(FSInit( ) == FALSE)
//
// Failed to initialize
```

The low-level function MediaIsPresent() can be used to check whether or not the card is present. If the card is removed, FSInit must be called again to remount the card.

8.4.2 FSfopen

This function opens a file on the SD card and associates a file structure with it. The function has two parameters: the filename and the mode.

The filename must be NULL terminated and must be less than eight characters, followed by a dot "." and a three-character file-extension name. The filename must be stored in RAM memory; thus, it should be declared as a character string.

The mode is a NULL-terminated one-character string that specifies the mode of access. The mode must be stored in RAM memory. Some of the valid modes are

r	Read-only
w	Write (a new file is created if it already exists)
w+	Create a new file (read and writes are enabled)
a	Append (if the file exists, any writing will be appended to the end of the file. If the file does not exist, a new file will be created)

The function returns a file structure or a NULL if the file could not be opened.

The following example shows how this function can be called to open a new file called "MYFILE.DAT":

```
FSFILE * MyFile;
char FileName[11] = "MYFILE.DAT";
char FileMode[2] = "w+";
MyFile = FSfopen(Filename, FileMode);
```

8.4.3 FSfopenpgm

This function opens a file on the SD card and associates a file structure with it. The function has two parameters: the filename and the mode.

The filename must be NULL terminated and must be less than eight characters, followed by a dot "." and a three-character file-extension name. The filename must be stored in ROM memory; thus, the filename can be entered directly as a string in the function.

The mode is a NULL-terminated one-character string that specifies the mode of access. The mode must be stored in the ROM memory. Valid modes are

r	Read-only
w	Write (a new file is created if it already exists)
w+	Create a new file (read and writes are enabled)
a	Append (if the file exists, any writing will be appended to the end of the File. If the file does not exist, a new file will be created)

The function returns a file structure or a NULL if the file could not be opened.

The following example shows how this function can be called to open a new file called "MYFILE.DAT":

```
FSFILE * MyFile;
MyFile = FSfopenpgm("MYFILE.DAT", "w+");
```

8.4.4 FSfclose

This function is called to close an opened file. The function returns an integer 0 if the file is closed successfully or a –1 (EOF) if the file failed to close. A pointer to the opened file (obtained from a previous call to FSopen) must be specified as the argument.

The following example shows how an opened file can be closed:

```
if(FSfclose(MyFile) == 0)
//
// File closed successfully
```

8.4.5 FSfeof

This function detects whether end-of-file is reached while reading from a file. A pointer to the opened file must be specified as the argument. The function returns an integer 1 (EOF) if the end-of-file is reached, otherwise 0 is returned. A pointer to the opened file must be specified as the argument.

The following example shows how the end-of-file can be detected:

```
if(FSfeof(MyFile) == EOF)
//
// end-of-file detected
```

8.4.6 FSfread

This function reads *n* bytes of data, each of length **size** bytes from the opened file and copies the data to the buffer pointed to by the buffer pointer.

The following parameters are required:

pntr	Pointer to the buffer that is to hold the data
size	Length of each item (bytes)
n	Number of items to read
ptr	Pointer to the opened file

The total number of bytes read is actually *n**size.

The function returns the number of items read or a 0 if there is an error in transferring *n**size bytes.

The following example shows how 20 packets of size 10 bytes each can be read and transferred to a buffer pointed to by pntr:

```
count = FSfread(pntr, 10, 20, MyFile);
```

In the above example, if the transfer is successful, **count** stores the total number of items actually read from the card. The possibilities are:

```
count == 0        // no data was read from the card
count < 20        // could not read 20 packets (EOF, or other error)
count == 20       // all 20 packets have been read
```

8.4.7 FSfwrite

This function writes *n* bytes of data, each of length **size** bytes, from a buffer pointed to by **pntr** to a previously opened file.

The following parameters are required:

pntr	Pointer to the buffer where data is to be written
size	Length of each item (bytes)
n	Number of items to read
ptr	Pointer to the opened file

The total number of bytes read is actually *n**size.

The function returns the number of items read or 0 if there is an error in transferring *n**size bytes.

The following example shows how 20 packets of size 10 bytes each can be written from a buffer pointed to by pntr to the card:

```
count = FSfwrite(pntr, 10, 20, MyFile);
```

In the above example, if the transfer is successful, **count** stores the total number of items actually written to the card. The possibilities are

```
count == 0          // no data was written to the card
count < 20          // could not write 20 packets
count == 20         // all 20 packets have been written
```

8.4.8 FSremove

This function deletes a file from the current directory. The filename must be specified in the RAM.

The function returns 0 if the deletion is successful or –1 on failure.

The following example shows how the file named "MYFILE.DAT" can be deleted:

```
char FileName[12] = "MYFILE.DAT";
if(FSremove(FileName) == 0)
//
// file deleted successfully
```

8.4.9 FSremovepgm

This function deletes a file from the current directory. The filename must be specified in the ROM, i.e., the filename can be directly entered as a string to the function.

The function returns 0 if the deletion is successful, or –1 on failure.

The following example shows how the file named "MYFILE.DAT" can be deleted:

```
if(FSremovepgm("MYFILE.DAT") == 0)
//
// file deleted successfully
```

8.4.10 FSrewind

This function sets the file pointer to the beginning of the file. The file structure opened earlier must be specified as an argument to the function. The function does not return anything.

A typical call to this function is

```
FSrewind(MyFile);
```

8.4.11 FSmkdir

This function creates a directory where the directory path string must be passed as an argument to the function. Directory names must be eight ASCII characters or less and must be delimited by the backslash character "\". The standard MSDOS "dot" formatting is used, where a dot "." accesses the current directory and two dots ".." access the previous directory. A directory in the "root directory" is created by specifying a backslash "\" before the directory name.

If the directory creation is successful, 0 is returned, otherwise –1 is returned.

In the following example, a directory called "COUNTS" will be created in the current directory:

```
char DirPath[ ] = "\COUNTS";
if(!FSmkdir(DirPath))
//
// directory created successfully
```

8.4.12 FSrmdir

This function deletes a directory where the directory path string must be passed as an argument to the function. Directory names must be eight ASCII characters or less and must be delimited by the backslash character "\". The standard MSDOS "dot" formatting is used, where a dot "." accesses the current directory and two dots ".." access the previous directory. A directory in the "root directory" is accessed by specifying a backslash "\" before the directory name.

The function requires two arguments: the directory path name and the mode. If the mode is TRUE, all subdirectories and files will be deleted. If the mode is FALSE, the directory will be deleted only if it is empty.

The function returns 0 on success and –1 if it fails to delete the directory.

In the following example, the directory called "COUNTS" in the current directory will be deleted. In addition, all the files within this directory will also be deleted:

```
char DirPath[ ] = "\COUNTS";
if(FSrmdir(DirPath, TRUE) == 0)
//
// directory and all its files deleted successfully
```

8.4.13 FSchdir

This function changes the current default directory. Directory names must be eight ASCII characters or less and must be delimited by the backslash character "\". The standard MSDOS "dot" formatting is used, where a dot "." accesses the current directory and two dots ".." access the previous directory. A directory in the "root directory" is accessed by specifying a backslash "\" before the directory name.

The path of the directory to be changed to must be specified as an argument to the function.

The function returns 0 on success and –1 if it fails to change the default working directory.

In the following example, the working directory is changed to "NUMBERS":

```
char NewDir[ ] = "\NUMBERS";
if(FSchdir(NewDir == 0)
//
// directory changed successfully
```

8.4.14 FSformat

This function deletes the FAT and root directory of a card. A new boot sector will be created if required. The function only supports FAT16 formatting.

The function has three arguments:

mode	0 erases the FAT and root directory
	1 creates a new master boot sector (MBR must be present)
serno	Serial number to write into the new boot sector
volume	Volume ID of the card (up to eight characters)

The function returns 0 if the formatting is successful, otherwise –1 is returned.

In the following example, an SD card is formatted and a new boot sector is created. The card volume name is set to "LOGS," and the card serial number is set to hexadecimal 0x11223344:

```
char CardVol[ ] = "LOGS";
if(FSformat(1, 0x11223344, CardVol) == 0)
//
// card formatted successfully
```

8.4.15 FSrename

This function changes the name of a file or a directory. The new filename and the name of the file to be changed must be specified as arguments to the function. If a NULL is passed as a pointer, then the name of the current working directory will be changed.

The function returns 0 if the filename is changed, otherwise –1 is returned.

In the following example, the name of the file pointed to by structure ptr will be changed to "NEW.DAT":

```
FSFILE *ptr;
if(FSrename("NEW.DAT", ptr) == 0)
//
// filename changed successfully
```

8.4.16 FindFirst

This function finds the first file in the current directory that matches the specified filename and file attribute criteria passed in the arguments.

The function has three arguments:

filename	The filename that must match (see Table 8.9)
attr	The file attribute that must match (see Table 8.10)
rec	Pointer to a structure of type **SearchRec** that will contain the file information if the file is found (see Table 8.11)

The function returns 0 if a match is found, otherwise –1 is returned.

Table 8.9: Filename Formats

Format	Description
.	Any file or directory
FILE.EXT	File with name FILE.EXT
FILE.*	Any file with name FILE and any extension
*.EXT	Any file with any name and extension EXT
*	Any directory
DIRS	Directory names DIRS

Table 8.10: File Attributes (Can Be Logically OR'ed)

Attribute	Description
ATTR_READ_ONLY	Files with read-only attribute
ATTR_HIDDEN	Hidden files
ATTR_SYSTEM	System files
ATTR_VOLUME	File may be a volume label
ATTR_DIRECTORY	File may be a directory
ATTR_ARCHIVE	File with archive attribute
ATTR_MASK	File with any attribute

Table 8.11: The SearchRec Structure

Member	Function
Char filename	Name of the file (NULL terminated)
Unsigned char attributes	File attributes
Unsigned long size	Size of files (bytes)
Unsigned long time-stamp	File creation date and time: 31:25 Year (0 = 1980, 1 = 1981,....) 24-21 Month (1 = Jan, 2 = Feb,...) 20:16 Day (1-31) 15:11 Hours (0-23) 10:05 Minutes (0-59) 04:00 Seconds/2 (0-29)
Unsigned int entry	Internal use only
Char search name	Internal use only
Unsigned char search attr	Internal use only
Unsigned int cwd clus	Internal use only
Unsigned char initialized	Internal use only

In the following example, files starting with "MY" and having extensions ".TXT" are searched in the current directory. The attribute field is set to hidden or system files:

```
SearchRec MyFile;
unsigned char attr = ATTR_HIDDEN | ATTR_SYSTEM;
char FileName[ ] = "MY*.TXT";
if(FindFirst(FileName, attr, &MyFile) == 0)
//
// file match found
```

Note that after a file match is found, we can get information about the file using the function SearchRec. For example, the full name of the file can be obtained using **MyFile.filename** or the file size can be obtained using **MyFile.size**.

8.4.17 FindFirstpgm

This function is very similar to **FindFirst,** except that the filename must be specified in ROM.

In the following example, files starting with "MY" and having extensions ".TXT" are searched in the current directory. The attribute is set to hidden or system files:

```
SearchRec MyFile;
unsigned char attr = ATTR_HIDDEN | ATTR_SYSTEM;
if(FindFirst("MY*.TXT", attr, &MyFile) == 0)
//
// file match found
```

8.4.18 FindNext

This function matches the next file in the current directory to the filename and attribute criteria specified by the last call to Findfirst (or FindFirstpgm).

The pointer to a SearchRec structure must be specified as an argument. The function returns 0 if a match is found, otherwise –1 is returned.

In the following example, a second match is found to a previously specified filename and attribute criteria (here, it is assumed that a **FindFirst** function call was made earlier):

```
if(FindNext(&MyFile) == 0)
//
// second match is found
```

8.4.19 SetClockVars

This function is used when the user-defined Clock mode is selected, and the function sets the timing variables used to set file creation and modification times.

The following arguments must be specified when the function is called:

Year	The current year (1980–2107)
Month	The current month (1–12)
Day	The current day (1–31)
Hour	The current hour (0–23)
Minute	The current minute (0–59)
Second	The current second (0–59)

The function returns 0 on success and –1 if one or more parameters are invalid.

In the example given below, the date and time are set to July 10, 2009, 11:30:10 A.M.

```
if(SetClockVars(2009, 7, 10, 11, 30, 10) == 0)
//
// date and time set successfully
```

8.4.20 FSfprintf

This function writes a formatted string to a file opened on a card. The function is similar to the standard **fprintf** statement. The first argument is a pointer to the file, the second argument is the string to write to (must be specified in ROM), and the other arguments are the format specifiers.

The function returns the number of bytes written on success or –1 if the write failed.

The format specifiers are similar to that used in standard fprintf statements and are normally written in the following order:

Flag characters, field width, field precision, size specification, and conversion specifiers.
Further details on format specifiers can be obtained from the Microchip Application note: AN1045 (*Implementing File I/O Functions Using Microchip's Memory Disk Drive File System Library*).

An example is given below:

```
FSfprintf(ptr, "Binary number=%#16b", 0x1eff);
```

The output will be:

```
Binary number=0b0001111011111111
```

8.5 Summary

In this chapter, the MDD functions available for reading and writing onto SD cards using the MPLAB C18 compiler are described. The use of SD cards in microcontroller-based projects is highly simplified when these functions are used.

8.6 Exercises

1. Explain the steps required to read and write onto an SD card using the MDD library functions.
2. Assume that a text file called "DATA.TXT" exists on an SD card. It is required to create a new file called "NEW.TXT" on the card and copy the first 100 bytes from "DATA.TXT" to "NEW.TXT." Show the steps required to perform this operation.
3. It is required to find out how many files with extensions ".TXT" are there in the current default working directory of an SD card. Show the steps required to perform this operation.
4. Assume that a file called "MONTHLY.DAT" exists on an SD card. Show the steps required to change the name of this file to "MONTH2.DAT."
5. Show the steps required to delete a file on an SD card.
6. Show the steps required to find out how many files with names "TEMP" there are in the working default directory of an SD card and show how this number can be displayed on an LCD.
7. It is required to find the names and sizes of all the files in the current working directory of an SD card. Show how this information can be found and stored in a buffer in RAM.

8. Assume that a text file called "MYDATA.TXT" exists in the current directory. Show the steps required to create a new directory called NEWD in the current directory, and also show how the file can be copied to this new directory.
9. Show the steps required to delete a directory called "DIRS" and all of its contents. Assume that this directory is in the current working directory.
10. It is required to find out the number of files in the current directory with the system attribute. Show the steps required to perform this operation, and also show how the number found can be displayed on an LCD.
11. Show the steps required to format and create a new boot sector on an SD card. Give the volume name "MYSD" and serial number 0x87654321 to the formatted card.
12. It is required to delete all the files in the current directory of an SD card. Show the steps required to perform this operation.
13. Explain what happens when an SD card is removed from its socket. How can we find out whether or not the card is mounted?

CHAPTER 9

Secure Digital Card Projects

The details of the Microchip memory disk drive (MDD) library, which consist of a large number of MPLAB C18 compiler–compatible functions that can be used in secure digital (SD) card-based projects, are given in Chapter 8.

In this chapter, we shall be looking at how these functions can actually be used in practical projects. The many simple-to-complex projects given in this chapter show how SD card-based projects can be built and how an SD card can be used as a large external storage medium.

The projects have been organized by increasing level of complexity. Thus, it is advised that the reader start from Project 1 and then move to more complex projects as experience is gained.

The following information is given for each project:

- Description of the project
- Aim of the project
- Block diagram of the project
- Circuit diagram of the project
- Operation of the project
- Program code of the project
- Description of the program code
- Suggestions for future work

The projects can be built using most of the commercially available PIC18 microcontroller development boards. Alternatively, complete projects can be built on a breadboard of suitable size. It is recommended to use an external +5-V power supply to provide power to the microcontroller and the associated circuitry used in the projects. Alternatively, a 9-V battery and a 7805 type +5-V regulator can be used to supply power to the projects.

In this book, the PICDEM PIC18 Explorer demonstration board (see Figure 9.1) is used for the projects. As described in Chapter 5, this development board has been specifically

D.O.I.: 10.1016/B978-1-85617-719-1.00013-0

Figure 9.1: PICDEM PIC18 Explorer Demonstration Board

designed for PIC18-based applications. The board incorporates a PIC18F8722-type microcontroller operating with a 10-MHz clock. The board has the following features:

- Connector for external daughter boards (e.g., SD card board)
- LCD display
- Eight LEDs
- Analog temperature sensor
- Push-button switches
- RS232 socket for serial communications
- External reset button
- Potentiometer for analog inputs

Figure 9.2: PICtail Daughter Board for SD and MMC Cards

- In-circuit-debugger connector
- Header pins to use different processors

A Microchip daughter SD card board (known as the PICtail daughter board for SD and MMC Cards, see Figure 9.2) is used as the SD card interface. This board directly plugs into the PICDEM PIC18 Explorer board (see Figure 9.3) and provides SD card interface to the demonstration board. (Note that there are minor design faults with the voltage-level conversion circuitry on some of the PICtail daughter boards for SD and MMC Cards. You can get around these problems by providing a 3.3-V supply for the daughter board directly from the PICDEM Explorer board. Cut short the power supply pin of the daughter board connector and connect this pin to the +3.3-V test point on the PICDEM Explorer board.)

The SD card daughter board has an on-board positive-regulated charge pump DC/DC converter chip (MCP1253) used to convert the +5-V supply to +3.3 V required for the SD

Figure 9.3: The Daughter Board Plugs onto the PICDEM Board

card. In addition, the board has buffers to provide correct voltages for the SD card inputs. Seven jumpers are provided on the board to select the SD card signal interface. The following jumpers should be selected:

Jumpers	Descriptions
JPl Pin 1-2	SCK connected to RC3
JP2 Pin 1-2	SDI connected to RC4
JP3 Pin 1-2	SDO connected to RC5
JP4 Pin 2-3 (default)	Card detect to RB4 (not used)
JP5 Pin 2-3 (default)	Write protect to RA4 (not used)
JP6 Pin 2-3 (default)	CS connected to RB3
JP7 Pin 2-3 (default)	Shutdown (not used)

The default jumper positions are connected by circuit tracks on the board, and these tracks should be cut to change the jumper positions if different connections are desired. Signals "card detect," "write protect," and "shutdown" are not used in this book, and the jumper settings can be left as they are.

Before starting the programming, make sure that you have a suitable programming device that can program the PIC18 series of microcontrollers. In addition, you will require a copy of the MPLAB C18 compiler and a copy of the Microchip MDD library. In this book, the ICD3 in-circuit-debugger device is used to program the PICDEM Explorer board.

9.1 Creating an MPLAB C18 Template

In this section, we shall be creating an MPLAB C18 template that can be used in all of our SD card projects. The template will be based on using the PIC18F8722 microcontroller with the MDD library. The steps are given below:

- Start the MPLAB and select Project -> Project Wizard (see Figure 9.4)
- Click Next and select the processor type as PIC18F8722 (see Figure 9.5)
- Click Next. Select Microchip C18 Toolsuite. Make sure that the tool suite components point to the correct directories (see Figure 9.6):

  MPASM assembler (mpasmwin.exe) -> C:\MCC18\mpasm\mpasmwin.exe

  MPLINK Object Linker (mplink.exe) -> C:\MCC18\bin\mplink.exe

  MPLAB C18 C Compiler (mcc18.exe)-> C:\MCC18\bin\mcc18.exe

  MPLIB Librarian (mplib.exe) -> C:\MCC18\bin\mplib.exe

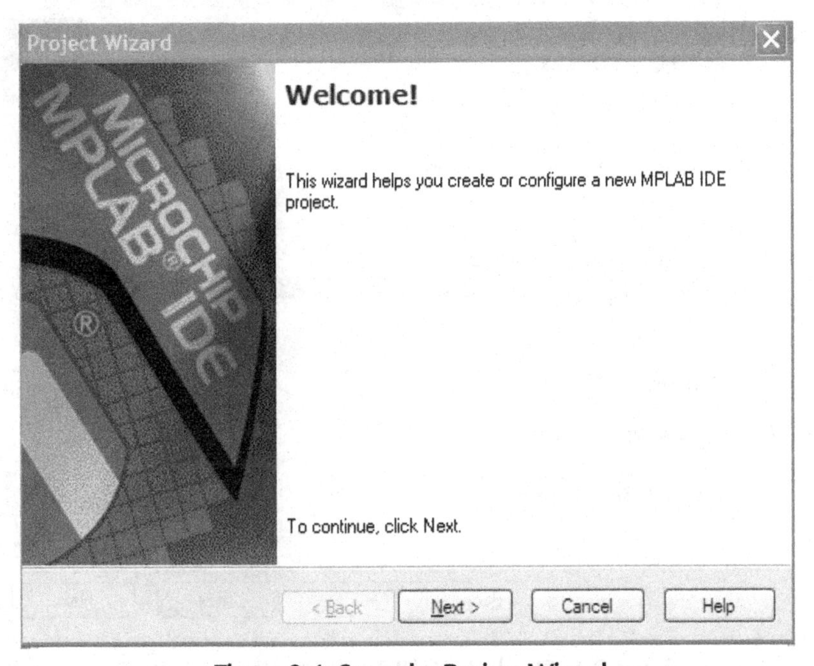

Figure 9.4: Start the Project Wizard

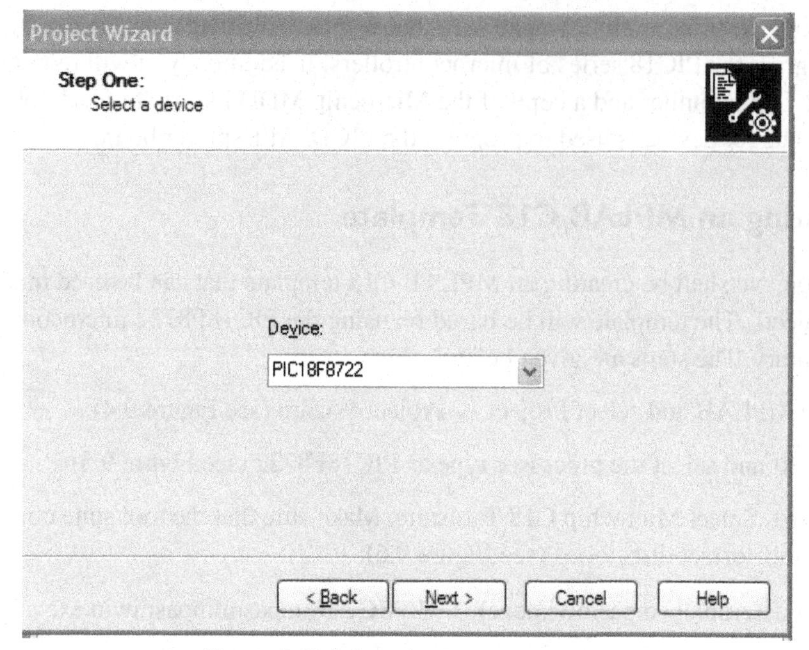

Figure 9.5: Select the Processor Type

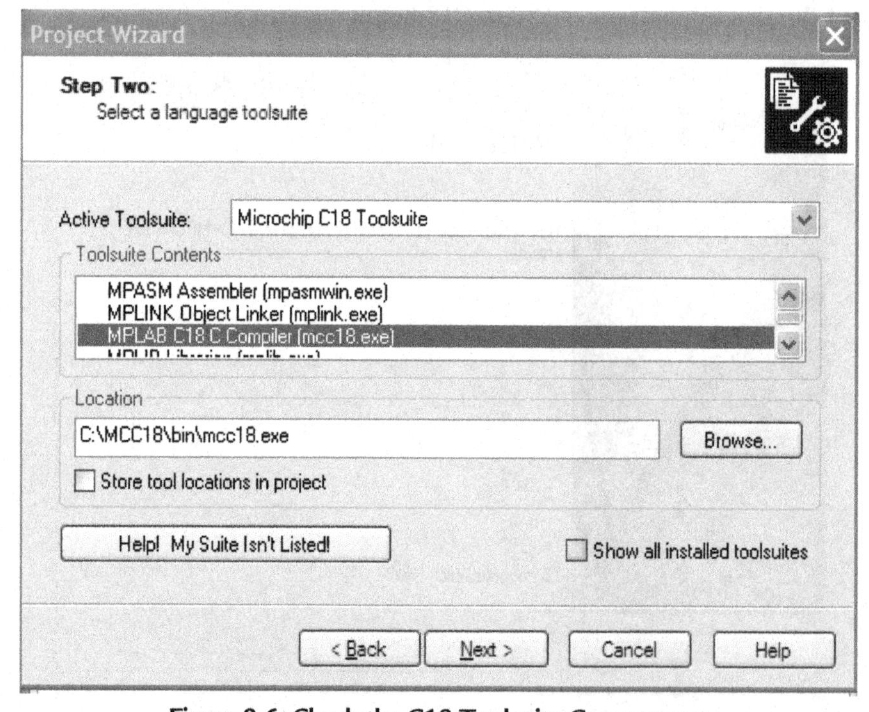

Figure 9.6: Check the C18 Toolsuite Components

- Click Next. Create a new project called SD_CARD_PROJECTS in directory C:\ Microchip Solutions\MDD File System-SD Card. That is, enter C:\Microchip Solutions\ MDD File System-SD Card\SD_CARD_PROJECTS (see Figure 9.7). You can click the Browse to find the directory, and then enter the filename and click Save button. Click Next to create the project file.
- Open the Notepad and create a text file called FIRST.C. Save this file in directory C:\ Microchip Solutions\MDD File System-SD Card. This will be your template source file. The contents of this file should be as shown in Figure 9.8.
- Click Next. Click the following files on the left-hand side and click Add to add them to the project (see Figure 9.9):

  C:\Microchip Solutions\Microchip\MDD File System\FSIO.c

  C:\Microchip Solutions\Microchip\MDD File System\SD-SPI.c

  C:\Microchip Solutions\Microchip\PIC18 salloc\salloc.c

  C:\Microchip Solutions\Microchip\Include\Compiler.h

  C:\Microchip Solutions\Microchip\Include\GenericTypeDefs.h

  C:\Microchip Solutions\MDD File System-SD Card\Pic18f\FSconfig.h

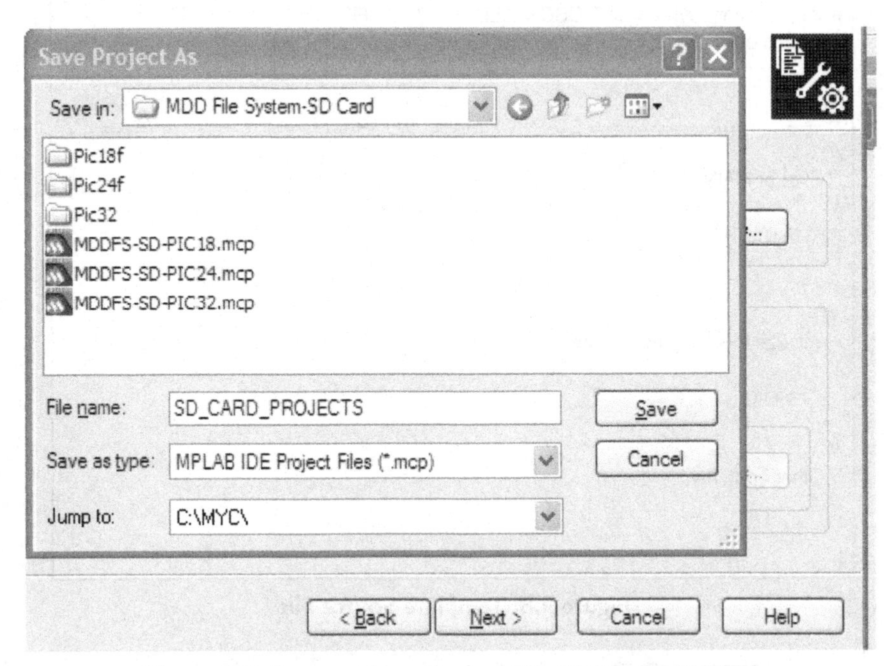

Figure 9.7: Create a New Project SD_CARD_PROJECTS

```
/*****************************************************************************************************

                    PROJECT TO WRITE SHORT TEXT TO AN SD CARD
                    ==========================================

In these projects, a PIC18F8722-type microcontroller is used. The microcontroller is
operated with a 10-MHz crystal.

An SD card is connected to the microcontroller as follows:

SD card                         microcontroller
CS                              RB3
CLK                             RC3
DO                              RC4
DI                              RC5

The program uses the Microchip MDD library functions to read and write to the SD card.

**************************** Insert other comments here ****************************




Author:             Dogan Ibrahim
Date:               July 2009
File:               write filename here
*****************************************************************************************************/
#include <p18f8722.h>
#include <FSIO.h>


#pragma config WDT = OFF, OSC = HSPLL, LVP = OFF
#pragma config MCLRE = ON, CCP2MX = PORTC, MODE = MC


/* =================== START OF MAIN PROGRAM =================== */
//
// Start of MAIN Program
//
void main(void)
{

//
// Initialize the SD card routines
//
    FSInit();

//
// Other code here
//

}
```

Figure 9.8: Template Source File

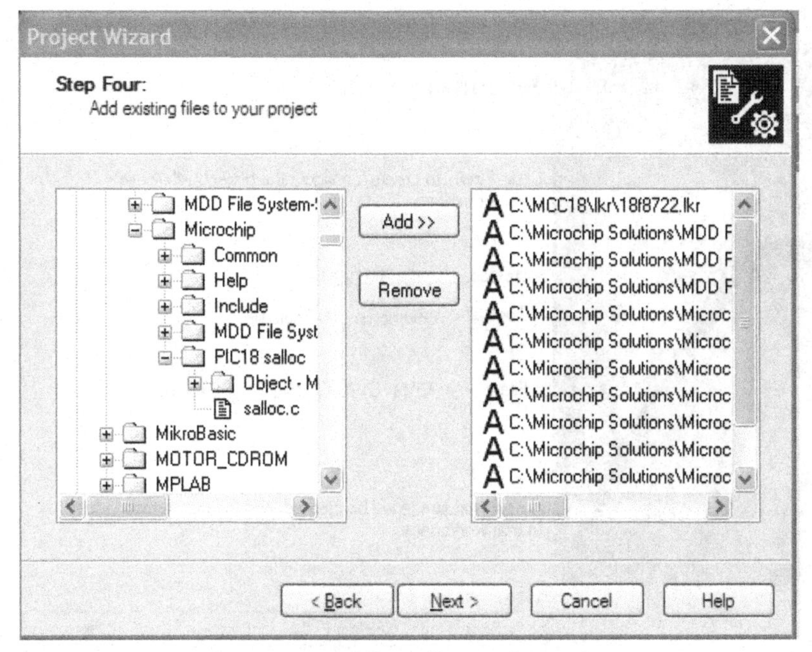

Figure 9.9: Adding Files to the Project

C:\Microchip Solutions\MDD File System-SD Card\Pic18f\HardwareProfile.h

C:\Microchip Solutions\Microchip\Include\MDD File System\FSDefs.h

C:\Microchip Solutions\Microchip\Include\MDD File System\SD-SPI.h

C:\Microchip Solutions\Microchip\Include\MDD File System\FSIO.h

C:\Microchip Solutions\Microchip\Include\PIC18 salloc\salloc.h

- Click Next. Click Finish to complete (see Figure 9.10)
- Specify the MDD header include files in the project. Click Project -> Build Options -> Project (see Figure 9.11)
- Select Include Search Path in Show directories for textbox and enter the following directory names (click New before entering a new set of data), see Figure 9.12:

```
C:\mcc18\h
.\Pic18f
..\Microchip\Include
..\Microchip\Include\MDD File System
.
..\Microchip\Include\PIC18 salloc
..\Microchip\PIC18 salloc
```

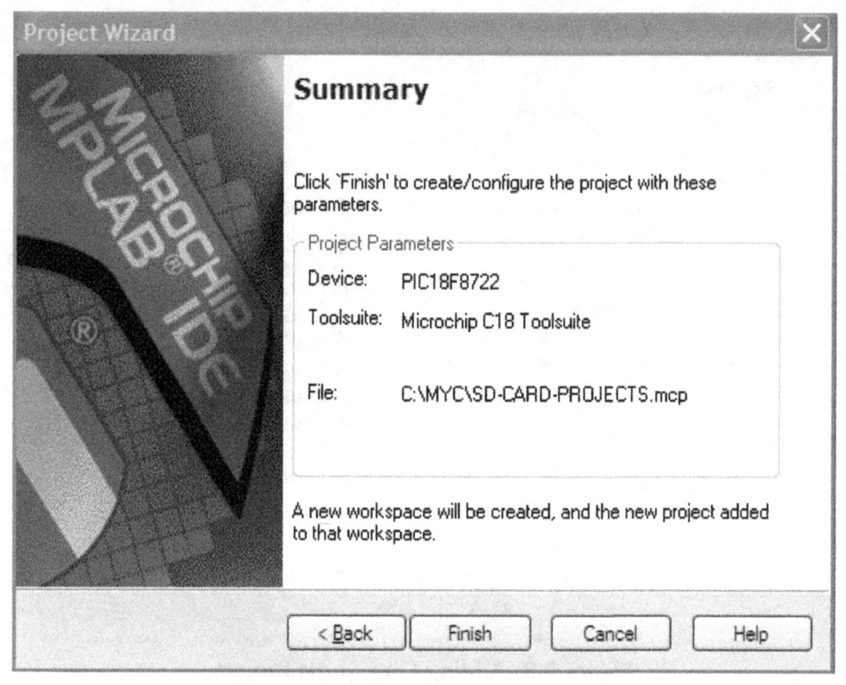

Figure 9.10: Click Finish to complete

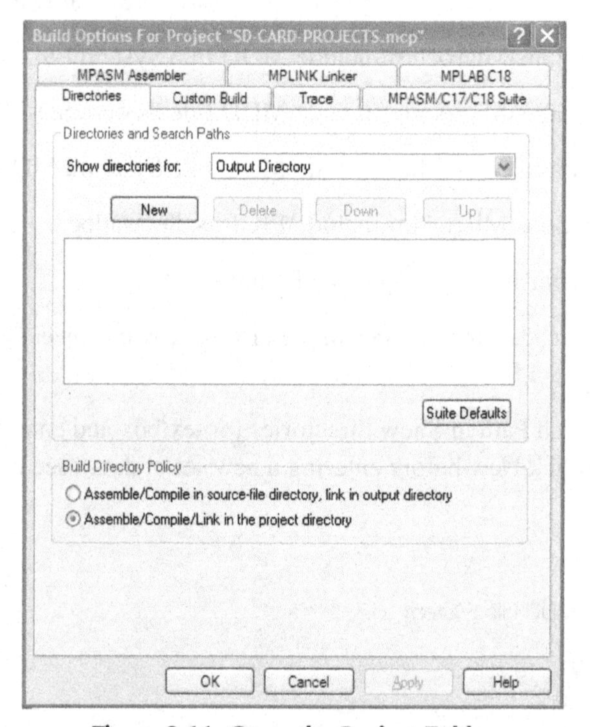

Figure 9.11: Open the Project Folder

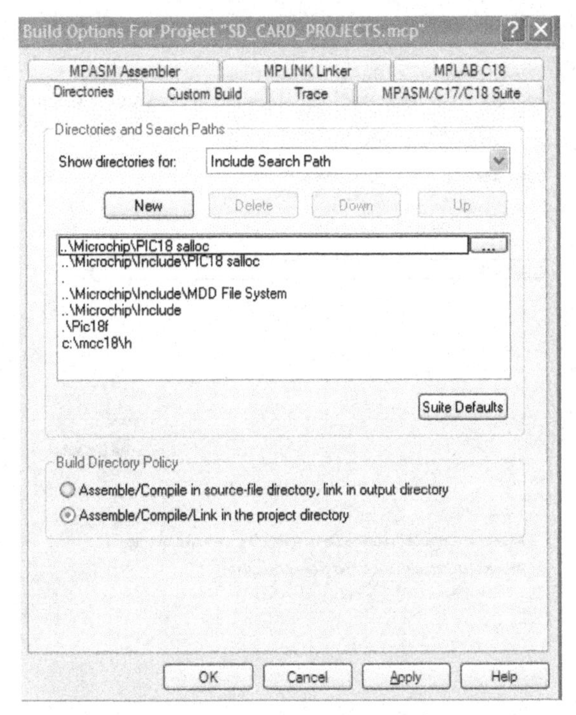

Figure 9.12: Enter the Include Search Path

- Select Library Search Path in Show directories for textbox; click New and enter the following directory name (see Figure 9.13):

  C:\mcc18\lib

- Click OK. Click File -> Save Workspace and then File -> Exit to exit from MPLAB after saving it.

The compiler linker file must be modified to include a 512-byte section for the data read-write and also a 512-byte section for the FAT allocation. This is done by editing the linker file **18f8722.lkr** in folder **c:\mcc18\lkr** and adding lines for a **dataBuffer** and an **FATBuffer**. In addition, it is required to add a section named **_SRAM_ALLOC_HEAP** to the linker file. The modified linker file is shown in Figure 9.14.

We can now verify if everything has been setup correctly.

- Restart MPLAB. Select Project -> Open and select SD_CARD_PROJECTS. Click Open. Double-click on program file FIRST.C. The project file should now open.

- Compile the program by clicking Build All. The program should compile and link with no errors.

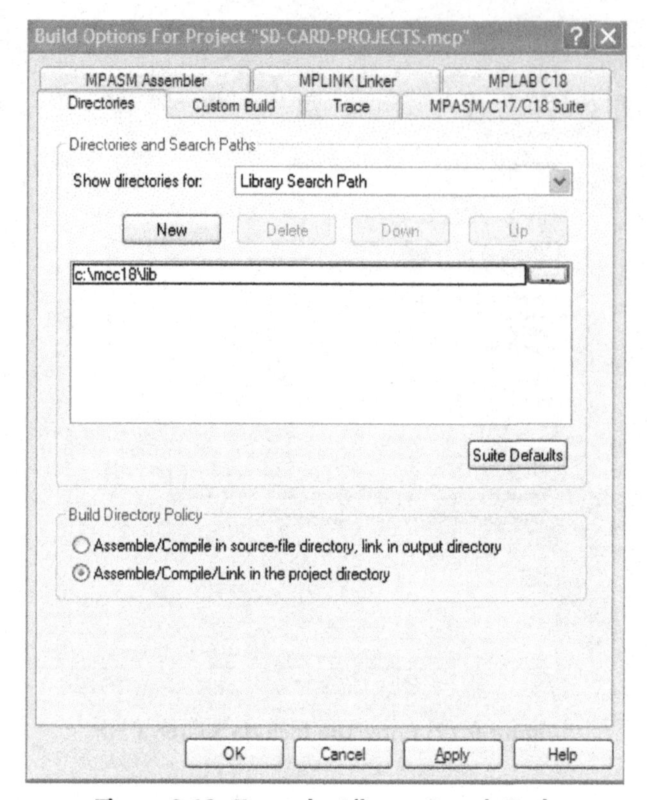

Figure 9.13: Enter the Library Search Path

9.1.1 Setting the Configuration Files

It is now necessary to customize some of the header files for our requirements. You should make the following modifications when using the PICDEM PIC18 Explorer demonstration board with the PICtail SD card daughter board (you are recommended to make copies of the original files before modifying them, in case you ever want to return to them):

- Modify the file C:\Microchip Solutions\MDD File System-SD Card\Pic18f\FSconfig.h and enable the following defines:

  1. #define FS_MAX_FILES_OPEN 2
  2. #define MEDIA_SECTOR_SIZE 512
  3. #define ALLOW_FILESEARCH
     #define ALLOW_WRITES

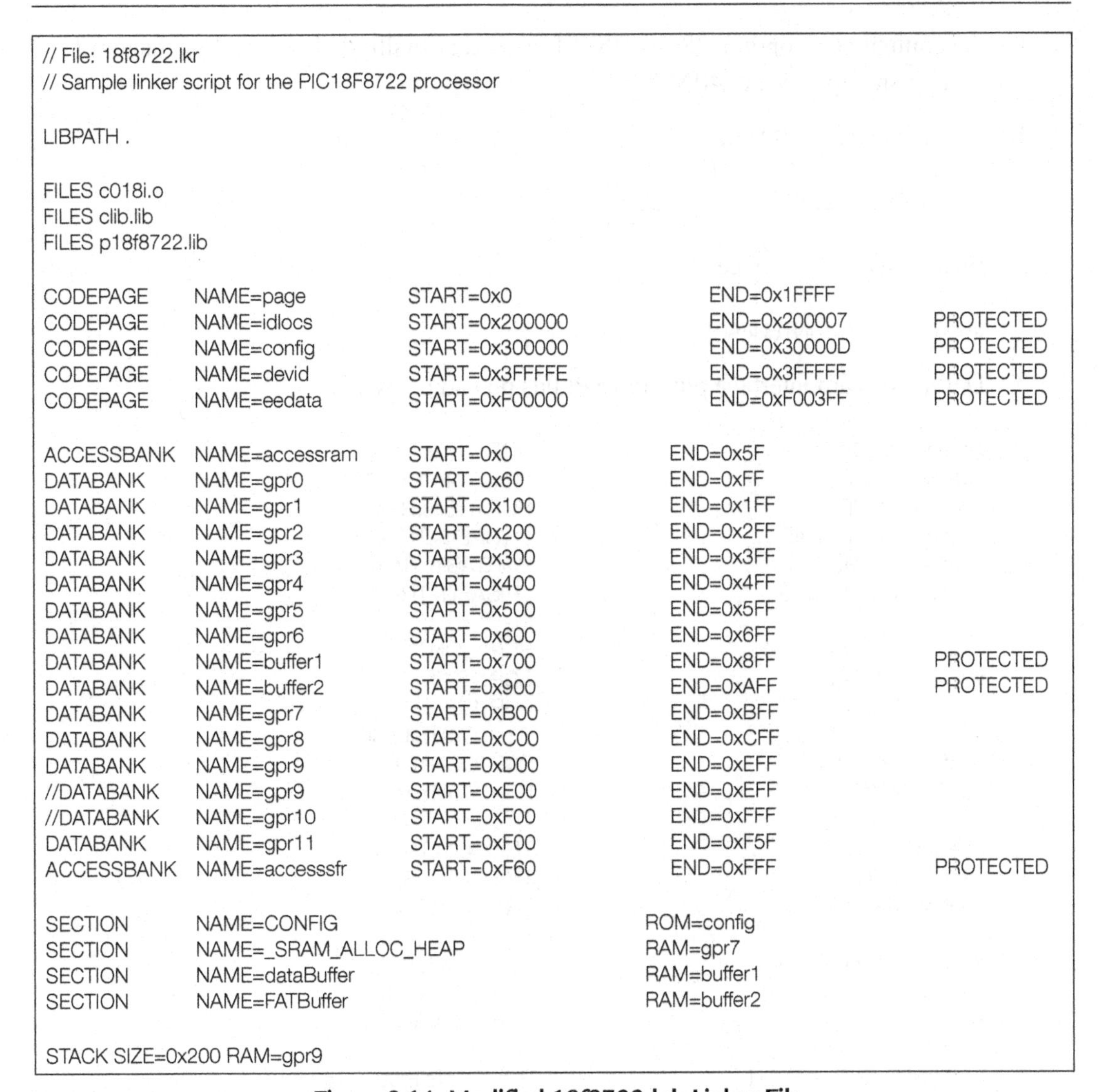

```
// File: 18f8722.lkr
// Sample linker script for the PIC18F8722 processor

LIBPATH .

FILES c018i.o
FILES clib.lib
FILES p18f8722.lib

CODEPAGE     NAME=page        START=0x0          END=0x1FFFF
CODEPAGE     NAME=idlocs      START=0x200000     END=0x200007    PROTECTED
CODEPAGE     NAME=config      START=0x300000     END=0x30000D    PROTECTED
CODEPAGE     NAME=devid       START=0x3FFFFE     END=0x3FFFFF    PROTECTED
CODEPAGE     NAME=eedata      START=0xF00000     END=0xF003FF    PROTECTED

ACCESSBANK   NAME=accessram   START=0x0        END=0x5F
DATABANK     NAME=gpr0        START=0x60       END=0xFF
DATABANK     NAME=gpr1        START=0x100      END=0x1FF
DATABANK     NAME=gpr2        START=0x200      END=0x2FF
DATABANK     NAME=gpr3        START=0x300      END=0x3FF
DATABANK     NAME=gpr4        START=0x400      END=0x4FF
DATABANK     NAME=gpr5        START=0x500      END=0x5FF
DATABANK     NAME=gpr6        START=0x600      END=0x6FF
DATABANK     NAME=buffer1     START=0x700      END=0x8FF       PROTECTED
DATABANK     NAME=buffer2     START=0x900      END=0xAFF       PROTECTED
DATABANK     NAME=gpr7        START=0xB00      END=0xBFF
DATABANK     NAME=gpr8        START=0xC00      END=0xCFF
DATABANK     NAME=gpr9        START=0xD00      END=0xEFF
//DATABANK   NAME=gpr9        START=0xE00      END=0xEFF
//DATABANK   NAME=gpr10       START=0xF00      END=0xFFF
DATABANK     NAME=gpr11       START=0xF00      END=0xF5F
ACCESSBANK   NAME=accesssfr   START=0xF60      END=0xFFF       PROTECTED

SECTION      NAME=CONFIG                   ROM=config
SECTION      NAME=_SRAM_ALLOC_HEAP         RAM=gpr7
SECTION      NAME=dataBuffer               RAM=buffer1
SECTION      NAME=FATBuffer                RAM=buffer2

STACK SIZE=0x200 RAM=gpr9
```

Figure 9.14: Modified 18f8722.lnk Linker File

```
#define ALLOW_DIRS
#define ALLOW_PGMFUNCTIONS
```

4. #define USERDEFINEDCLOCK

5. Make sure that the file object allocation is dynamic. i.e.
   #if 1

- Modify the file C:\Microchip Solutions\MDD File System-SD Card\Pic18f\HardwareProfile.h and set the following options (notice that the system clock is 10 MHz, but the configuration option OSC = HSPLL is used to multiply the clock by a factor of four, and it should be set to 40 MHz):

1. Set clock rate to 40 MHz:

```
#define GetSystemClock() 40000000
```

2. Enable SD-SPI interface.

```
#define USE_SD_INTERFACE_WITH_SPI
```

3. Define SD card interface pins and SPI bus pins to be used:

```
#define SD_CS                    PORTBBits.RB3
#define SD_CS_TRIS               TRISBBits.TRISB3
#define SD_CD                    PORTBBits.RB4
#define SD_CD_TRIS               TRISBBits.TRISB4
#define SD_WE                    PORTABits.RA4
#define SD_WE_TRIS               TRISABits.TRISA4

#define SPICON1                  SSP1CON1
#define SPISTAT                  SSP1STAT
#define SPIBUF                   SSP1BUF
#define SPISTAT_RBF              SSP1STATbits.BF
#define SPICON1bits              SSP1CON1bits
#define SPISTATbits              SSP1STATbits

#define SPICLOCK                 TRISCbits.TRISC3
#define SPIIN                    TRISCbits.TRISC4
#define SPIOUT                   TRISCbits.TRISC5

#define SPICLOCKLAT              LATCbits.LATC3
#define SPIINLAT                 LATCbits.LATC4
#define SPIOUTLAT                LATCbits.LATC5

#define SPICLOCKPORT             PORTCbits.RC3
#define SPIINPORT                PORTCbits.RC4
#define SPIOUTPORT               PORTCbits.RC5
```

9.1.2 The Memory Model

The memory model should now be selected correctly. Select **Project -> Build Option -> Project**; then click the **MPLAB C18** tab and select **Memory Model** in **Categories**. Set the following options (see Figure 9.15):

Code Model: Large code model

Data Model: Large data model

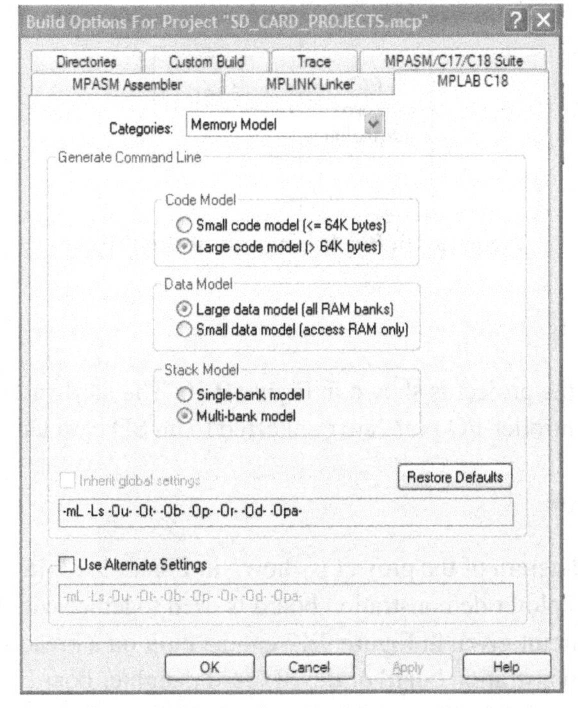

Figure 9.15: Setting the Memory Model

Stack Model: Multibank model

We are now ready to develop projects using the created template file and the working environment.

9.2 PROJECT 1 – Writing a Short Text Message to an SD Card

9.2.1 *Description*

In this project, a file called "MESSAGE.TXT" is created on an SD card and the following short text message is written to this file:

"This is a TEXT message"

9.2.2 *Aim*

The aim of this project is to familiarize the reader with the minimum hardware required to build an SD card-based project. In addition, the configuration of the MDD library and the MPLAB C18 compiler are described so that the reader can compile and build the basic software for an SD card-based project. With the knowledge gained in this project, the reader should be able to move to more complex SD card-based projects.

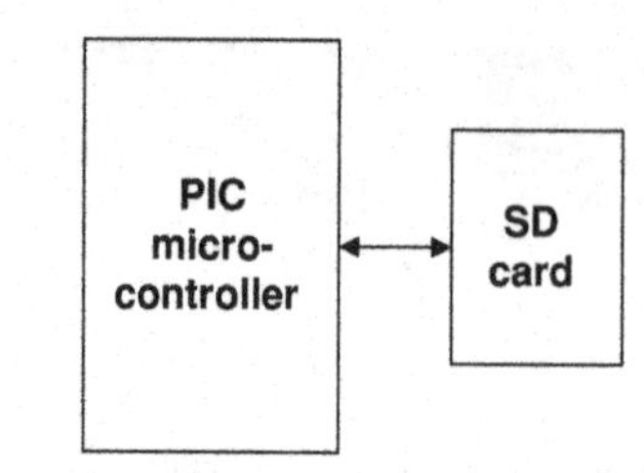

Figure 9.16: Block Diagram of the Project

9.2.3 Block Diagram

The block diagram of the project is shown in Figure 9.16. The hardware setup is very simple. Basically, the microcontroller I/O ports are connected to an SD card using a card holder.

9.2.4 Circuit Diagram

The complete circuit diagram of the project is shown in Figure 9.17. In the actual implementation, the PICDEM Explorer demonstration board is used together with the PICtail SD card daughter board. The circuit given in Figure 9.17 can be built on a breadboard if you do not have the PICDEM demonstration board or the SD card daughter board.

The circuit is built around a PIC18F8722-type microcontroller operated from a 10-MHz crystal. The MCLR input is connected to an external push-button switch for external reset of the microcontroller.

The interface between the microcontroller and the SD card pins is as follows. (The card adapter on the PICtail daughter board provides two additional signals: card detect [CD] and write enable [WE].)

SD Card Pins	Microcontroller Pins
CS	RB3
CLK	RC3
DO	RC4
DI	RC5

The maximum allowable input voltage at the inputs of an SD card is +3.6 V. The voltage at the outputs of the microcontroller is about +4.3 V, which is too high for the inputs (CS, DO, CLK) of the SD card. As a result, potential divider resistors are used to lower the voltage to acceptable levels (on the PICDEM board, buffers are used to lower the voltage levels). With the 2.2- and 3.3-K resistors, the voltage at the inputs of the SD card will be

$$\text{SD card input voltage} = 4.3\,\text{V} \times 3.3\,\text{K}/(2.2\,\text{K} + 3.3\,\text{K}) = 2.48\,\text{V}$$

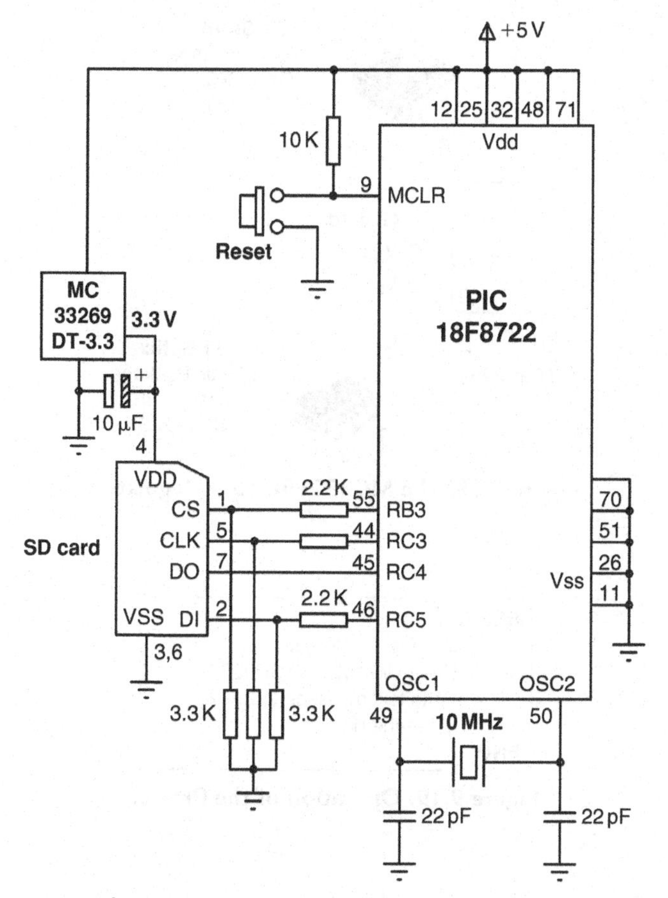

Figure 9.17: Circuit Diagram of the Project

In Figure 9.17, the +3.3-V power for the SD card is derived from a MC33269DT-3.3 type regulator (see Figure 9.18) powered from the +5-V power. (On the PICtail SD card daughter board, an MCP1253 type DC/DC converter is used to provide the +3.3-V supply for the SD card.)

9.2.5 Operation of the Project

The operation of the project is very simple and can be described by the program description language (PDL) given in Figure 9.19.

9.2.6 Program Code

The program code (file WRITE1.C) is shown in Figure 9.20.

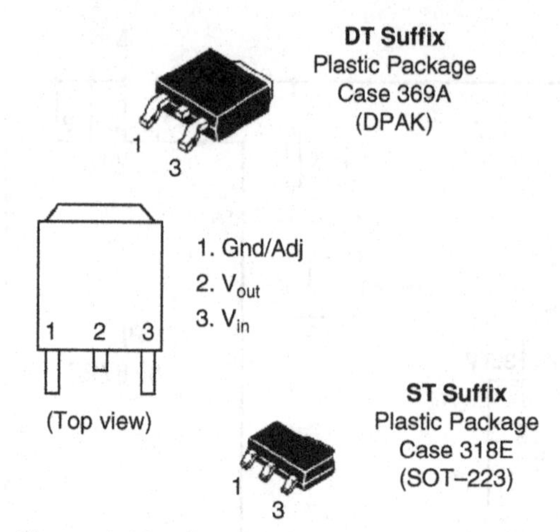

Figure 9.18: The MC33269DT3-3 Regulator

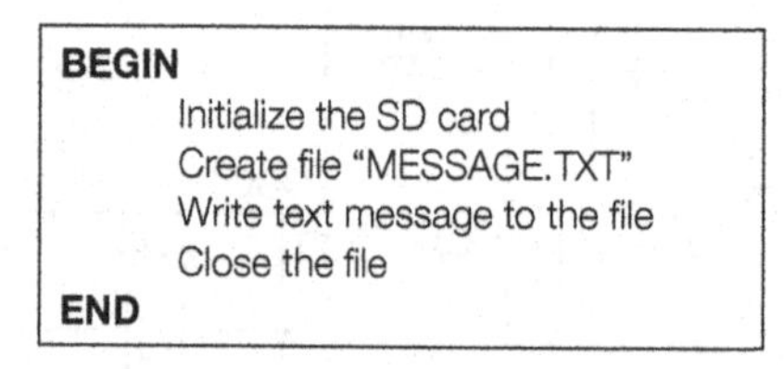

Figure 9.19: Operation of the Project

9.2.7 Description of the Program Code

At the beginning of the program, file pointer **MyFile** is declared and the text to be written to the card is assigned to character array **txt**. The MDD file system is initialized by calling function FSInit, and file called MESSAGE.TXT is opened on the SD card using function call FSfopenpgm. The text message is then written to the file by calling function FSfwrite. Finally, the file is closed by calling function FSfclose.

The program given in Figure 9.20 works, but there is no indication as to whether or not all the function calls returned success or as to when the program is terminated and the SD card removed. The program can be made more user friendly by testing the return of each function call for success. In addition, an LED can be connected to the RD0 pin of the microcontroller, and this LED can be turned ON to indicate the successful termination of the program. The modified program listing (file WRITE2.C) is shown in Figure 9.21.

```
/*************************************************************************************************
                PROJECT TO WRITE SHORT TEXT TO AN SD CARD
                ========================================

In these projects, a PIC18F8722-type microcontroller is used. The microcontroller
is operated with a 10-MHz crystal.

An SD card is connected to the microcontroller as follows:

SD card          microcontroller
CS               RB3
CLK              RC3
DO               RC4
DI               RC5

The program uses the Microchip MDD library functions to read and write to
the SD card.


Author:   Dogan Ibrahim
Date:     July 2009
File:     WRITE1.C
*************************************************************************************************/
#include <p18f8722.h>
#include <FSIO.h>

#pragma config WDT = OFF, OSC = HSPLL, LVP = OFF
#pragma config MCLRE = ON, CCP2MX = PORTC, MODE = MC

/* ================= START OF MAIN PROGRAM ================= */
//
// Start of MAIN Program
//
void main(void)
{
    FSFILE *MyFile;
    unsigned char txt[]="This is a TEXT message";
//
// Initialize the SD card routines
//
    FSInit();
//
// Create a new file called MESSAGE.TXT
//
    MyFile = FSfopenpgm("MESSAGE.TXT", "w+");
//
// Write message to the file
//
    FSfwrite(txt, 1, 22, MyFile);
//
```

Figure 9.20: The Program Code

```
// Close the file
//
    FSfclose(MyFile);

while(1);
}
```

Figure 9.20: *Cont'd*

```
/**************************************************************************************************
                    PROJECT TO WRITE SHORT TEXT TO AN SD CARD
                    =========================================

In these projects, a PIC18F8722-type microcontroller is used. The microcontroller
is operated with a 10-MHz crystal.

An SD card is connected to the microcontroller as follows:

SD card                     microcontroller
CS                          RB3
CLK                         RC3
DO                          RC4
DI                          RC5

The program uses the Microchip MDD library functions to read and write to
the SD card.

In this version of the program an LED is connected to port RD0 and
the LED is turned ON when the program is terminated successfully.


Author:    Dogan Ibrahim
Date:      July 2009
File:      WRITE2.C
**************************************************************************************************/
#include <p18f8722.h>
#include <FSIO.h>

#pragma config WDT = OFF, OSC = HSPLL, LVP = OFF
#pragma config MCLRE = ON, CCP2MX = PORTC, MODE = MC

#define LED PORTDbits.RD0
#define ON 1
#define OFF 0

/* ================ START OF MAIN PROGRAM ================ */
//
// Start of MAIN Program
//
```

Figure 9.21: Modified Program Code

```
void main(void)
{
    FSFILE *MyFile;
    unsigned char txt[]="This is a TEXT message";

    TRISD = 0;
    PORTD = 0;
//
// Initialize the SD card routines
//
    while(!FSInit());
//
// Create a new file called MESSAGE.TXT
//
    MyFile = FSfopenpgm("MESSAGE.TXT", "w+");
    if(MyFile == NULL)while(1);
//
// Write message to the file
//
    if(FSfwrite((void *)txt, 1, 22, MyFile) != 22)while(1);
//
// Close the file
//
    if(FSfclose(MyFile) != 0)while(1);
//
// Success. Turn ON the LED
//
    LED = ON;

while(1);
}
```

Figure 9.21: *Cont'd*

9.2.8 Suggestions for Future Work

The programs given in Figures 9.20 and 9.21 can be improved using several LEDs to indicate the cause of the error if the program does not terminate. Alternatively, an LCD can be used to show the status of the program and the cause of any errors.

9.3 PROJECT 2 – Time Stamping a File

9.3.1 Description

In this project, a file called TIME.TXT is created and the following text is written into the file:

The date is 10-July-2009, time is 10:12:05.

The file creation date is set to July 10, 2009, time 10:12:05.

Turn ON the LED connected to port RD0 when the program terminates.

9.3.2 Aim

The aim of this project is to show how the creation (or modification) date of a file can be set.

9.3.3 Block Diagram

The block diagram of this project is as in Figure 9.16.

9.3.4 Circuit Diagram

The circuit diagram of this project is the same as in Figure 9.17.

9.3.5 Operation of the Project

The operation of the project is shown in Figure 9.22.

9.3.6 Program Code

The program code (file WRITE3.C) is shown in Figure 9.23.

9.3.7 Description of the Program Code

The program is similar to the program given in Project 1 except that the function **SetClockVars** is called in this program to set the file creation date and time. Figure 9.24 shows the card directory listing (obtained on a PC) with file TEXT.TXT having the creation date of 10/07/2009 10:12.

9.3.8 Suggestions for Future Work

The program given in Figure 9.23 can be extended by opening several files with different creation dates.

```
BEGIN
    Turn OFF the LED
    Initialize the SD card
    Set date to 10th July 2009, time 10:12:05
    Create file "MESSAGE.TXT"
    Write text message to the file
    Close the file
    Turn ON the LED
END
```

Figure 9.22: Operation of the Project

```
/*****************************************************************************************************
                          PROJECT TO TIME STAMP A FILE
                          ==========================

In these projects, a PIC18F8722-type microcontroller is used. The microcontroller
is operated with a 10-MHz crystal.

An SD card is connected to the microcontroller as follows:

SD card                       microcontroller
CS                            RB3
CLK                           RC3
DO                            RC4
DI                            RC5

The program uses the Microchip MDD library functions to read and write to
the SD card.

In this program a new file called TIME.TXT is created and its creation date is
set to 10th of July 2009, 10:12:05


Author:   Dogan Ibrahim
Date:     July 2009
File:     WRITE3.C
*****************************************************************************************************/
#include <p18f8722.h>
#include <FSIO.h>

#pragma config WDT = OFF, OSC = HSPLL,LVP = OFF
#pragma config MCLRE = ON,CCP2MX = PORTC, MODE = MC

#define LED PORTDbits.RD0
#define ON 1
#define OFF 0

/* ================= START OF MAIN PROGRAM =============== */
//
// Start of MAIN Program
//
void main(void)
{
    FSFILE *MyFile;
    unsigned char txt[]="The date is 10-July-2009, time is 10:12:05";

    TRISD = 0;
    PORTD = 0;
//
// Initialize the SD card routines
//
    while(!FSInit());
//
```

Figure 9.23: Program Code

```
// Set the date and time values
//
    SetClockVars(2009, 7, 10, 10, 12, 5);
//
// Create a new file called MESSAGE.TXT
//
    MyFile = FSfopenpgm("TIME.TXT", "w+");
//
// Write message to the file
//
    FSfwrite(txt, 1, 42, MyFile);
//
// Close the file
//
    FSfclose(MyFile);
//
// Success. Turn ON the LED
//
    LED = ON;

while(1);
}
```

Figure 9.23: *Cont'd*

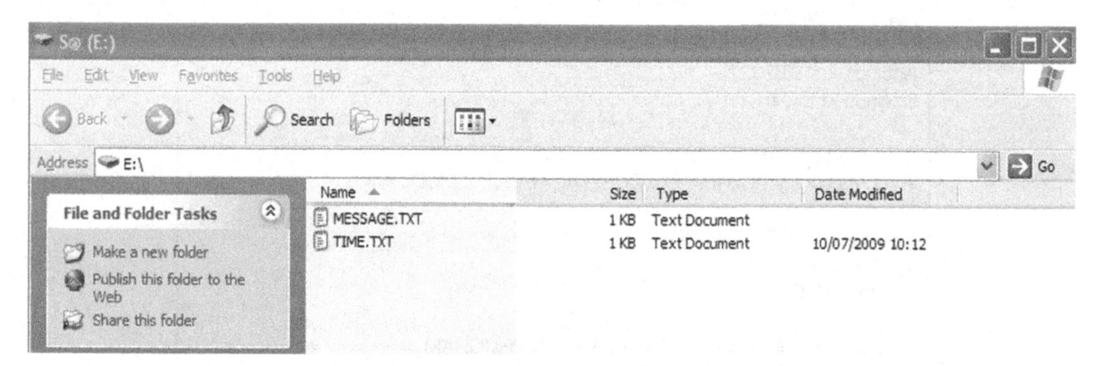

Figure 9.24: Directory Listing of the SD Card

9.4 PROJECT 3 – Formatting a Card

9.4.1 Description

In this project, an SD card is formatted with the volume name MYSDCRD and the serial number is set to hexadecimal 0x11223344. An LED connected to port RD0 is turned ON to indicate the end of formatting.

9.4.2 Aim

The aim of this project is to show how an SD card can be formatted with a given volume name and serial number using the MDD library functions.

9.4.3 Block Diagram

The block diagram of this project is as in Figure 9.16.

9.4.4 Circuit Diagram

The circuit diagram of this project is the same as in Figure 9.17.

9.4.5 Operation of the Project

The operation of the project is shown in Figure 9.25.

9.4.6 Program Code

The program code is shown in Figure 9.26.

```
BEGIN
    Turn OFF the LED
    Format the card
    Turn ON the LED
END
```

Figure 9.25: Operation of the Project

```
/*********************************************************************************************
                    PROJECT TO FORMAT A CARD
                    ========================

In these projects, a PIC18F8722-type microcontroller is used. The microcontroller
is operated with a 10-MHz crystal.

An SD card is connected to the microcontroller as follows:

SD card                     microcontroller
CS                          RB3
CLK                         RC3
DO                          RC4
DI                          RC5

The program uses the Microchip MDD library functions to read and write to
the SD card.

This program formats a file to volume name MYSDCRD. A new boot-sector
is also created on the card.


Author:   Dogan Ibrahim
Date:     July 2009
```

Figure 9.26: The Program Code

```
File:      FORMAT.C
**************************************************************************************************/
#include <p18f8722.h>
#include <FSIO.h>

#pragma config WDT = OFF, OSC = HSPLL,LVP = OFF
#pragma config MCLRE = ON,CCP2MX = PORTC, MODE = MC

#define LED PORTDbits.RD0
#define ON 1
#define OFF 0

/* =============== START OF MAIN PROGRAM ================== */
//
// Start of MAIN Program
//
void main(void)
{
    char VolName[] = "MYSDCRD";

    TRISD = 0;
    PORTD = 0;
//
// Format the card
//
    FSformat(0, 0x11223344, VolName);
//
// Success. Turn ON the LED
//
    LED = ON;

while(1);
}
```

Figure 9.26: *Cont'd*

9.4.7 Description of the Program Code

The program simply calls function FSformat to format the card. Notice that the entries #define ALLOW_FORMATS, #define ALLOW_DIRS, and #define ALLOW_WRITES must be enabled in configuration file FSconfig.h

9.4.8 Suggestions for Future Work

Try formatting different cards with different volume names.

9.5 PROJECT 4 – Deleting a File

9.5.1 Description

In this project, a file called TEMP.TXT that was created earlier on the SD card is deleted. An LED connected to port RD0 is turned ON to indicate the end of the program.

9.5.2 Aim

The aim of this project is to show how a file can be deleted using the MDD library functions.

9.5.3 Block Diagram

The block diagram of this project is as in Figure 9.16.

9.5.4 Circuit Diagram

The circuit diagram of this project is the same as in Figure 9.17.

9.5.5 Operation of the Project

The operation of the project is shown in Figure 9.27.

9.5.6 Program Code

The program code is shown in Figure 9.28.

9.5.7 Description of the Program Code

The program (DELETE.C) simply calls the function FSInit to initialize the MDD library and then calls the function FSremovepgm to delete the file.

```
BEGIN
    Turn OFF the LED
    Initialise the MDD library
    Delete the file
    Turn ON the LED
END
```

Figure 9.27: Operation of the Project

```
/*****************************************************************************************
                    PROJECT TO DELETE A FILE
                    ========================

In these projects, a PIC18F8722-type microcontroller is used. The microcontroller
is operated with a 10-MHz crystal.

An SD card is connected to the microcontroller as follows:

SD card                     microcontroller
CS                          RB3
CLK                         RC3
DO                          RC4
DI                          RC5

The program uses the Microchip MDD library functions to read and write to
the SD card.

This program deletes file called TEMP.TXT


Author:   Dogan Ibrahim
Date:     July 2009
File:     DELETE.C
*****************************************************************************************/
#include <p18f8722.h>
#include <FSIO.h>

#pragma config WDT = OFF, OSC = HSPLL,LVP = OFF
#pragma config MCLRE = ON,CCP2MX = PORTC, MODE = MC

#define LED PORTDbits.RD0
#define ON 1
#define OFF 0

/* =============== START OF MAIN PROGRAM =================== */
//
// Start of MAIN Program
//
void main(void)
{
    TRISD = 0;
    PORTD = 0;
//
// Initialize MDD library
//
    while(!FSInit());
//
// Delete the file
//
```

Figure 9.28: The Program Code

```
    if(FSremovepgm("TEMP.TXT") != 0)while(1);
//
// Success. Turn ON the LED
//
    LED = ON;

while(1);
}
```

Figure 9.28: *Cont'd*

9.5.8 Suggestions for Future Work

Try creating two files, check the card directory on a PC, then delete one of the files and check the directory again to make sure that the correct file is deleted.

9.6 PROJECT 5 – Renaming a File

9.6.1 Description

In this project, a file called TEMP.TXT is created on the SD card and then the name of this file is changed to MYTEMP.TXT.

9.6.2 Aim

The aim of this project is to show how a file can be renamed using the MDD library functions.

9.6.3 Block Diagram

The block diagram of this project is as in Figure 9.16.

9.6.4 Circuit Diagram

The circuit diagram of this project is the same as in Figure 9.17.

9.6.5 Operation of the Project

The operation of the project is shown in Figure 9.29.

9.6.6 Program Code

The program code is shown in Figure 9.30.

9.6.7 Description of the Program Code

The program (RENAME.C) first creates a file called TEMP.TXT and then calls the function FSrenamepgm to change the name of this file to MYTEMP.TXT. Note that the file must be closed before its name can be changed.

```
BEGIN
      Turn OFF the LED
      Initialise the MDD library
      Create a file
      Close the file
      Rename the file
      Turn ON the LED
END
```

Figure 9.29: Operation of the Project

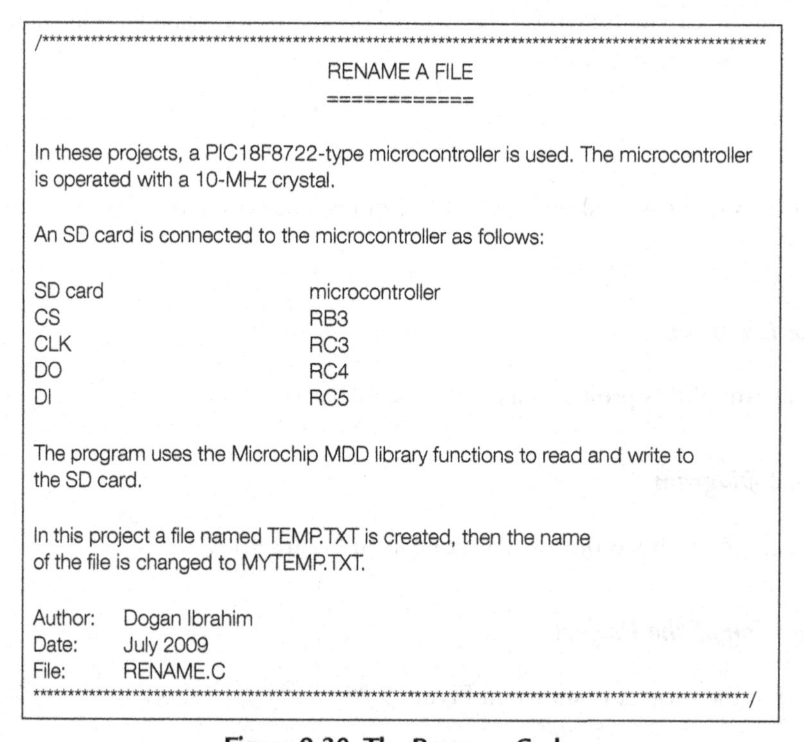

```
/*****************************************************************************************
                              RENAME A FILE
                              ============

In these projects, a PIC18F8722-type microcontroller is used. The microcontroller
is operated with a 10-MHz crystal.

An SD card is connected to the microcontroller as follows:

SD card                    microcontroller
CS                         RB3
CLK                        RC3
DO                         RC4
DI                         RC5

The program uses the Microchip MDD library functions to read and write to
the SD card.

In this project a file named TEMP.TXT is created, then the name
of the file is changed to MYTEMP.TXT.

Author:   Dogan Ibrahim
Date:     July 2009
File:     RENAME.C
*****************************************************************************************/
```

Figure 9.30: The Program Code

```
#include <p18f8722.h>
#include <FSIO.h>

#pragma config WDT = OFF, OSC = HSPLL,LVP = OFF
#pragma config MCLRE = ON,CCP2MX = PORTC, MODE = MC


/* ================= START OF MAIN PROGRAM ================= */
//
// Start of MAIN Program
//
void main(void)
{
    FSFILE * MyFile;

//
// Initialize MDD library
//
    while(!FSInit());
//
// Create file TEMP.TXT
//
    MyFile =FSfopenpgm("TEMP.TXT", "w+");
//
// Close the file
//
    FSfclose(MyFile);
//
// Rename the file to MYTEMP.TXT
//
    FSrenamepgm("MYTEMP.TXT", MyFile);

while(1);
}
```

Figure 9.30: *Cont'd*

9.6.8 Suggestions for Future Work

Try creating two files, check the card directory on a PC, then delete one of the files and rename the other file, and check the directory again to make sure that the correct files are deleted and renamed.

9.7 PROJECT 6 – Creating a Directory

9.7.1 Description

In this project, a directory called MYDATA is created in the current directory.

9.7.2 Aim

The aim of this project is to show how a directory can be created using the MDD library functions.

9.7.3 Block Diagram

The block diagram of this project is as in Figure 9.16.

9.7.4 Circuit Diagram

The circuit diagram of this project is the same as in Figure 9.17.

9.7.5 Operation of the Project

The operation of the project is shown in Figure 9.31.

9.7.6 Program Code

The program code is shown in Figure 9.32.

9.7.7 Description of the Program Code

The program (DIR1.C) initializes the MDD library and calls the function FSmkdir to create a directory within the default working directory.

9.7.8 Suggestions for Future Work

Create two directories called DATA1 and DATA2 in the current working directory. Then, create a directory called DATA1-1 inside the first directory. Check the card directory using a PC.

```
BEGIN
    Initialise the MDD library
    Create a directory
END
```

Figure 9.31: Operation of the Project

```
/*****************************************************************************************
                         CREATE A DIRECTORY
                         =================

In these projects, a PIC18F8722-type microcontroller is used. The microcontroller
is operated with a 10-MHz crystal.

An SD card is connected to the microcontroller as follows:

SD card                          microcontroller
CS                               RB3
CLK                              RC3
DO                               RC4
DI                               RC5

The program uses the Microchip MDD library functions to read and write to
the SD card.

In this project a directory called MYDATA is created within the
current default working directory

Author:   Dogan Ibrahim
Date:     July 2009
File:     DIR1.C
*****************************************************************************************/
#include <p18f8722.h>
#include <FSIO.h>

#pragma config WDT = OFF, OSC = HSPLL,LVP = OFF
#pragma config MCLRE = ON,CCP2MX = PORTC, MODE = MC


/* ================= START OF MAIN PROGRAM ================= */
//
// Start of MAIN Program
//
void main(void)
{
    char DirPath[] = "\MYDATA";
//
// Initialize MDD library
//
    while(!FSInit());
//
// Create directory MYDATA within the current working directory
//
    FSmkdir(DirPath);

while(1);
}
```

Figure 9.32: The Program Code

9.8 PROJECT 7 – Create a Directory and a File

9.8.1 Description

In this project, a directory called MYDATA is created in the current directory and then a file called RESULTS.DAT is created inside this directory. The following numbers are written inside this file: 24, 45, 22, 10, 28, 30.

9.8.2 Aim

The aim of this project is to show how a directory and a file inside this directory can be created using the MDD library functions.

9.8.3 Block Diagram

The block diagram of this project is as in Figure 9.16.

9.8.4 Circuit Diagram

The circuit diagram of this project is the same as in Figure 9.17.

9.8.5 Operation of the Project

The operation of the project is shown in Figure 9.33.

9.8.6 Program Code

The program code is shown in Figure 9.34.

9.8.7 Description of the Program Code

The program (DIR2.C) initializes the MDD library and calls the function FSmkdir to create a directory MYDATA within the default working directory. Then the function FSchdir is used to change the working directory to MYDATA. The file RESULTS.DAT is then created inside

```
BEGIN
    Initialise the MDD library
    Create a directory
    Create a file inside this directory
    Write numbers inside the file
    Close the file
END
```

Figure 9.33: Operation of the Project

```
/**************************************************************************************************
                        CREATE A DIRECTORY AND A FILE
                        ===========================

In these projects, a PIC18F8722-type microcontroller is used. The microcontroller
is operated with a 10-MHz crystal.

An SD card is connected to the microcontroller as follows:

SD card                         microcontroller
CS                              RB3
CLK                             RC3
DO                              RC4
DI                              RC5

The program uses the Microchip MDD library functions to read and write to
the SD card.

In this project a directory called MYDATA is created within the current default
working directory and then a file called RESULTS.DAT is created inside this
directory. Some numbers are then written into this file

Author:   Dogan Ibrahim
Date:     July 2009
File:     DIR2.C
**************************************************************************************************/
#include <p18f8722.h>
#include <FSIO.h>

#pragma config WDT = OFF, OSC = HSPLL,LVP = OFF
#pragma config MCLRE = ON,CCP2MX = PORTC, MODE = MC


/* ================ START OF MAIN PROGRAM ================== */
//
// Start of MAIN Program
//
void main(void)
{
    FSFILE * MyFile;

    char DirPath[] = "\MYDATA";
    char txt[] = "24, 45, 22,10, 28, 30";
//
// Initialize MDD library
//
    while(!FSInit());
//
// Create directory MYDATA within the current working directory
//
    FSmkdir(DirPath);
```

Figure 9.34: The Program Code

```
//
// Change default working directory to MYDATA
//
     FSchdir(DirPath);
//
// Create a file in this directory
//
     MyFile = FSfopenpgm("RESULTS.DAT", "w+");
//
// Write numbers into this file
//
     FSfwrite((void *)txt, 1, 17, MyFile);
//
// Close the file
//
     FSfclose(MyFile);

while(1);
}
```

Figure 9.34: *Cont'd*

this directory and the required numbers are written to this file. The contents of the card can be verified using a PC to list the card directory.

9.8.8 Suggestions for Future Work

Create a directory called MYTEXT inside the current default directory. Then, create two files inside this directory and write some text to both the files. Verify the card directory and contents of the files using a PC.

9.9 PROJECT 8 – File Copying

9.9.1 Description

In this project, the contents of a file are copied to another file. The source file called SRC. TXT is loaded with some text using a PC. The contents of this file are then copied to a file called DST.TXT. The success of the copy operation is verified by reading the destination file on a PC.

9.9.2 Aim

The aim of this project is to show how multiple files can be handled using the MDD library. In addition, the steps to read and write to a card are described in this project.

9.9.3 Block Diagram

The block diagram of this project is as in Figure 9.16.

9.9.4 Circuit Diagram

The circuit diagram of this project is the same as in Figure 9.17.

9.9.5 Operation of the Project

The operation of the project is shown in Figure 9.35.

9.9.6 Program Code

The program code (COPY.C) is shown in Figure 9.36.

9.9.7 Description of the Program Code

After initializing the MDD library, file SRC.TXT is opened in read mode and file DST.TXT is created as a new file. Then a **while** loop is formed and the statements inside this loop are executed until the end of the source file is reached. Inside this loop, 10 items of 1 byte each are read (other sizes could also be used). Variable **ReadCnt** actually stores the number of bytes read from SRC.TXT. The data is then written to file DST.TXT. Both the source and the destination files are closed at the end of the copy operation.

9.9.8 Suggestions for Future Work

Modify the program given in Figure 9.36 to copy a file using different numbers of items and different lengths for each item.

```
BEGIN
    Initialise the MDD library
    Open source file
    Open destination file
    WHILE not end of source file
        Read from source file
        Write to destination file
    WEND
    Close source file
    Close destination file
END
```

Figure 9.35: Operation of the Project

```
/*********************************************************************************************************
                                FILE COPY
                                =========

In these projects, a PIC18F8722-type microcontroller is used. The microcontroller
is operated with a 10-MHz crystal.

An SD card is connected to the microcontroller as follows:

SD card                         microcontroller
CS                              RB3
CLK                             RC3
DO                              RC4
DI                              RC5

The program uses the Microchip MDD library functions to read and write to
the SD card.

In this project a file called SRC.TXT is copied to another file
called DST.TXT. It is assumed that SRC.TXT had some text data in
it before the copy operation.


Author:   Dogan Ibrahim
Date:     July 2009
File:     COPY.C
*********************************************************************************************************/
#include <p18f8722.h>
#include <FSIO.h>

#pragma config WDT = OFF, OSC = HSPLL,LVP = OFF
#pragma config MCLRE = ON,CCP2MX = PORTC, MODE = MC


/* ================== START OF MAIN PROGRAM ================== */
//
// Start of MAIN Program
//
void main(void)
{
     FSFILE *MySrcFile, *MyDstFile;
     char bufr[10];
     char ReadCnt;
//
// Initialize MDD library
//
     while(!FSInit());
//
// Open the source file
//
     MySrcFile = FSfopenpgm("SRC.TXT", "r");
```

Figure 9.36: The Program Code

```
//
// Create the destination file
//
            MyDstFile = FSfopenpgm("DST.TXT", "w+");
//
// Read from SRC.TXT and write to DST.TXT until the end-of-file
//
     while(FSfeof(MySrcFile) == 0)
     {
          ReadCnt = FSfread((void *)bufr,1, 10, MySrcFile);
          FSfwrite((void *)bufr, 1, ReadCnt, MyDstFile);
     }
//
// Close the files
//
     FSfclose(MySrcFile);
     FSfclose(MyDstFile);

while(1);
}
```

Figure 9.36: *Cont'd*

9.10 PROJECT 9 – Displaying File on a PC

9.10.1 Description

In this project, the contents of a file are displayed on a PC connected to the microcontroller. Before starting the project, a file called MYTEST.TXT is created on the SD card using a PC and this file is loaded with some text. The card is then connected to the microcontroller, the file is opened and its contents are displayed on the PC screen. The message "Displaying the file…" is sent to the PC before displaying contents of the file.

9.10.2 Aim

The aim of this project is to show how the serial port of the microcontroller can be connected to a PC and how the contents of a file can be displayed on the PC screen using the MDD library functions.

9.10.3 Block Diagram

The block diagram of this project is as in Figure 9.37.

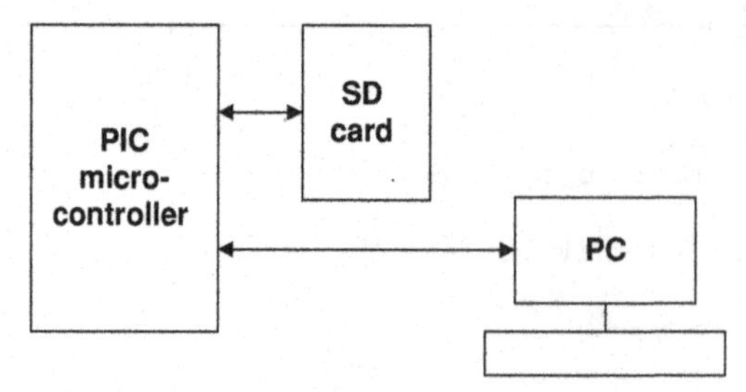

Figure 9.37: Block Diagram of the Project

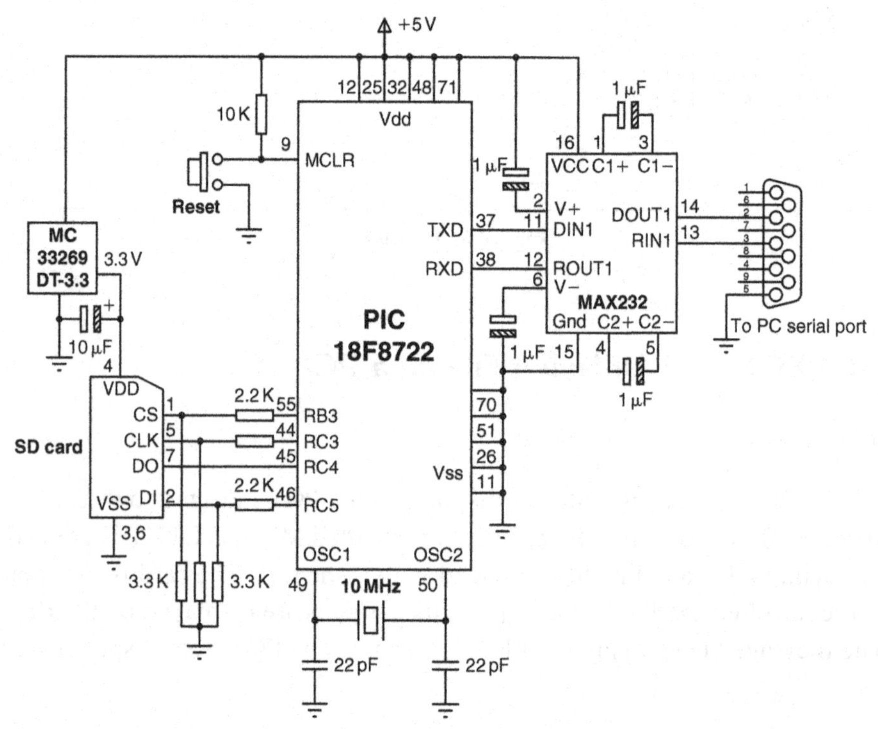

Figure 9.38: Circuit Diagram of the Project

9.10.4 Circuit Diagram

The circuit diagram of the project is shown in Figure 9.38. As in the other projects in this book, the PICDEM PIC18 Explorer board and the PICtail SD card daughter board are used in this project, but the circuit can easily be built on a breadboard if desired (the RS232 port jumper J13 should be set as described in the Explorer Demonstration board if this board is used for RS232 communication). PIC18F8722 contains two UART-type serial hardware modules. In this project, UART 1 (pin RX1 = RC7 and TX1 = RC6) is used for serial communication and is connected to a MAX232-type level converter chip. This chip converts the 0 to +5-V TTL level output voltage of the microcontroller to ±12-V RS232 levels and also the ±12 V RS232 levels to 0 to +5 V required for the microcontroller inputs. The MAX232 is connected to the serial port of a PC via a 9-pin D-type connector.

There are two types of RS232 connector: 9-pin and 25-pin. The required pins in each type are as follows:

Pins	Signals
9-pin	
2	Transmit (TX)
3	Receive (RX)
5	Ground (GND)
25-pin	
2	Transmit (TX)
3	Receive (RX)
7	Ground (GND)

9.10.5 Operation of the Project

The operation of the project is shown in Figure 9.39.

9.10.6 The Program Code

The program code (RS232.C) is shown in Figure 9.40.

9.10.7 Description of the Program Code

At the beginning of the program, the USART module is initialized by calling the C18 library function Open1USART (since there are two USART modules on the PIC18F8722 microcontroller, we have to specify which module we shall be using). Serial port interrupts are disabled, USART is set to asynchronous mode, and the baud-rate clock is set to low

```
BEGIN
    Initialize USART module
    Initialize the MDD library
    Open file on SD card
    WHILE not end of source file
        Read from source file
        Send to USART
    WEND
    Close file
    Close USART
END
```

Figure 9.39: Operation of the Project

```
/*******************************************************************************************
                         SEND FILE CONTENTS TO THE PC
                         ==========================

In these projects, a PIC18F8722-type microcontroller is used. The microcontroller
is operated with a 10-MHz crystal.

An SD card is connected to the microcontroller as follows:

SD card                         microcontroller
CS                              RB3
CLK                             RC3
DO                              RC4
DI                              RC5

The program uses the Microchip MDD library functions to read and write to
the SD card.

In this project the microcontroller is connected to a PC via a MAX232-type
level converter chip. USART 1 hardware module of the microcontroller is used
for serial communication. The pin configuration of the RS232 connector is as
follows:

          Pin 2     TX
          Pin 3     RX
          Pin 3     GND

The communication is established using the C18 USART library functions. The USART
is configured as follows:

          4800 Baud
          8 data bits
          No parity
          1 stop bit

In this project a text file called MYTEST.TXT is opened on the SD card and the
contents of this file are displayed on the PC.



Author:   Dogan Ibrahim
Date:     August 2009
File:     RS232.C
*******************************************************************************************/
#include <p18f8722.h>
#include <usart.h>
#include <FSIO.h>

#pragma config WDT = OFF, OSC = HSPLL, LVP = OFF
#pragma config MCLRE = ON, CCP2MX = PORTC, MODE = MC
```

Figure 9.40: The Program Code

```
/* ==================START OF MAIN PROGRAM ================= */
//
// Start of MAIN Program
//
void main(void)
{
    FSFILE *MySrcFile;
    char bufr[1];

//
// Initialize the USART
//
    Open1USART(    USART_TX_INT_OFF  &
                   USART_RX_INT_OFF  &
                   USART_ASYNCH_MODE &
                   USART_EIGHT_BIT   &
                   USART_CONT_RX     &
                   USART_BRGH_LOW,
                   129);
//
// Send a message to the PC
//
    while(Busy1USART());
    putrs1USART(" Displaying the file...\n\r");
//
// Initialize MDD library
//
    while(!FSInit());
//
// Open the source file
//
    MySrcFile = FSfopenpgm("MYTEST.TXT", "r");
//
// Read from the file and send to the PC until the end-of-file
//
    while(FSfeof(MySrcFile) == 0)
    {
        FSfread((void *)bufr,1, 1, MySrcFile);
        while(Busy1USART());
        putc1USART(bufr[0]);
    }
//
// Close the file
//
    FSfclose(MySrcFile);
//
// Close USART
//
    Close1USART();

while(1);
}
```

Figure 9.40: *Cont'd*

speed. In this project, the serial communication baud rate is set to 4800. The last argument of openUSART (spbrg) sets the baud rate according to the formula:

$$\text{Baud rate} = \text{Fosc}/(64 \times (\text{spbrg} + 1)),$$

where Fosc is the microcontroller operating clock frequency. Here, the value of spbrg is calculated as

$$\text{spbrg} = \text{Fosc}/(64 \times \text{baud rate}) - 1$$

giving

$$\text{spbrg} = 40 \times 10^6/(64 \times 4800) - 1 = 129$$

Then the message "Displaying the file…," followed by the characters carriage-return and line feed are sent to the PC. File MYTEST.TXT is then opened in read mode on the SD card. A **while** loop is formed that executes as long as the end of file is not detected. Inside this loop, a byte is read from the file using the function FSfread and this byte is sent to the USART by calling the function putc1USART. Note that we should make sure that the USART is ready to receive a character before we send the next character. After all the data in the file is sent, the file is closed and the USART is disabled by closing it.

The operation of the program can be tested using a serial communications program on the PC, such as the HyperTerminal. The steps in using this program are given below:

- Start the HyperTerminal program by selecting it from Start -> Programs -> Accessories -> Communications -> HyperTerminal
- Enter a name for the connection (e.g., TEST) and click OK (see Figure 9.41)

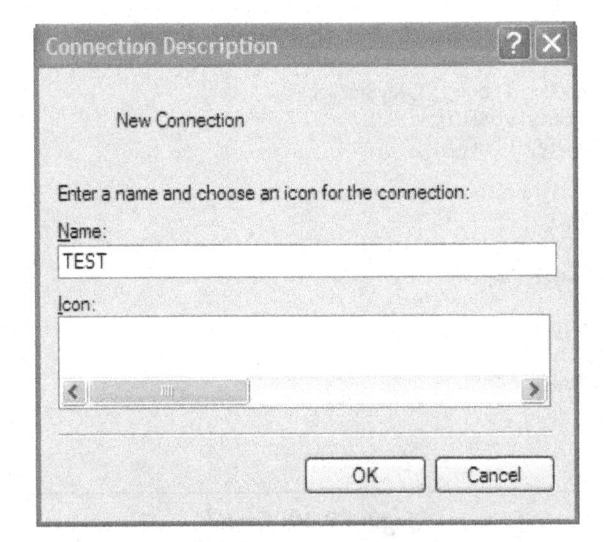

Figure 9.41: Create a New Serial Connection

Figure 9.42: Select the Serial Port Number

- Select the serial port the microcontroller is connected to. In this example, COM2 is used, as shown in Figure 9.42. (If you are not sure which serial ports are available in your PC, check the Device Manager in Control Panel. Most modern PCs do not have any serial ports and you may have to use a serial to USB connector device to provide serial ports to your PC.) Click OK.
- Select the serial port parameters as (see Figure 9.43)

Bits per second:	4800
Data bits:	8
Parity:	None
Stop bits:	1
Flow control:	None

Click **Apply** and **OK**.

- Insert the SD card into its holder and make sure that there is a file called MYTEST.TXT on the card.
- Press the Reset button. The contents of the file should be displayed on the PC (see the example in Figure 9.44).

9.10.8 Suggestions for Future Work

Write a program to read a file name from the PC and then open this file on the SD card and display its contents on the PC screen.

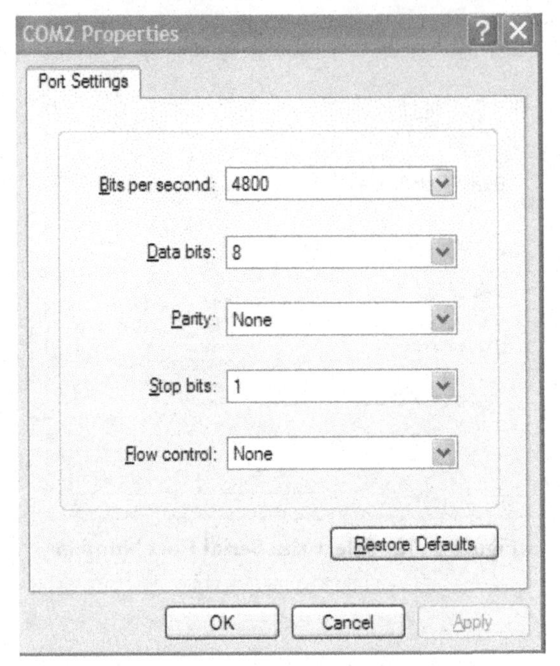

Figure 9.43: Select the Serial Port Parameters

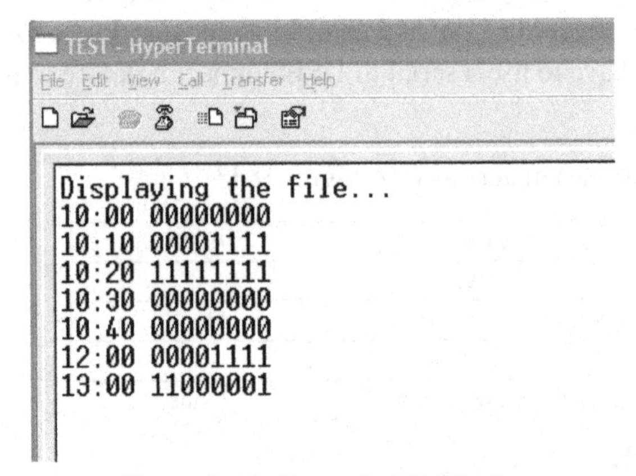

Figure 9.44: Example PC Display

9.11 PROJECT 10 – Reading a Filename from the PC and Displaying the File

9.11.1 Description

In this project, the name of a file is received from the PC and then contents of this file are displayed on the PC. Before starting the project, it is assumed that a file called MYTEST. TXT exists on the SD card. The following message is displayed on the PC screen by the microcontroller (the filename string is terminated when the **Enter** key is pressed on the PC keyboard):

Enter the Filename:

9.11.2 Aim

The aim of this project is to show how a string of data can be received from a PC and how a file can be opened on the SD card and its contents displayed on the PC screen.

9.11.3 Block Diagram

The block diagram of this project is as in Figure 9.37.

9.11.4 Circuit Diagram

The circuit diagram of this project is as in Figure 9.38.

9.11.5 Operation of the Project

The operation of the project is shown in Figure 9.45.

9.11.6 Program Code

The program code (RS232-2.C) is shown in Figure 9.46.

9.11.7 Description of the Program Code

The USART module is initialized at the beginning of the program (as in the earlier project) by calling to C18 library function Open1USART (since there are two USART modules on the PIC18F8722 microcontroller, we have to specify which module we shall be using).

```
BEGIN
    Initialize USART module
    Initialize the MDD library
    Display a heading message
    Read the filename
    Open the file on SD card
    WHILE not end of source file
        Read from the file
        Send to USART
    WEND
    Close the file
    Close USART
END
```

Figure 9.45: Operation of the Project

Serial port interrupts are disabled, USART is set to asynchronous mode, the baud-rate clock is set to low speed, and the baud rate is set to 4800. The message "Enter the Filename:" is displayed on the PC screen and the user is requested to enter the name of the file. The filename is stored in a character array called FileName. This array is terminated with a NULL character as the function FSfopen requires the name to be a NULL-terminated string. The required text file is then opened on the SD card and its contents are displayed on the PC screen as in the earlier project.

Figure 9.47 shows a typical output from the project.

```
/**************************************************************************************************
        READ FILENAME AND SEND FILE CONTENTS TO THE PC
        ==========================================

In these projects, a PIC18F8722-type microcontroller is used. The microcontroller
is operated with a 10-MHz crystal.

An SD card is connected to the microcontroller as follows:

SD card                     microcontroller
CS                          RB3
CLK                         RC3
DO                          RC4
DI                          RC5

The program uses the Microchip MDD library functions to read and write to
the SD card.

In this project the microcontroller is connected to a PC via
a MAX232-type level converter chip. USART 1 hardware module
of the microcontroller is used for serial communication. The
pin configuration of the RS232 connector is as follows:

        Pin 2     TX
        Pin 3     RX
        Pin 3     GND

The communication is established using the C18 USART library
functions. The USART is configured as follows:

        4800 Baud
        8 data bits
        No parity
        1 stop bit

In this project the filename is read from the PC keyboard and the content of
the file is displayed on the PC screen (it is assumed that the requested file exists
on the SD card).
```

Figure 9.46: The Program Code

```
Author:   Dogan Ibrahim
Date:     August 2009
File:     RS232-2.C
*******************************************************************************************************************
#include <p18f8722.h>
#include <usart.h>
#include <FSIO.h>

#pragma config WDT = OFF, OSC = HSPLL, LVP = OFF
#pragma config MCLRE = ON, CCP2MX = PORTC, MODE = MC


/* =================== START OF MAIN PROGRAM =================== */
//
// Start of MAIN Program
//
void main(void)
{
    FSFILE *MySrcFile;
    char bufr[1];
    char FileName[20];
    char FileLen = 0, itm = 0;
    char mode[2] = "r";
//
// Initialize the USART
//
    Open1USART(USART_TX_INT_OFF  &
               USART_RX_INT_OFF  &
               USART_ASYNCH_MODE &
               USART_EIGHT_BIT   &
               USART_CONT_RX     &
               USART_BRGH_LOW,
               129);
//
// Send a message to the PC
//
    while(Busy1USART());
    putrs1USART(" Enter the Filename:");
//
// Read the filename (until the Enter key is pressed)
//
    while(itm != 0x0D)
    {
    while(!DataRdy1USART());
    itm = getc1USART();
    putc1USART(itm);
    FileName[FileLen] = itm;
    FileLen++;
    }
//
// Terminate the Filename with a NULL character
//
```

Figure 9.46: *Cont'd*

```
        FileLen--;
        FileName[FileLen] = '\0';
//
// Insert a new line
//
        putrs1USART("\n\r");
//
// Initialize MDD library
//
        while(!FSInit());
//
// Open the file
//
        MySrcFile = FSfopen(FileName, mode);
//
// Read from the file and send to the PC until the end-of-file
//
        while(FSfeof(MySrcFile) == 0)
        {
            FSfread((void *)bufr,1, 1, MySrcFile);
            while(Busy1USART());
            putc1USART(bufr[0]);
        }
//
// Close the file
//
        FSfclose(MySrcFile);
//
// Close USART
//
        Close1USART();

while(1);
}
```

Figure 9.46: *Cont'd*

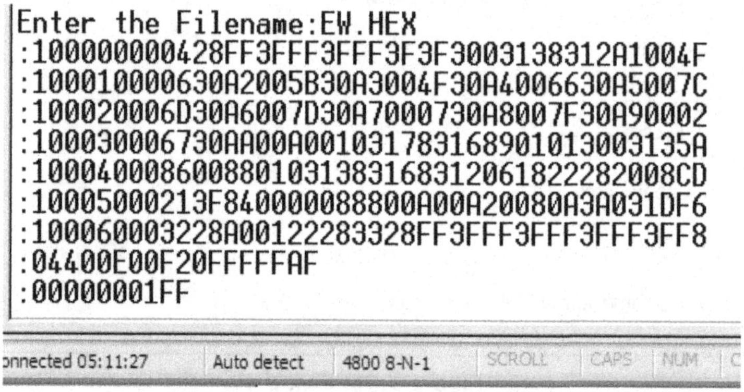

```
Enter the Filename:EW.HEX
:100000000428FF3FFF3FFF3F3F3003138312A1004F
:100010000630A2005B30A3004F30A4006630A5007C
:100020006D30A6007D30A7000730A8007F30A90002
:100030006730AA00A00103178316890101300313 5A
:10004000860088010313831683120618222820 08CD
:1000500213F840000088800A00A20080A3A031DF6
:100060003228A0012228332 8FF3FFF3FFF3FFF3FF8
:04400E00F20FFFFFAF
:00000001FF
```

onnected 05:11:27 | Auto detect | 4800 8-N-1 | SCROLL | CAPS | NUM | C

Figure 9.47: Example PC Display

9.11.8 Suggestions for Future Work

Store several files on the SD card with the extension ".DAT." Write a program to read a file name from the keyboard without the file extension. Append the extension ".DAT" to the file name and display the contents of the file on the PC screen.

9.12 PROJECT 11 – Looking for a File

9.12.1 Description

In this project, the current directory is searched for a given filename with any attribute value. The name of the file to be searched is entered from the PC keyboard as in Project 10. If the file is not found, the message NOT FOUND is displayed; if the file is found, the size of the file is displayed in bytes on the PC screen.

9.12.2 Aim

The aim of this project is to show how a file can be searched in the current directory using the MDD library functions.

9.12.3 Block Diagram

The block diagram of this project is as in Figure 9.37.

9.12.4 Circuit Diagram

The circuit diagram of this project is as in Figure 9.38.

9.12.5 Operation of the Project

The operation of the project is shown in Figure 9.48.

```
BEGIN
    Initialize USART module
    Initialize the MDD library
    Read the filename
    Search for the file in current directory
    IF file not found
       Display NOT FOUND
    ELSE
       Display size of the file
    ENDIF
    Close USART
END
```

Figure 9.48: Operation of the Project

9.12.6 Program Code

The program code (FIND.C) is shown in Figure 9.49.

9.12.7 Description of the Program Code

At the beginning of the program, variable **Attribute** is set to ATTR_MASK so that the file attribute is not considered while finding the specified file. As in the previous project, the USART module is initialized by calling to C18 library function Open1USART (since there are two USART modules on the PIC18F8722 microcontroller, we have to specify which

```
/*****************************************************************************************************
                LOOK FOR A FILE IN CURRENT DIRECTORY
                ===================================

In these projects, a PIC18F8722-type microcontroller is used. The microcontroller
is operated with a 10-MHz crystal.

An SD card is connected to the microcontroller as follows:

SD card                     microcontroller
CS                          RB3
CLK                         RC3
DO                          RC4
DI                          RC5

The program uses the Microchip MDD library functions to read and write to
the SD card.

In this project the microcontroller is connected to a PC via a MAX232-type level
converter chip. USART 1 hardware module of the microcontroller is used for serial
communication. The pin configuration of the RS232 connector is as follows:

          Pin 2     TX
          Pin 3     RX
          Pin 3     GND

The communication is established using the C18 USART library functions. The
USART is configured as follows:

          4800 Baud
          8 data bits
          No parity
          1 stop bit
```

Figure 9.49: The Program Code

```
In this project a given file is searched in the current default directory and the message
NOT FOUND is displayed if the file is not found; otherwise, the size of the file is
displayed on the PC screen.


Author:   Dogan Ibrahim
Date:     August 2009
File:     FIND.C
*************************************************************************************************/
#include <p18f8722.h>
#include <usart.h>
#include <stdlib.h>
#include <FSIO.h>

#pragma config WDT = OFF, OSC = HSPLL,LVP = OFF
#pragma config MCLRE = ON,CCP2MX = PORTC, MODE = MC


/* ================= START OF MAIN PROGRAM ================== */
//
// Start of MAIN Program
//
void main(void)
{
    FSFILE *MySrcFile;
    SearchRec File;
    char bufr[1];
    char FileName[20];
    char FileLen = 0, itm = 0;
    unsigned char Attribute = ATTR_MASK;
    unsigned long FileSize;
    unsigned char FileSizeStr[10];
//
// Initialize the USART
//
    Open1USART(USART_TX_INT_OFF  &
               USART_RX_INT_OFF  &
               USART_ASYNCH_MODE &
               USART_EIGHT_BIT   &
               USART_CONT_RX     &
               USART_BRGH_LOW,
               129);
//
// Send a message to the PC
//
    while(Busy1USART());
    putrs1USART(" Enter the Filename:");
//
// Read the filename (until the Enter key is pressed)
//
```

Figure 9.49: *Cont'd*

```
    while(itm != 0x0D)
    {
        while(!DataRdy1USART());
        itm = getc1USART();
        putc1USART(itm);
        FileName[FileLen] = itm;
        FileLen++;
    }
//
// Terminate the Filename with a NULL character
//
    FileLen—;
    FileName[FileLen] = '\0';
//
// Insert a new line
//
    putrs1USART("\n\r");
//
// Initialize MDD library
//
    while(!FSInit());
//
// Look for the file
//
    if(FindFirst(FileName, Attribute, &File) != 0)
        putrs1USART("NOT FOUND");
    else
    {
        FileSize = File.filesize;
        ltoa(FileSize, ( void*)FileSizeStr);
        puts1USART((void *)FileSizeStr);
    }
//
// Close USART
//
    Close1USART();

while(1);
}
```

Figure 9.49: *Cont'd*

module we shall be using). Serial port interrupts are disabled, USART is set to asynchronous mode, the baud-rate clock is set to low speed, and the baud rate is set to 4800. The message "Enter the Filename:" is displayed on the PC screen, and the user is requested to enter the name of the file. The filename is stored in a character array called FileName and is terminated with a NULL character.

The MDD library function FindFirst is called to search for the specified file in the current default directory. If the file is not found, the function returns a nonzero value and the message "NOT FOUND" is displayed on the PC screen. If the file is found, its size is

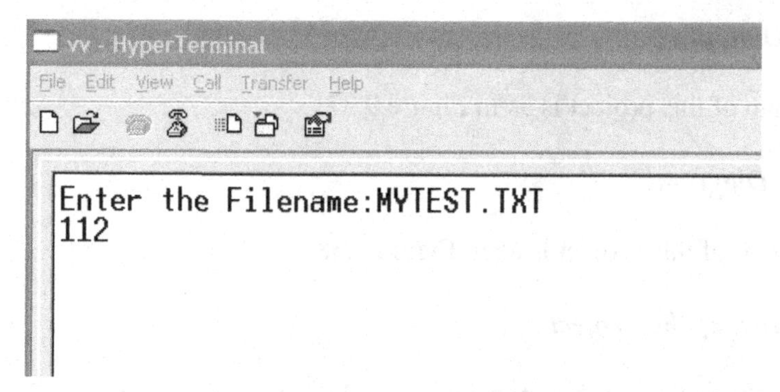

Figure 9.50: Example PC Display

extracted from the **filesize** member of structure SearchRec. The file size is a long variable; it is converted into a string using the C18 library function **ltoa** and is stored in character array FileSizeStr. USART function puts1USART is then called to display the file size (in bytes).

Figure 9.50 shows a typical output from the project where the size of the file was 112 bytes.

9.12.8 Suggestions for Future Work

Write a program to look for a file on the SD card, and if the file is found, display the creation date of the file; otherwise, display a message "NOT FOUND" on the PC screen.

9.13 PROJECT 12 – Looking for a Number of Files with a Given File Extension

9.13.1 Description

In this project, a file extension is given and then all the files in the current directory with the given extension name are found and listed on the PC screen. If no files are found, the message "NO SUCH FILES" is displayed. For example, if the file extension is specified as

*.TXT,

then all the files having extensions "TXT" will be searched for in the current directory.

9.13.2 Aim

The aim of this project is to show how multiple files can be searched in the current directory using the MDD library functions.

9.13.3 Block Diagram

The block diagram of this project is as in Figure 9.37.

9.13.4 Circuit Diagram

The circuit diagram of this project is as in Figure 9.38.

9.13.5 Operation of the Project

The operation of the project is shown in Figure 9.51.

9.13.6 Program Code

The program code (FINDEXT.C) is shown in Figure 9.52.

9.13.7 Description of the Program Code

At the beginning of the program, variable **Attribute** is set to ATTR_MASK so that the file attribute is not considered while finding the specified files. As in the previous project, the USART module is initialized by calling to C18 library function Open1USART (since there are two USART modules on the PIC18F8722 microcontroller, we have to specify which module we shall be using). Serial port interrupts are disabled, USART is set to asynchronous mode, the baud-rate clock is set to low speed, and the baud rate is set to 4800. The message "Enter the File extension:" is displayed on the PC screen and the user is requested to enter the file extension name. The file extension is stored in a character array called FileName and is terminated with a NULL character.

The MDD library function FindFirst is called to search for the specified file in the current default directory. If the file is not found, the function returns a nonzero value and the message "NO SUCH FILES" is displayed on the PC screen. If a file is found matching the specified

```
BEGIN
    Initialize USART module
    Initialize the MDD library
    Read the file extension
    Search for all files with the given extension
    IF file not found
        Display NO SUCH FILES
    ELSE
        Display names of all found files
    ENDIF
    Close USART
END
```

Figure 9.51: Operation of the Project

```
/**************************************************************************************************
                LOOK FOR A NUMBER OF FILES WITH A GIVEN EXTENSION
                ================================================

In these projects, a PIC18F8722-type microcontroller is used. The microcontroller
is operated with a 10-MHz crystal.

An SD card is connected to the microcontroller as follows:

SD card                         microcontroller
CS                              RB3
CLK                             RC3
DO                              RC4
DI                              RC5

The program uses the Microchip MDD library functions to read and write to
the SD card.

In this project the microcontroller is connected to a PC via a MAX232-type level
converter chip. The USART 1 hardware module of the microcontroller is used for serial
communication. The pin configuration of the RS232 connector is as follows:

          Pin 2     TX
          Pin 3     RX
          Pin 3     GND

The communication is established using the C18 USART library functions. The USART
is configured as follows:

          4800 Baud
          8 data bits
          No parity
          1 stop bit

In this project a file extension is given and then names of all the files in the current
directory having the given extension are displayed. If there are no files with the given
extension then message NO SUCH FILES is displayed. For example, the following file
specification:

          *.TXT

displays all the files having extensions "TXT"


Author:   Dogan Ibrahim
Date:     August 2009
File:     FINDEXT.C
**************************************************************************************************/
#include <p18f8722.h>
#include <usart.h>
#include <stdlib.h>
#include <FSIO.h>
```

Figure 9.52: The Program Code

```
#pragma config WDT = OFF, OSC = HSPLL,LVP = OFF
#pragma config MCLRE = ON,CCP2MX = PORTC, MODE = MC


/* =============== START OF MAIN PROGRAM ================= */
//
// Start of MAIN Program
//
void main(void)
{
    FSFILE *MySrcFile;
    SearchRec File;
    char bufr[1];
    char FileName[12];
    char FileLen = 0, itm = 0;
    unsigned char Attribute = ATTR_MASK;
    char Flag = 0;
//
// Initialize the USART
//
    Open1USART(USART_TX_INT_OFF  &
               USART_RX_INT_OFF  &
               USART_ASYNCH_MODE &
               USART_EIGHT_BIT   &
               USART_CONT_RX     &
               USART_BRGH_LOW,
               129);
//
// Send a message to the PC
//
    while(Busy1USART());
    putrs1USART(" Enter the File extension: ");
//
// Read the filename (until the Enter key is pressed)
//
    while(itm != 0x0D)
    {
        while(!DataRdy1USART());
        itm = getc1USART();
        putc1USART(itm);
        FileName[FileLen] = itm;
        FileLen++;
    }
//
// Terminate the Filename with a NULL character
//
    FileLen--;
    FileName[FileLen] = '\0';
//
// Insert a new line
//
    putrs1USART("\n\r");
```

Figure 9.52: *Cont'd*

```
//
// Initialize MDD library
//
    while(!FSInit());
//
// Look for the files with given extension
//
    if(FindFirst(FileName, Attribute, &File) != 0)
            putrs1USART("NO SUCH FILES");
        else
        {
            puts1USART(File.filename);
            putrs1USART("\n\r");

            while(Flag == 0)
            {
                if(FindNext(&File) != 0)
                    Flag = 1;
                else
                {
                    puts1USART(File.filename);
                    putrs1USART("\n\r");
                }
            }
        }
//
// Close USART
//
    Close1USART();

while(1);
}
```

Figure 9.52: *Cont'd*

extension name, then the full name of this file is extracted from the **filename** member of structure SearchRec and the filename is displayed on the PC screen using USART function puts1USART. The MDD function FindNext is then called in a **while** loop to check if there are any more files with the specified extension name. The return of function FindNext is zero if another file is found, and the name of the found file is displayed, followed by a new-line character. When there are no more files matching the specified extension name, then function FindNext returns a nonzero value and the variable **Flag** is set to 1 to terminate the **while** loop.

Figure 9.53 shows a typical output from the project. In this example, there are six files on the SD card with the following names:

TEST1.TXT

TEST2.TXT

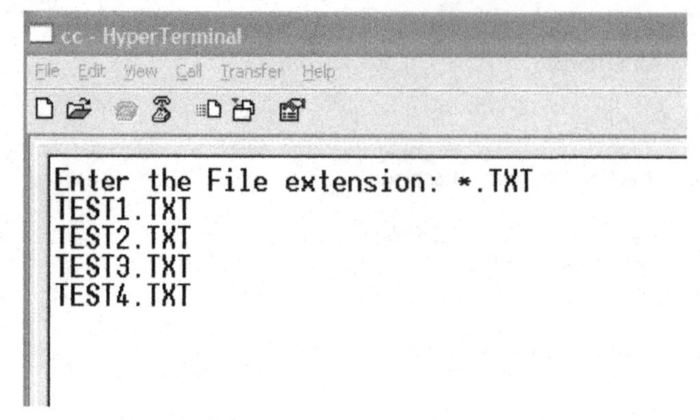

Figure 9.53: Example PC Display

TEST3.TXT

TEST4.TXT

TEST5.DAT

TEST6.DAT

Only the files with extensions "TXT" are displayed.

Note that the program given in this project can be used to look for other types of files. Some examples are given below:

.	All files in the current directory
*.ext	Files with extension "ext"
File.*	Files with name "File"
*	Any directory

9.13.8 Suggestions for Future Work

Write a program to look for a file in the current directory and in all the subdirectories of the SD card.

9.14 PROJECT 13 – Displaying the Attributes of a File

9.14.1 Description

In this project, a file name is read from the PC keyboard and then the specified file is searched on the SD card and the attributes of the file are displayed on the PC screen. If the file does not exist, then the message "NO SUCH FILE" is displayed.

9.14.2 Aim

The aim of this project is to show how the attributes of a given file can be determined.

9.14.3 Block Diagram

The block diagram of this project is as in Figure 9.37.

9.14.4 Circuit Diagram

The circuit diagram of this project is as in Figure 9.38.

9.14.5 Operation of the Project

The operation of the project is shown in Figure 9.54.

9.14.6 Program Code

The program code (ATTR.C) is shown in Figure 9.55.

9.14.7 Description of the Program Code

At the beginning of the program, the variable **Attribute** is set to ATTR_MASK so that the file attribute is not considered while finding the specified file. As in the previous project, the USART module is initialized by calling to C18 library function Open1USART (since there are two USART modules on the PIC18F8722 microcontroller, we have to specify which module we shall be using). Serial port interrupts are disabled, USART is set to asynchronous mode, the baud-rate clock is set to low speed, and the baud rate is set to 4800. The message "Enter the Filename:" is displayed on the PC screen and the user is requested to enter the file name. The file name is stored in a character array called FileName and is terminated with a NULL character.

```
BEGIN
    Initialize USART module
    Initialize the MDD library
    Read the filename
    Search for the file
    IF file not found
        Display NO SUCH FILE
    ELSE
        Display attributes of the file
    ENDIF
    Close USART
END
```

Figure 9.54: Operation of the Project

```
/*****************************************************************************************************************
                              DISPLAY ATTRIBUTES OF A FILE
                              ============================

In these projects, a PIC18F8722-type microcontroller is used. The microcontroller is operated
with a 10-MHz crystal.

An SD card is connected to the microcontroller as follows:

SD card                       microcontroller
CS                            RB3
CLK                           RC3
DO                            RC4
DI                            RC5

The program uses the Microchip MDD library functions to read and write to
the SD card.

In this project the microcontroller is connected to a PC via a MAX232-type level converter chip.
The USART 1 hardware module of the microcontroller is used for serial communication. The pin
configuration of the RS232 connector is as follows:

          Pin 2     TX
          Pin 3     RX
          Pin 3     GND

The communication is established using the C18 USART library functions. The
USART is configured as follows:

          4800 Baud
          8 data bits
          No parity
          1 stop bit

In this project a file name is read from the PC keyboard and then the attributes of this file are
found and displayed on the PC screen.


Author:   Dogan Ibrahim
Date:     August 2009
File:     ATTR.C
*****************************************************************************************************************/
#include <p18f8722.h>
#include <usart.h>
#include <stdlib.h>
#include <FSIO.h>

#pragma config WDT = OFF, OSC = HSPLL,LVP = OFF
#pragma config MCLRE = ON,CCP2MX = PORTC, MODE = MC
```

Figure 9.55: The Program Code

```
/* ================ START OF MAIN PROGRAM ================
*/
//
// Start of MAIN Program
//
void main(void)
{
    SearchRec File;
    char bufr[1];
    char FileName[12];
    char FileLen = 0, itm = 0;
    unsigned char Attribute = ATTR_MASK;
    unsigned char Attributes;
//
// Initialize the USART
//
    Open1USART(USART_TX_INT_OFF  &
               USART_RX_INT_OFF  &
               USART_ASYNCH_MODE &
               USART_EIGHT_BIT   &
               USART_CONT_RX     &
               USART_BRGH_LOW,
               129);
//
// Send a message to the PC
//
    while(Busy1USART());
    putrs1USART(" Enter the Filename: ");
//
// Read the filename (until the Enter key is pressed)
//
    while(itm != 0x0D)
    {
        while(!DataRdy1USART());
        itm = getc1USART();
        putc1USART(itm);
        FileName[FileLen] = itm;
        FileLen++;
    }
//
// Terminate the Filename with a NULL character
//
    FileLen--;
    FileName[FileLen] = '\0';
//
// Insert a new line
//
    putrs1USART("\n\r");
//
// Initialize MDD library
```

Figure 9.55: *Cont'd*

```
//
    while(!FSInit());
//
// Look for the files with given extension
//
    if(FindFirst(FileName, Attribute, &File) != 0)
        putrs1USART("NO SUCH FILES");
    else
    {
        Attributes = File.attributes;
        if(Attributes & 0x01)putrs1USART("Read-only file\n\r");
        if(Attributes & 0x02)putrs1USART("Hidden file\n\r");
        if(Attributes & 0x04)putrs1USART("System file\n\r");
        if(Attributes & 0x08)putrs1USART("Volume label\n\r");
        if(Attributes & 0x20)putrs1USART("Archived File\n\r");
    }
//
// Close USART
//
    Close1USART();

while(1);
}
```

Figure 9.55: *Cont'd*

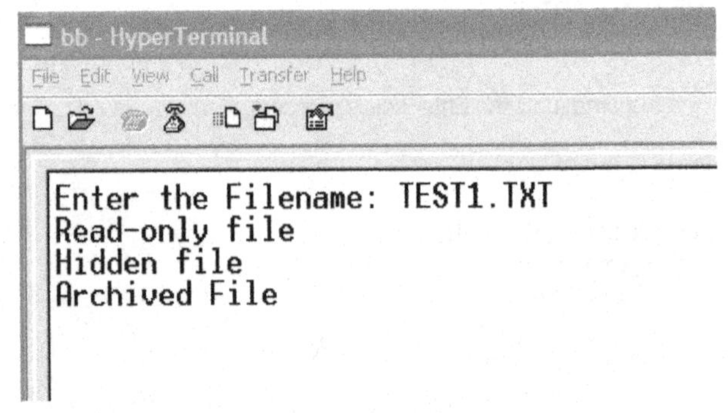

Figure 9.56: Example for PC Display

The MDD library function FindFirst is called to search for the specified file in the current default directory. If the file is not found, the function returns a nonzero value and the message "NO SUCH FILE" is displayed on the PC screen. If a file is found matching the specified extension name, then the attributes of this file is extracted from the **attributes** member of structure SearchRec and the attributes of the file are displayed on the PC screen.

Figure 9.56 shows typical output from the project. In this example, the file TEST1.TXT had read-only, archive, and hidden attributes set.

9.14.8 Suggestions for Future Work

Write a program to display a list of all the files in the current default working directory with their sizes and attributes.

9.15 PROJECT 14 – SD Card File Handling

9.15.1 Description

In this project, a program is developed to handle a number of SD card operations. The following operations are performed:

CD	Change directory
DEL	Delete a file
DIR	Directory listing
FORM	Format SD card
HELP	Help on these commands
MD	Create a directory
RD	Delete a directory
REN	Rename a file
TYPE	Display the contents of a file
COPY	Copy a file

The command mode is identified by character "$" where any of the above commands can be entered. After entering a command, the user should press the **Enter** key and then enter the appropriate command option. The following command options are available for each command:

CD
Enter directory name:

DEL
Enter filename:

DIR

FORM
Enter volume name:

HELP

MD
Enter directory name:

RD
Enter directory name:

REN
Source filename:
Destination filename:

TYPE
Enter filename:

COPY
Source filename:
Destination filename:

9.15.2 Aim

The aim of this project is to show how several file-handling operations can be performed in a program.

9.15.3 Block Diagram

The block diagram of this project is as in Figure 9.37.

9.15.4 Circuit Diagram

The circuit diagram of this project is as in Figure 9.38.

9.15.5 Operation of the Project

The operation of the project is shown in Figure 9.57.

9.15.6 Program Code

The program code (FILES.C) is shown in Figure 9.58.

9.15.7 Description of the Program Code

The following functions are used in the program:

Read_Filename: This function reads the filename, directory name, or the volume name from the keyboard. The argument specifies the text string to be displayed before accepting the user input. The function is terminated when the **Enter** key is pressed and a NULL character is added to the end of the name.

Read_SrcDst: This function reads the source filename and the destination filename in commands Copy and Rename.

As in the previous project, at the beginning of the program, the USART module is initialized by calling to C18 library function Open1USART (since there are two USART modules on the PIC18F8722 microcontroller, we have to specify which module we shall be using). Serial port interrupts are disabled, USART is set to asynchronous mode, the baud-rate clock is set to low speed, and the baud rate is set to 4800.

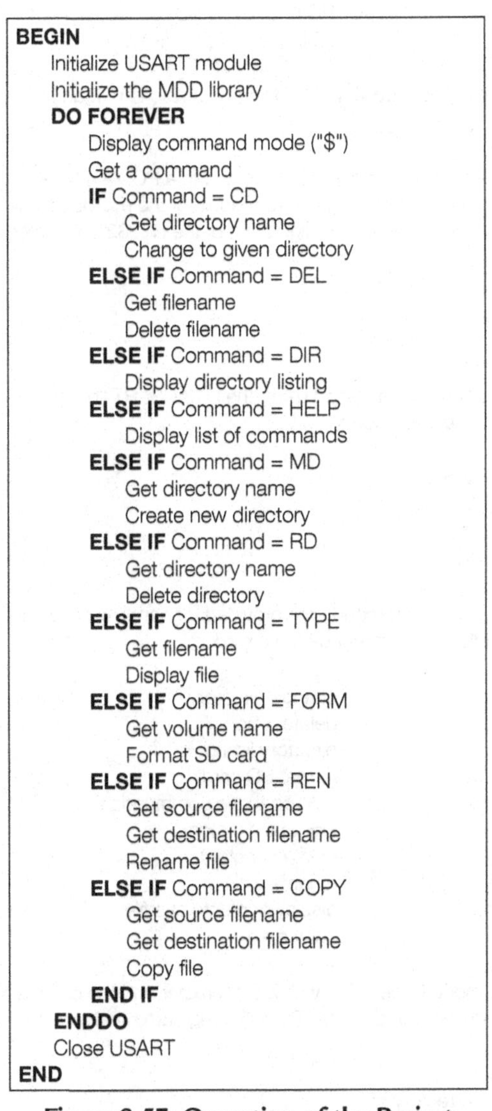

```
BEGIN
    Initialize USART module
    Initialize the MDD library
    DO FOREVER
        Display command mode ("$")
        Get a command
        IF Command = CD
            Get directory name
            Change to given directory
        ELSE IF Command = DEL
            Get filename
            Delete filename
        ELSE IF Command = DIR
            Display directory listing
        ELSE IF Command = HELP
            Display list of commands
        ELSE IF Command = MD
            Get directory name
            Create new directory
        ELSE IF Command = RD
            Get directory name
            Delete directory
        ELSE IF Command = TYPE
            Get filename
            Display file
        ELSE IF Command = FORM
            Get volume name
            Format SD card
        ELSE IF Command = REN
            Get source filename
            Get destination filename
            Rename file
        ELSE IF Command = COPY
            Get source filename
            Get destination filename
            Copy file
        END IF
    ENDDO
    Close USART
END
```

Figure 9.57: Operation of the Project

```
/**************************************************************************************************
                    SD CARD FILE HANDLING COMMANDS
                    ==============================

In these projects, a PIC18F8722-type microcontroller is used. The microcontroller
is operated with a 10-MHz crystal.

An SD card is connected to the microcontroller as follows:

SD card                     microcontroller
CS                          RB3
CLK                         RC3
DO                          RC4
DI                          RC5

The program uses the Microchip MDD library functions to read and write to the
SD card.

In this project the microcontroller is connected to a PC via a MAX232-type level
converter chip. The USART 1 hardware module of the microcontroller is used for
serial communication. The pin configuration of the RS232 connector is as follows:

        Pin 2     TX
        Pin 3     RX
        Pin 3     GND

The communication is established using the C18 USART library functions. The
USART is configured as follows:

        4800 Baud
        8 data bits
        No parity
        1 stop bit

In this project various commands are provided for handling file operations on an
SD card. The following commands are provided:

        CD                  change directory
        DEL                 delete a file
        DIR                 directory listing
        FORM                format SD card
        HELP                help on these commands
        MD                  create a directory
        RD                  delete directory
        REN                 rename a file
        TYPE                display contents of a file
        COPY                copy a file

The command mode is identified with a $ character and the command should be
entered after the command mode. The following command tails are available:

        CD
        Enter Directory name:
```

Figure 9.58: The Program Code

```
            DEL
            Enter Filename:

            DIR

            FORM
            Enter Volume name

            HELP

            MD
            Enter Directory name:

            RD
            Enter Directory name:

            REN
            Source Filename:
            Destination Filename:

            TYPE
            Enter Filename:

            COPY
            Source Filename:
            Destination Filename:


Author:     Dogan Ibrahim
Date:       August 2009
File:       FILES.C
****************************************************************************************/
#include <p18f8722.h>
#include <usart.h>
#include <stdlib.h>
#include <string.h>
#include <FSIO.h>

#pragma config WDT = OFF, OSC = HSPLL,LVP = OFF
#pragma config MCLRE = ON,CCP2MX = PORTC, MODE = MC

SearchRec File;
char bufr[10];
char FileName[13];
char SourceFileName[12];
char DestinationFileName[12];
char DirectoryName[12];
char FileLen = 0, itm;
unsigned char Attribute = ATTR_MASK;

//
// Read a filename
//
```

Figure 9.58: *Cont'd*

```
void Read_Filename(const rom char *msg)
{
    itm = 0;
    FileLen = 0;
    putrs1USART(msg);
    while(itm != 0x0D)
    {
        while(!DataRdy1USART());
        itm = getc1USART();
        putc1USART(itm);
        FileName[FileLen] = itm;
        FileLen++;
    }
//
// Terminate the Filename with a NULL character
//
    FileLen--;
    FileName[FileLen] = '\0';
}


//
// Read source filename and destination filename
//
void Read_SrcDst(void)
{
    char i = 0;
    Read_Filename("    Enter source filename: ");
    for(i=0; i<12; i++) SourceFileName[i] = FileName[i];
    Read_Filename("\n\rEnter destination filename: ");
    for(i=0; i<12; i++) DestinationFileName[i] = FileName[i];
}

/* ================== START OF MAIN PROGRAM ====================== */
//
// Start of MAIN Program
//
void main(void)
{
    FSFILE *pntrs, *pntrd;
    char Command[6];
    char CmdLen, Flag, i;
    char mode_read[2] = "r";
    char mode_write[3] = "w+";
    int ReadCnt;
//
// Initialize the USART
//
    Open1USART(USART_TX_INT_OFF  &
               USART_RX_INT_OFF  &
               USART_ASYNCH_MODE &
               USART_EIGHT_BIT   &
               USART_CONT_RX     &
```

Figure 9.58: *Cont'd*

```
                    USART_BRGH_LOW,
                    129);
//
// Display command prompt character $
//
    while(Busy1USART());
//
// ================== START OF LOOP ==================
//
while(1)
{
    putrs1USART("\n\r$ ");
//
// Read a command (until the Enter key is pressed)
//
    itm = 0;
    CmdLen = 0;
    while(itm != 0x0D)
    {
        while(!DataRdy1USART());
        itm = getc1USART();
        putc1USART(itm);
        Command[CmdLen] = itm;
        CmdLen++;
    }
//
// Terminate the Command with a NULL character
//
    CmdLen--;
    Command[CmdLen] = '\0';
//
// Insert a new line
//
    putrs1USART("\n\r");
//
// Initialize MDD library
//
    while(!FSInit());
//
// Determine what type of command it is and process the command
//
    if(strcmppgm2ram(Command, "HELP") == 0)
    {
        putrs1USART("CD\tchange directory\n\r");
        putrs1USART("DEL\tdelete a file\n\r");
        putrs1USART("DIR\tdirectory listing\n\r");
        putrs1USART("FORM\tformat SD card\n\r");
        putrs1USART("HELP\tthis display\n\r");
        putrs1USART("MD\tcreate directory\n\r");
        putrs1USART("RD\tdelete directory\n\r");
        putrs1USART("REN\trename a file\n\r");
        putrs1USART("TYPE\tdisplay a file\n\r");
        putrs1USART("COPY\tcopy a file\n\r");
    }
```

Figure 9.58: *Cont'd*

```
else if(strcmppgm2ram(Command, "CD") == 0)
{
    Read_Filename("Enter directory name: ");
    FSchdir(FileName);
}
else if(strcmppgm2ram(Command, "DEL") == 0)
{
    Read_Filename("Enter filename: ");
    FSremove(FileName);
}
else if(strcmppgm2ram(Command, "DIR") == 0)
{
    FileName[0] = '*';
    FileName[1] = '.';
    FileName[2] = '*';
    FileName[3] = '\0';
    FindFirst(FileName, Attribute, &File);
    puts1USART(File.filename);
    putrs1USART("\n\r");

    Flag = 0;
    while(Flag == 0)
    {
        if(FindNext(&File) == 0)
        {
            puts1USART(File.filename);
            putrs1USART("\n\r");
        }
        else Flag = 1;
    }
}
else if(strcmppgm2ram(Command, "FORM") == 0)
{
    Read_Filename("Enter Volume name: ");
    FSformat(1, 0x12345678, FileName);
}
else if(strcmppgm2ram(Command, "MD") == 0)
{
    Read_Filename("Enter directory name: ");
    FSmkdir(FileName);
}
else if(strcmppgm2ram(Command, "RD") == 0)
{
    Read_Filename("Enter directory name: ");
    FSrmdir(FileName, 1);
}
else if(strcmppgm2ram(Command, "REN") == 0)
{
    Read_SrcDst();
    pntrs = FSfopen(SourceFileName, mode_read);
    FSrename(DestinationFileName, pntrs);
    FSfclose(pntrs);
}
```

Figure 9.58: *Cont'd*

```
        else if(strcmppgm2ram(Command, "TYPE") == 0)
        {
            Read_Filename("Enter filename: ");
            pntrs = FSfopen(FileName, mode_read);

            ReadCnt = 1;
            while(ReadCnt != 0)
            {
                ReadCnt = FSfread(bufr, 1, 10, pntrs);
                for(i=0; i<ReadCnt; i++)
                {
                    while(Busy1USART());
                    putc1USART(bufr[i]);
                }
            }
            FSfclose(pntrs);
        }
        else if(strcmppgm2ram(Command, "COPY") == 0)
        {
            Read_SrcDst();
            pntrs = FSfopen(SourceFileName, mode_read);
            pntrd = FSfopen(DestinationFileName, mode_write);

            ReadCnt = 1;
            while(ReadCnt != 0)
            {
                ReadCnt = FSfread(bufr, 1, 10, pntrs);
                FSfwrite((void *)bufr, 1, ReadCnt, pntrd);
            }
            FSfclose(pntrs);
            FSfclose(pntrd);
        }
}

//
// Close USART
//
    Close1USART();

while(1);
}
```

Figure 9.58: *Cont'd*

When the microcontroller is reset, the command mode identifier “$” character is displayed and the user is expected to enter a command. The user command is stored in a character array **Command** and the requested command is determined and processed. The C18 string comparison function strcmppgm2ram is used to determine the command entered by the user. The commands are processed as follows:

HELP

A list of the available commands is displayed using the USART function putrs1USART. The command names and command descriptions are separated by a tab character "\t":

```
putrs1USART("CD\tchange directory\n\r");
putrs1USART("DEL\tdelete a file\n\r");
putrs1USART("DIR\tdirectory listing\n\r");
putrs1USART("FORM\tformat SD card\n\r");
putrs1USART("HELP\tthis display\n\r");
putrs1USART("MD\tcreate directory\n\r");
putrs1USART("RD\tdelete directory\n\r");
putrs1USART("REN\trename a file\n\r");
putrs1USART("TYPE\tdisplay a file\n\r");
putrs1USART("COPY\tcopy a file\n\r");
```

CD

The directory name to be changed is read from the keyboard and the MDD function FSchdir is used to change this directory:

```
Read_Filename("Enter directory name: ");
FSchdir(FileName);
```

DEL

The name of the file to be deleted is read from the keyboard and the MDD function FSremove is used to delete the file:

```
Read_Filename("Enter filename: ");
FSremove(FileName);
```

DIR

The filename is set to wild character string "*.*" and the MDD function FindFirst is called to find any file in the default directory. The name of the found file is then displayed. Then, the MDD function FindNext is used in a **while** loop to look for more files in the directory and the names of all found files are displayed:

```
FileName[0] = '*';
FileName[1] = '.';
FileName[2] = '*';
FileName[3] = '\0';
FindFirst(FileName, Attribute, &File);
puts1USART(File.filename);
putrs1USART("\n\r");

Flag = 0;
while(Flag == 0)
```

```
    {
        if(FindNext(&File) == 0)
        {
            puts1USART(File.filename);
            putrs1USART("\n\r");
        }
        else Flag = 1;
    }
```

FORM

The volume name to be given to the SD card is read from the keyboard and the MDD function FSformat is used to format the card. The card is given the serial number 0x12345678:

```
Read_Filename("Enter Volume name: ");
FSformat(1, 0x12345678, FileName);
```

MD

The name of the directory to be created is read from the keyboard and the MDD function FSmkdir is used to create a new directory:

```
Read_Filename("Enter directory name: ");
FSmkdir(FileName);
```

FS

The name of the directory to be deleted is read from the keyboard and the MDD function FSrmdir is used to delete the directory:

```
Read_Filename("Enter directory name: ");
FSrmdir(FileName, 1);
```

REN

The function Read_SrcDst is called to read the source and the destination filenames. Then, the source file is opened in read mode with pointer pntrs. The MDD function FSrename is called to change the name of the source file. The source file is closed after the rename operation:

```
Read_SrcDst();
pntrs = FSfopen(SourceFileName, mode_read);
FSrename(DestinationFileName, pntrs);
FSfclose(pntrs);
```

TYPE

The name of the file to be displayed is read from the keyboard and the MDD function FSfopen is used to open the file in read mode. The contents of the file are read into a buffer called **bufr** in a **while** loop using the function FSfread until the end of file is detected. The contents of this buffer are then sent to the PC screen using USART function putc1USART:

```
    Read_Filename("Enter filename: ");
    pntrs = FSfopen(FileName, mode_read);

    ReadCnt = 1;
    while(ReadCnt != 0)
    {
        ReadCnt = FSfread(bufr, 1, 10, pntrs);
        for(i=0; i<ReadCnt; i++)
        {
            while(Busy1USART());
            putc1USART(bufr[i]);
        }
    }
    FSfclose(pntrs);
}
```

COPY

The function Read_SrcDst is called to read the source and the destination filenames. The source file is opened in read mode and the destination file is created in write mode. Then, a **while** loop is formed and data is read from the source file and copied to the destination file until the end of file is reached. At the end of the copy operation, both the source file and the destination files are closed:

```
Read_SrcDst();
pntrs = FSfopen(SourceFileName, mode_read);
pntrd = FSfopen(DestinationFileName, mode_write);

ReadCnt = 1;
while(ReadCnt != 0)
{
    ReadCnt = FSfread(bufr, 1, 10, pntrs);
    FSfwrite((void *)bufr, 1, ReadCnt, pntrd);
}
FSfclose(pntrs);
FSfclose(pntrd);
```

Figures 9.59 and 9.60 show typical outputs from the project.

9.15.8 Suggestions for Future Work

Write a file-handling program similar to the program given in Figure 9.58 but include appropriate error messages in your program to inform the user of any error conditions.

```
cc - HyperTerminal
File Edit View Call Transfer Help

$ HELP
CD      change directory
DEL     delete a file
DIR     directory listing
FORM    format SD card
HELP    this display
MD      create directory
RD      delete directory
REN     rename a file
TYPE    display a file
COPY    copy a file

$ DIR
SD
SEN.HEX
MONDAY.TXT

$
```

Figure 9.59: Example for PC Display

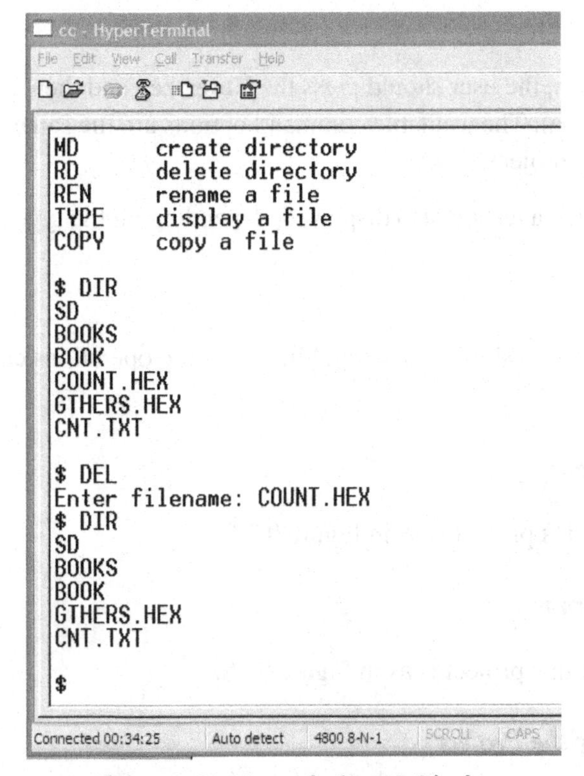
cc - HyperTerminal
File Edit View Call Transfer Help
MD create directory
RD delete directory
REN rename a file
TYPE display a file
COPY copy a file
$ DIR
SD
BOOKS
BOOK
COUNT.HEX
GTHERS.HEX
CNT.TXT
$ DEL
Enter filename: COUNT.HEX
$ DIR
SD
BOOKS
BOOK
GTHERS.HEX
CNT.TXT
$
Connected 00:34:25
Auto detect
4800 8-N-1
SCROLL
CAPS

Figure 9.60: Example for PC Display

9.16 PROJECT 15 – MENU-Based SD Card File Handling

9.16.1 Description

This project is similar to Project 14, but here the SD card file handling is MENU based. When the program is activated, the following MENU is displayed on the PC screen:

MENU
1. Change directory
2. Delete a file
3. Directory listing
4. Format SD card
5. Help
6. Create a directory
7. Delete a directory
8. Rename a file
9. Display a file
10. Copy a file

Option:

After making a selection, the user should press the **Enter** key and then enter the command option (e.g., the filename). The available command options are the same as the command options in the previous project.

In this project, error messages are also displayed, depending on the type of error.

9.16.2 Aim

The aim of this project is to show how several file-handling operations can be performed in a program.

9.16.3 Block Diagram

The block diagram of this project is as in Figure 9.37.

9.16.4 Circuit Diagram

The circuit diagram of this project is as in Figure 9.38.

9.16.5 Operation of the Project

The operation of the project is shown in Figure 9.61.

```
BEGIN
    Initialize USART module
    Initialize the MDD library
    DO FOREVER
        Display MENU
        Get an option
        IF Command = CD
            Get directory name
            Change to given directory
        ELSE IF Command = DEL
            Get filename
            Delete filename
        ELSE IF Command = DIR
            Display directory listing
        ELSE IF Command = HELP
            Display list of commands
        ELSE IF Command = MD
            Get directory name
            Create new directory
        ELSE IF Command = RD
            Get directory name
            Delete directory
        ELSE IF Command = TYPE
            Get filename
            Display file
        ELSE IF Command = FORM
            Get volume name
            Format SD card
        ELSE IF Command = REN
            Get source filename
            Get destination filename
            Rename file
        ELSE IF Command = COPY
            Get source filename
            Get destination filename
            Copy file
        END IF
    ENDDO
    Close USART
END
```

Figure 9.61: Operation of the Project

9.16.6 Program Code

The program code (MENU.C) is shown in Figure 9.62.

9.16.7 Description of the Program Code

As in the previous project, at the beginning of the program, the USART module is initialized by calling to C18 library function Open1USART (since there are two USART modules on the PIC18F8722 microcontroller, we have to specify which module we shall be using). Serial port

```
/****************************************************************************************************
                MENU-BASED SD CARD FILE-HANDLING COMMANDS
                =========================================

In these projects, a PIC18F8722-type microcontroller is used. The microcontroller is
operated with a 10-MHz crystal.

An SD card is connected to the microcontroller as follows:

SD card                 microcontroller
CS                      RB3
CLK                     RC3
DO                      RC4
DI                      RC5

The program uses the Microchip MDD library functions to read and write to the SD card.

In this project the microcontroller is connected to a PC via a MAX232-type level converter
chip. The USART 1 hardware module of the microcontroller is used for serial communication.
The pin configuration of the RS232 connector is as follows:

        Pin 2   TX
        Pin 3   RX
        Pin 3   GND

The communication is established using the C18 USART library functions. The USART is
configured as follows:

        4800 Baud
        8 data bits
        No parity
        1 stop bit

In this project a MENU system is designed to handle the various file operations. The
MENU consists of following options:

        MENU
        ====

 1. Change directory
 2. Delete a file
 3. Directory listing
 4. Format SD card
 5. Help
 6. Create a directory
 7. Delete a directory
 8. Rename a file
 9. Display a file
10. Copy a file

Option:
```

Figure 9.62: The Program Code

```
Author:    Dogan Ibrahim
Date:      August 2009
File:      MENU.C
****************************************************************************************************************/
#include <p18f8722.h>
#include <usart.h>
#include <stdlib.h>
#include <string.h>
#include <FSIO.h>

#pragma config WDT = OFF, OSC = HSPLL,LVP = OFF
#pragma config MCLRE = ON,CCP2MX = PORTC, MODE = MC

SearchRec File;
char bufr[10];
char FileName[13];
char SourceFileName[12];
char DestinationFileName[12];
char DirectoryName[12];
char FileLen = 0, itm;
unsigned char Attribute = ATTR_MASK;

//
// Read a filename
//
void Read_Filename(const rom char *msg)
{
    itm = 0;
    FileLen = 0;
    putrs1USART(msg);
    while(itm != 0x0D)
    {
        while(!DataRdy1USART());
        itm = getc1USART();
        putc1USART(itm);
        FileName[FileLen] = itm;
        FileLen++;
    }
//
// Terminate the Filename with a NULL character
//
    FileLen--;
    FileName[FileLen] = '\0';
}


//
// Read source filename and destination filename
//
void Read_SrcDst(void)
{
    char i = 0;
    Read_Filename("    Enter source filename: ");
    for(i=0; i<12; i++) SourceFileName[i] = FileName[i];
```

Figure 9.62: *Cont'd*

```
    Read_Filename("\n\rEnter destination filename: ");
    for(i=0; i<12; i++) DestinationFileName[i] = FileName[i];
}

/* ==================== START OF MAIN PROGRAM ================== */
//
// Start of MAIN Program
//
void main(void)
{
    FSFILE *pntrs, *pntrd;
    char Command[6];
    char CmdLen, Flag, i;
    char mode_read[2] = "r";
    char mode_write[3] = "w+";
    int ReadCnt, Cmd;
//
// Initialize the USART
//
    Open1USART(USART_TX_INT_OFF  &
               USART_RX_INT_OFF  &
               USART_ASYNCH_MODE &
               USART_EIGHT_BIT   &
               USART_CONT_RX     &
               USART_BRGH_LOW,
               129);
//
// Display command prompt character $
//
    while(Busy1USART());
//
// ====================== START OF LOOP =======================
//
while(1)
{
    putrs1USART("\n\rOption: ");

//
// Read an Option (until the Enter key is pressed)
//
    itm = 0;
    CmdLen = 0;
    while(itm != 0x0D)
    {
        while(!DataRdy1USART());
        itm = getc1USART();
        putc1USART(itm);
        Command[CmdLen] = itm;
        CmdLen++;
    }
//
// Terminate the Command with a NULL character
//
```

Figure 9.62: *Cont'd*

```
    CmdLen--;
    Command[CmdLen] = '\0';
//
// Insert a new line
//
    putrs1USART("\n\r");
//
// Initialize MDD library
//
    while(!FSInit());
//
// Determine what type of command it is and process the command
//
    Cmd = atoi(Command);                                    // convert to integer
//
// Process the command
//
    switch(Cmd)
    {
        case 1:
            Read_Filename("Enter directory name: ");
            if(FSchdir(FileName) != 0)
                putrs1USART("Error to change directory...\n\r");
            break;
        case 2:
            Read_Filename("Enter filename: ");
            if(FSremove(FileName) != 0)
                putrs1USART("Error to delete the file...\n\r");
            break;
        case 3:
            FileName[0] = '*';
            FileName[1] = '.';
            FileName[2] = '*';
            FileName[3] = '\0';
            FindFirst(FileName, Attribute, &File);
            puts1USART(File.filename);
            putrs1USART("\n\r");

            Flag = 0;
            while(Flag == 0)
            {
                if(FindNext(&File) == 0)
                {
                    puts1USART(File.filename);
                    putrs1USART("\n\r");
                }
                else Flag = 1;
            }
            break;
        case 4:
            Read_Filename("Enter Volume name: ");
            if(FSformat(1, 0x12345678, FileName) != 0)
                putrs1USART("Error to format...\n\r");
            break;
```

Figure 9.62: *Cont'd*

```
case 5:
    putrs1USART("\n\r       MENU\n\r");
    putrs1USART(   "        ====\n\r");
    putrs1USART("1. Change directory\n\r");
    putrs1USART("2. Delete a file\n\r");
    putrs1USART("3. Directory listing\n\r");
    putrs1USART("4. Format SD card\n\r");
    putrs1USART("5. Help\n\r");
    putrs1USART("6. Create a directory\n\r");
    putrs1USART("7. Delete a directory\n\r");
    putrs1USART("8. Rename a file\n\r");
    putrs1USART("9. Display a file\n\r");
    putrs1USART("10.Copy a file\n\r");
    break;
case 6:
    Read_Filename("Enter directory name: ");
    if(FSmkdir(FileName) != 0)
        putrs1USART("Error to create directory...\n\r");
    break;
case 7:
    Read_Filename("Enter directory name: ");
    if(FSrmdir(FileName, 1) != 0)
        putrs1USART("Error to delete directory...\n\r");
    break;
case 8:
    Read_SrcDst();
    pntrs = FSfopen(SourceFileName, mode_read);
    if(FSrename(DestinationFileName, pntrs) != 0)
        putrs1USART("Error to rename file...\n\r");
    FSfclose(pntrs);
    break;
case 9:
    Read_Filename("Enter filename: ");
    pntrs = FSfopen(FileName, mode_read);

    ReadCnt = 1;
    while(ReadCnt != 0)
    {
        ReadCnt = FSfread(bufr, 1, 10, pntrs);
        for(i=0; i<ReadCnt; i++)
        {
            while(Busy1USART());
            putc1USART(bufr[i]);
        }
    }
    FSfclose(pntrs);
    break;
case 10:
    Read_SrcDst();
    pntrs = FSfopen(SourceFileName, mode_read);
    pntrd = FSfopen(DestinationFileName, mode_write);

    ReadCnt = 1;
    while(ReadCnt != 0)
```

Figure 9.62: *Cont'd*

```
                {
                    ReadCnt = FSfread(bufr, 1, 10, pntrs);
                    FSfwrite((void *)bufr, 1, ReadCnt, pntrd);
                }
                FSfclose(pntrs);
                FSfclose(pntrd);
                break;
            default:
                putrs1USART("Wrong choice...\n\r");
        }
}

//
// Close USART
//
    Close1USART();

while(1);
}
```

Figure 9.62: *Cont'd*

interrupts are disabled, USART is set to asynchronous mode, the baud-rate clock is set to low speed, and the baud rate is set to 4800.

At the beginning of the program, string "Option:" is displayed and the user is expected to choose an option between 1 and 10. The received user option is saved in a character array **Command**. This option is then converted into an integer number called **Cmd** using built-in MPLAB C18 function **atoi** and a switch statement is used to process the user commands as described below:

Case 1:

The new directory name is read from the keyboard and the function FSchdir is used to change to the new directory. The error message "Error to change directory…" is displayed if an error is detected:

```
Read_Filename("Enter directory name: ");
if(FSchdir(FileName) != 0)
    putrs1USART("Error to change directory...\n\r");
break;
```

Case 2:

The name of the file to be deleted is read from the keyboard and the function FSremove is used to delete the file. The error message "Error to delete the file…" is displayed if an error is detected:

```
Read_Filename("Enter filename: ");
if(FSremove(FileName) != 0)
    putrs1USART("Error to delete the file...\n\r");
break;
```

Case 3:

The filename is set to "*.*" and names of all the files (including directory names) in the current working directory are displayed. The function FindFirst is used to find the first file in the directory and the function FindNext is used to find other files in the directory:

```
FileName[0] = '*';
FileName[1] = '.';
FileName[2] = '*';
FileName[3] = '\0';
FindFirst(FileName, Attribute, &File);
puts1USART(File.filename);
putrs1USART("\n\r");

Flag = 0;
while(Flag == 0)
{
    if(FindNext(&File) == 0)
    {
        puts1USART(File.filename);
        putrs1USART("\n\r");
    }
    else Flag = 1;
}
break;
```

Case 4:

This option reads the required volume name from the keyboard and formats the SD card by giving the card this volume name. The error message "Error to format..." is displayed if an error is detected:

```
Read_Filename("Enter Volume name: ");
if(FSformat(1, 0x12345678, FileName) != 0)
        putrs1USART("Error to format...\n\r");
break;
```

Case 5:

This option displays the MENU items:

```
putrs1USART("\n\r       MENU\n\r");
putrs1USART(   "       ====\n\r");
putrs1USART("1. Change directory\n\r");
putrs1USART("2. Delete a file\n\r");
putrs1USART("3. Directory listing\n\r");
putrs1USART("4. Format SD card\n\r");
putrs1USART("5. Help\n\r");
putrs1USART("6. Create a directory\n\r");
putrs1USART("7. Delete a directory\n\r");
putrs1USART("8. Rename a file\n\r");
```

```
putrs1USART("9. Display a file\n\r");
putrs1USART("10.Copy a file\n\r");
break;
```

Case 6:

This function reads a directory name and creates a directory with this name using the function FSmkdir. The error message "Error to create directory…" is displayed if an error is detected:

```
Read_Filename("Enter directory name: ");
if(FSmkdir(FileName) != 0)
    putrs1USART("Error to create directory...\n\r");
break;
```

Case 7:

This option reads a directory name and then this directory is deleted from the current working directory using the function FSrmdir. The error message "Error to delete directory…" is displayed if an error is detected:

```
Read_Filename("Enter directory name: ");
if(FSrmdir(FileName, 1) != 0)
    putrs1USART("Error to delete directory...\n\r");
break;
```

Case 8:

This option changes the name of a file. The user function Read_SrcDst is called to read the destination and the source filenames, and then the MDD function FSrename is used to rename the file. Note that the source file must be opened with read attribute and its file pointer must be used in the FSrename function. The error message "Error to rename file…" is displayed if an error is detected:

```
Read_SrcDst();
pntrs = FSfopen(SourceFileName, mode_read);
if(FSrename(DestinationFileName, pntrs) != 0)
    putrs1USART("Error to rename file...\n\r");
FSfclose(pntrs);
break;
```

Case 9:

This option displays a file on the PC screen. The filename is read from the keyboard and the file is opened in read mode. Then, a **while** loop is formed to read and display the contents of the file until the end of file is reached:

```
Read_Filename("Enter filename: ");
pntrs = FSfopen(FileName, mode_read);

ReadCnt = 1;
```

```
while(ReadCnt != 0)
{
    ReadCnt = FSfread(bufr, 1, 10, pntrs);
    for(i=0; i<ReadCnt; i++)
    {
        while(Busy1USART());
        putc1USART(bufr[i]);
    }
}
FSfclose(pntrs);
break;
```

Case 10:

This option copies one file to another file. The user function Read_SrcDst is called to read the source and the destination filenames. Then the source file is opened in read mode and the destination file is opened in write mode. A **while** loop is formed, and the data is read from the source file and written into the destination file until the end of the source file is reached. At the end, both the source file and the destination file are closed:

```
Read_SrcDst();
pntrs = FSfopen(SourceFileName, mode_read);
pntrd = FSfopen(DestinationFileName, mode_write);

ReadCnt = 1;
while(ReadCnt != 0)
{
    ReadCnt = FSfread(bufr, 1, 10, pntrs);
    FSfwrite((void *)bufr, 1, ReadCnt, pntrd);
}
FSfclose(pntrs);
FSfclose(pntrd);
break;
```

Default:

This option of the switch statement is executed if the user option is not between 1 and 10:

```
putrs1USART("Wrong choice...\n\r");
```

Figures 9.63 and 9.64 show typical outputs from the project.

In the program code in Figure 9.62, the error messages are displayed by checking the return values of the MDD functions. The MDD library provides a function called FSerror that can be used to find the exact cause of a problem when an error occurs in a previous MDD function call. FSerror returns an integer where the actual error messages can be found in the MDD documentation (e.g., the *Release Notes for Microchip Memory Disk Drive File System, Version 1.2.0, August 20, 2008*).

An example for the use of function FSerror is shown in Figure 9.65, where the MDD function FSremove is used to delete a file from the current working directory. For example,

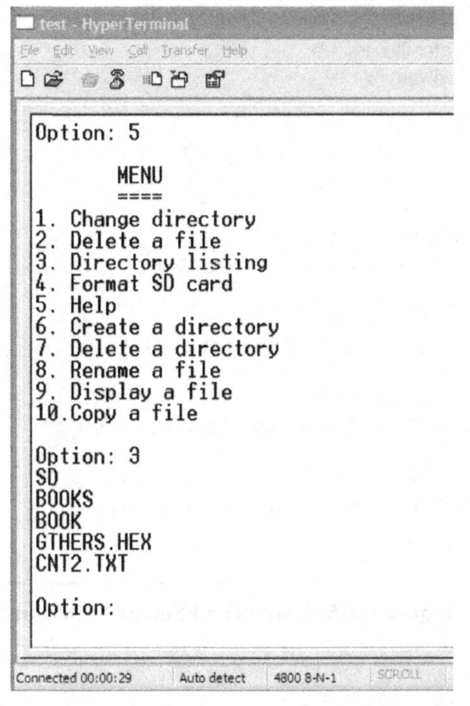

Figure 9.63: Typical Output from the Project

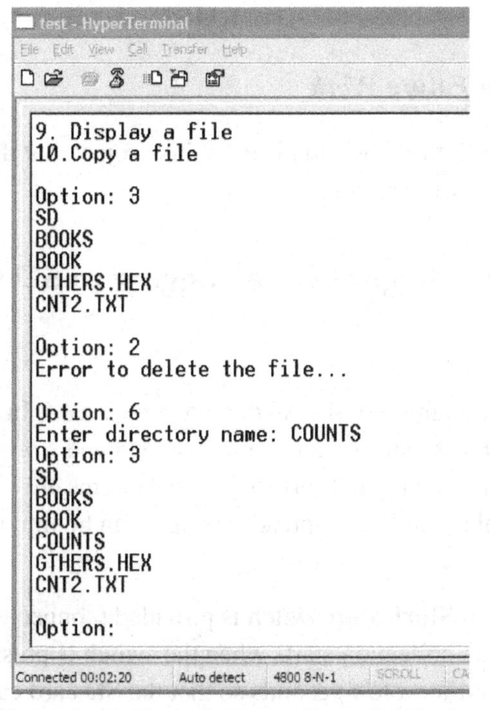

Figure 9.64: Typical Output from the Project

```
case 2:
    Read_Filename("Enter filename: ");
    Ret = FSremove(FileName);
    if(Ret != 0)
        Error = FSerror();

    switch(Error)
    {
        case CE_WRITE_PROTECTED:
            putrs1USART("Devide write protected...\n\r");
            break;
        case CE_FILE_NOT_FOUND:
            putrs1USART("The specified file is not found...\n\r");
            break;
        case CE_INVALID_FILENAME:
            putrs1USART("Specified filename was invalid...\n\r");
            break;
        case CE_ERASE_FAIL:
            putrs1USART("The file could not be erased...\n\r");
            break;
    }
```

Figure 9.65: Using the FSerror Function

if the file to be deleted is not found in the directory, the error message "The specified file could not be found" will be displayed (see the MDD documentation for a full list of error messages).

9.16.8 Suggestions for Future Work

Write a program, similar to the program given in this project, but include all the FSerror-based error messages in your program.

9.17 PROJECT 16 – Digital Data Logging to SD Card

9.17.1 Description

This project is about data logging of digital data to the SD card. In this project, microcontroller's PORTE and PORTF data are collected every second and stored on the SD card in a file called DIGITAL.TXT in hexadecimal format. The data items are separated with a comma so that the data is compatible with Excel spreadsheet and can be imported into Excel, if required for further processing.

An active low push-button **Start/Stop** switch is provided (connected to RB0) to control operation of the program. Data collection starts when the switch is pressed. The program stops when the switch is pressed for a few seconds so that the SD card can be removed safely.

9.17.2 Aim

The aim of this project is to show how digital data can be collected and stored on the SD card every second.

9.17.3 Block Diagram

The block diagram of this project is shown in Figure 9.66.

9.17.4 Circuit Diagram

The circuit diagram of this project is shown in Figure 9.67.

9.17.5 Operation of the Project

The operation of the project is shown in Figure 9.68.

9.17.6 Program Code

The program code (DIGLOG.C) is shown in Figure 9.69.

9.17.7 Description of the Program Code

At the beginning of the program, the port directions are configured and the program waits until the **Start/Stop** button is pressed. After the button is pressed and released, the MDD library is initialized and a new file called DIGITAL.TXT is created on the SD card. The main part of the program is in a **while** loop. Inside this loop, data from PORTE and PORTF are read and stored in the file using the MDD function FSfprint. Data is written to the file in the hexadecimal format and is separated with a comma. The program then checks whether or not the **Start/Stop** button is pressed to stop the program, and if so, the file is closed and the program stops waiting in a **while** loop forever.

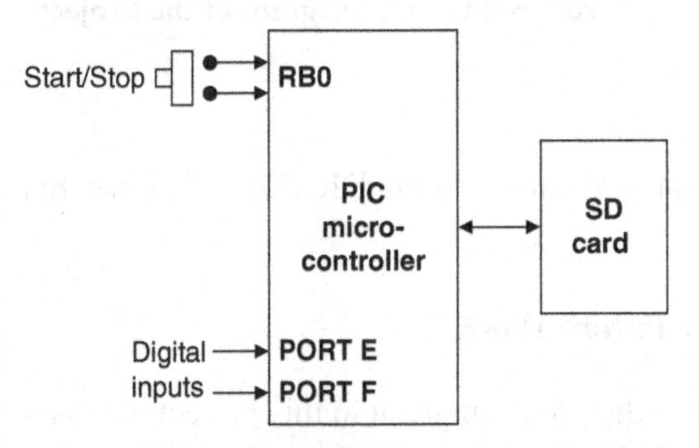

Figure 9.66: Block Diagram of the Project

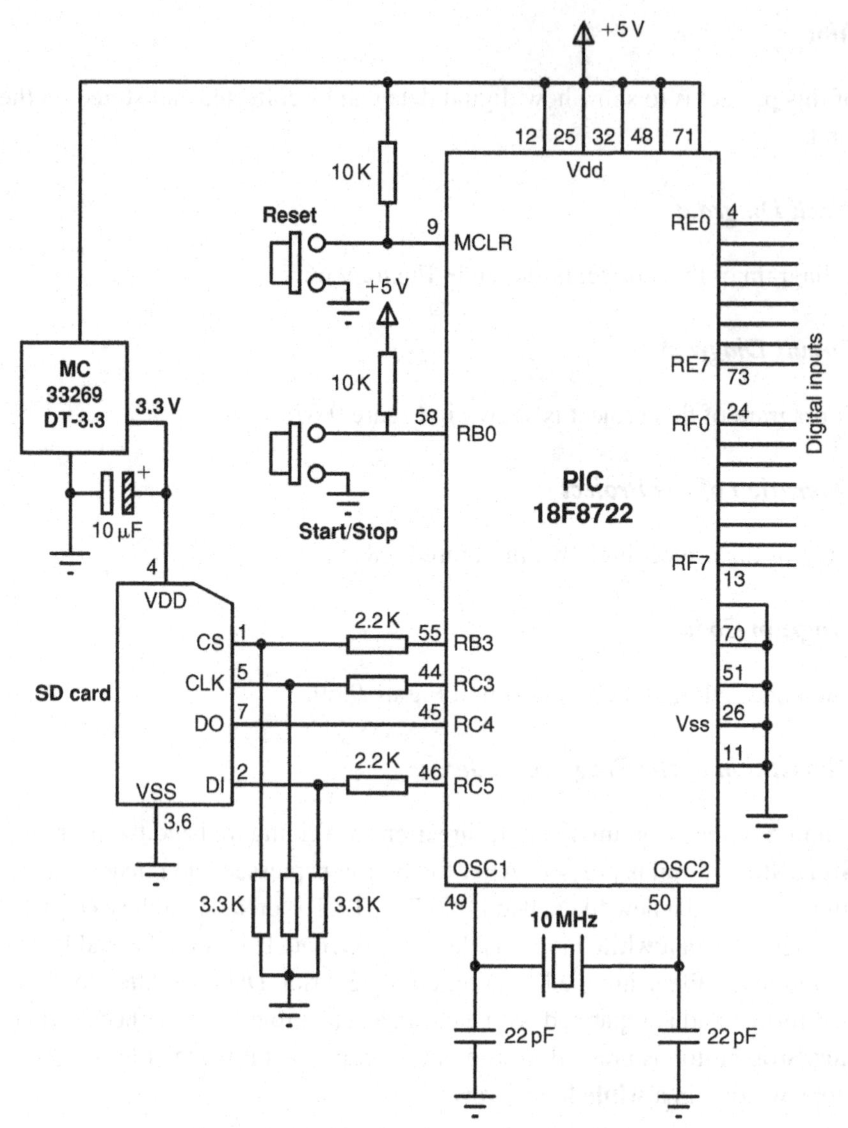

Figure 9.67: Circuit Diagram of the Project

Figure 9.70 shows the typical contents of file DIGITAL.TXT where data was collected for a short period of time.

9.17.8 Suggestions for Future Work

Write a program similar to the program given in this project, but save the data from all the free ports of the PIC18F8722 microcontroller and also save the data every minute.

```
BEGIN
    Initialise the MDD library
    Open new file on SD card
    Wait until START/STOP button is pressed
    DO FOREVER
        Read PORT B and PORT D data
        Store data in file (separated with a comma)
        IF Start/Stop button is pressed
            Close file
            Wait here forever
        ENDIF
        Wait a second
    ENDDO
END
```

Figure 9.68: Operation of the Project

```
/****************************************************************************************************
                    DIGITAL DATA LOGGING USING SD CARD
                    ==================================

In these projects, a PIC18F8722-type microcontroller is used. The microcontroller
is operated with a 10-MHz crystal.

An SD card is connected to the microcontroller as follows:

SD card                     microcontroller
CS                          RB3
CLK                         RC3
DO                          RC4
DI                          RC5

The program uses the Microchip MDD library functions to read and write to the SD card.

This is a digital data logging project where digital data from ports PORT E and PORT F
are read and stored on the SD card in a file called DIGITAL.TXT every second. The data
items are separated with a comms so that the data is compatible with most spreadsheet
programs.


Author:   Dogan Ibrahim
Date:     August 2009
File:     DIGLOG.C
****************************************************************************************************/
#include <p18f8722.h>
#include <stdlib.h>
#include <string.h>
#include <delays.h>
#include <FSIO.h>

#pragma config WDT = OFF, OSC = HSPLL,LVP = OFF
#pragma config MCLRE = ON,CCP2MX = PORTC, MODE = MC

#define STRT PORTBbits.RB0
```

Figure 9.69: The Program Code

```
void main(void)
{
    FSFILE *pntr;
    int PortEData, PortFData,i;
//
// Configure PORTS E and PORT F to digital inputs
//
    TRISE = 0xFF;
    TRISF = 0xFF;
    TRISB = 1;
    ADCON1 = 0xFF;
//
// Wait until START BUTTON is pressed and released
//
    while(STRT);
    while(!STRT);
//
// Initialize MDD library
//
    while(!FSInit());
//
// Open file DIGITAL.TXT in write mode (create it)
//
    pntr = FSfopenpgm("DIGITAL.TXT", "w+");
//
// ====================== START OF LOOP ======================
//
// Read PORT E and PORT F data and store in file
//
while(1)
{
    PortEData = PORTE;
    PortFData = PORTF;
//
// Write to SD card
//
    FSfprintf(pntr, "%02X, %02X\n", PortEData, PortFData);
//
// Check if BUTTON is pressed and if so, close the file and stop the program
//
    if(STRT == 0)
    {
        FSfclose(pntr);
        while(1);
    }
//
// Wait for a second. Clock freq = 40MHz, clock cycle = 0.1 microsecond
// The basic delay with Delay10KTCYx is 1ms. With a count of 1000 // the delay is 1 second.
A loop is formed to iterate 4 times // for the required delay of 1 second
//
    for(i=0; i< 4; i++)Delay10KTCYx(250);
}

}
```

Figure 9.69: *Cont'd*

```
20, 4F
20, 4E
20, 4E
23, 40
20, 4C
22, 4D
21, 42
23, 4A
0E, 44
0F, 44
1F, 42
```

Figure 9.70: Typical DIGITAL.TXT File

9.18 PROJECT 17 – Temperature Data Logging

9.18.1 Description

This project is about data logging of the ambient temperature data on the SD card every second. An analog temperature sensor IC (MCP9701A) is used to sense the ambient temperature and is connected to port RA1 of the PIC18F8722 microcontroller. A file called TEMP.TXT is created on the SD card, and the temperature is stored in this file every second, one data item on each line, with each data item followed by a new-line character.

An active low push-button **Start/Stop** switch is provided (connected to RB0) to control operation of the program. Data collection starts when the switch is pressed. The program stops when the switch is pressed for a few seconds so that the SD card can be removed safely.

9.18.2 Aim

The aim of this project is to show how analog data can be collected and stored on the SD card every second.

9.18.3 Block Diagram

The block diagram of this project is shown in Figure 9.71.

9.18.4 Circuit Diagram

The circuit diagram of this project is shown in Figure 9.72. The PICDEM PIC18 Explorer board is equipped with an MCP9701A temperature sensor IC connected to port RA1 of the microcontroller. Thus, the board can be used with this project without any hardware changes.

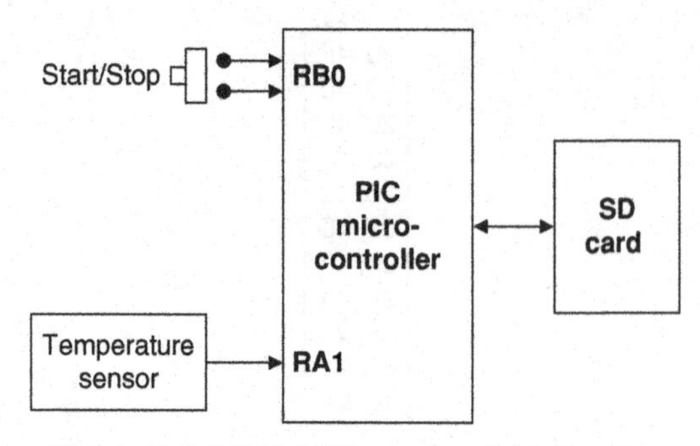

Figure 9.71: Block Diagram of the Project

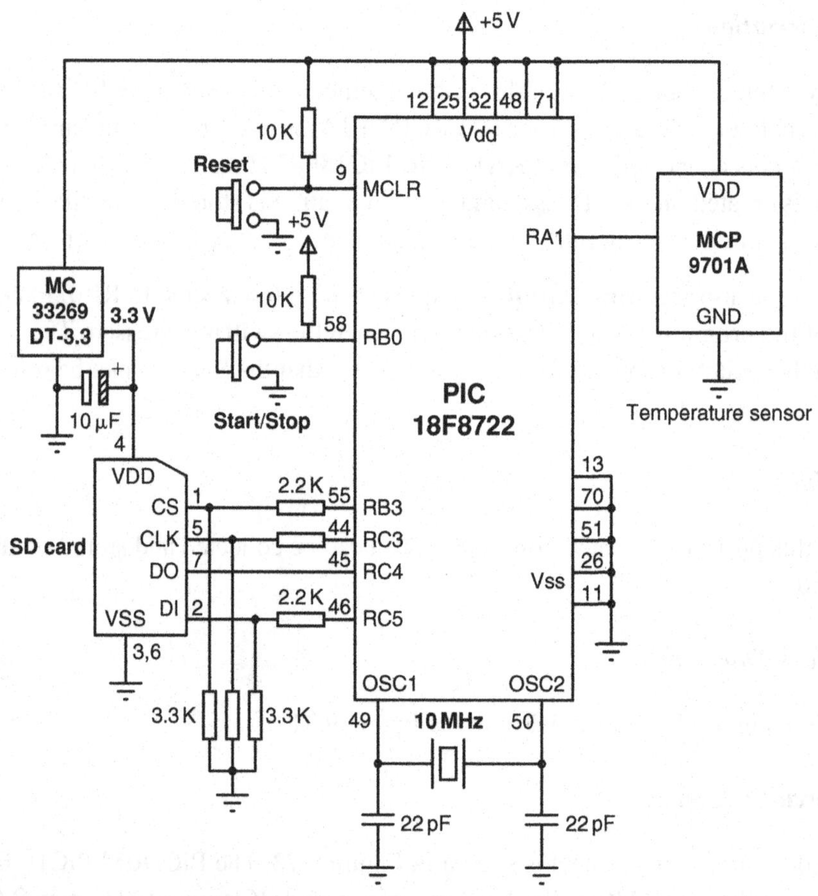

Figure 9.72: Circuit Diagram of the Project

9.18.5 Operation of the Project

The operation of the project is shown in Figure 9.73.

9.18.6 Program Code

The program code (ANALOG.C) is shown in Figure 9.74.

9.18.7 Description of the Program Code

Before looking at the code in detail, it is worthwhile to see how the MPC9701A temperature sensor IC operates.

MCP9701A is a linear active analog thermistor IC that senses and converts ambient temperature into analog voltage. The IC is a low-power, low-cost sensor with an accuracy of ±1°C from 0°C to +70°C and operating with a current as low as 6 μA. Unlike resistive thermistors, this sensor does not require an additional signal-conditioning circuit. The voltage output pin of the device can be directly connected to the analog input of the microcontroller. An advantage of this sensor is that it can drive large capacitive loads and thus can be located away from the microcontroller.

MCP9701A has three pins:

- Pin 1 is the supply voltage (+5 V)
- Pin 2 is the output voltage
- Pin 3 is the ground

```
BEGIN
   Initialise the MDD library
   Open new file on SD card
   Wait until START/STOP button is pressed
   DO FOREVER
      Read temperature from analog port
      Format to find temperature in Celsius
      Store temperature in file
      IF Start/Stop button is pressed
         Close file
         Wait here forever
      ENDIF
      Wait a second
   ENDDO
END
```

Figure 9.73: Operation of the Project

```
/*****************************************************************************************************************
                          ANALOG TEMPERATURE DATA LOGGING
                          ================================

In these projects, a PIC18F8722-type microcontroller is used. The microcontroller
is operated with a 10-MHz crystal.

An SD card is connected to the microcontroller as follows:

SD card                    microcontroller
CS                         RB3
CLK                        RC3
DO                         RC4
DI                         RC5

The program uses the Microchip MDD library functions to read and write to
the SD card.
This is an analog temperature data logger project. The temperature is read
every second and stored in a filed called ANALOG.TXT on an SD card.

The temperature is sensed using the analog sensor chip MCP9701A. This chip
is a 3-pin device where pin 1 the supply (+5V), pin 2 is the Vout and pin 3 is the ground.

MCP9701A is connected to analog port RA1 of the PIC18F877 microcontroller.

The output voltage of the sensor is proportional to the temperature and the
temperature is given by:
                                C = (Vout – 400) / 19.5

where, Vout is the sensor output voltage in mV

Data logging is started when button START/STOP (connected to RB0) is pressed.
During the data collection, pressing this button for a few seconds stops the data collection.
The SD card should only be removed after the data collection is stopped


Author:   Dogan Ibrahim
Date:     August 2009
File:     ANALOG.C
*****************************************************************************************************************/
#include <p18f8722.h>
#include <stdlib.h>
#include <stdio.h>
#include <string.h>
#include <delays.h>
#include <adc.h>
#include <FSIO.h>

#pragma config WDT = OFF, OSC = HSPLL, LVP = OFF
#pragma config MCLRE = ON,MODE = MC, CCP2MX = PORTC

#define STRT PORTBbits.RB0

//
// This function opens the A/D converter, reads analog data, converts into
// digital and then returns. The required channel number is passed as an
// argument to the function
//
int Read_ADC_Chan(unsigned char chan)
{
```

Figure 9.74: Program Code

```
    int Res;
    OpenADC(ADC_FOSC_64             &           // Open the A/D
            ADC_RIGHT_JUST          &
            ADC_0_TAD,
            ADC_CH1                 &
            ADC_INT_OFF             &
            ADC_VREFPLUS_VDD &
            ADC_VREFMINUS_VSS,
            13);
    Delay10TCYx(10);                    // Delay for 100 cycles
    SetChanADC(chan);                   // Select channel
    ConvertADC();                       // Convert
    while(BusyADC());                   // Wait until complete
    Res = ReadADC();                    // Read analog data
    CloseADC();                         // Close A/D
    return Res;                         // Return the result
}

void main(void)
{
    FSFILE *pntr;
    int result, i, j;
    float mV, Temp;
//
// Configure PORT RB0 is inout (START/STOP switch and RA1 is the
// analog input where the MCP9701A is connected to
//
    TRISAbits.TRISA1 = 1;
    TRISB = 1;
//
// Wait until BUTTON is pressed and relesed
//
    while(STRT);
    while(!STRT);
//
// Initialize MDD library
//
    while(!FSInit());
//
// Open file ANALOG.TXT in write mode (create it)
//
    pntr = FSfopenpgm("ANALOG.TXT", "w+");
//
// ======================== START OF LOOP ==========================
//
// Read the temperature and store in the file
//

while(1)
{
    result = Read_ADC_Chan(ADC_CH1);               // Get the data
    mV = result * 5000.0 / 1024.0;          // Convert to mV
    Temp = (mV – 400.0) / 19.5;             // Calculate temperature
    i = (int)Temp;                          // Integer part
    j = (int)((Temp-i)*10);                 // Fractional part
```

Figure 9.74: *Cont'd*

```
//
// Write to SD card
//
        FSfprintf(pntr, "%d.%d\n", i, j);
//
// Check if BUTTON is pressed and if so, close the file and stop the program
//
        if(STRT == 0)
        {
            FSfclose(pntr);
            while(1);
        }
//
// Wait for a second. Clock freq = 40MHz, clock cycle = 0.1 microsecond
// The basic delay with Delay10KTCYx is 1ms. With a count of 1000, the
// delay is 1 second. A loop is formed to iterate 4 times for the required delay
// of 1 second
//

        for(i=0; i< 4; i++)Delay10KTCYx(250);
}

}
```

Figure 9.74: *Cont'd*

The output voltage V_o of MCP9701A is proportional to the ambient temperature and the relationship is given by

$$V_o = T_C \times T_A + V_{0C},$$

where V_o is the sensor output voltage, T_C is the sensor temperature coefficient, T_A is the ambient temperature, and V_{0C} is the sensor output voltage at 0°C.

From the manufacturer's specifications,

$$T_C = 19.5\,\text{mV/°C}$$

$$V_{0C} = 400\,\text{mV}$$

Thus, we can write the equation as

$$V_o = 400 + 19.5\text{C},$$

where V_o is the sensor output voltage (mV) and C is the ambient temperature in °C.

Rearranging the formula, we get

$$C = \frac{V_o - 400}{19.5}$$

Figure 9.75 shows change of the output voltage with the ambient temperature.

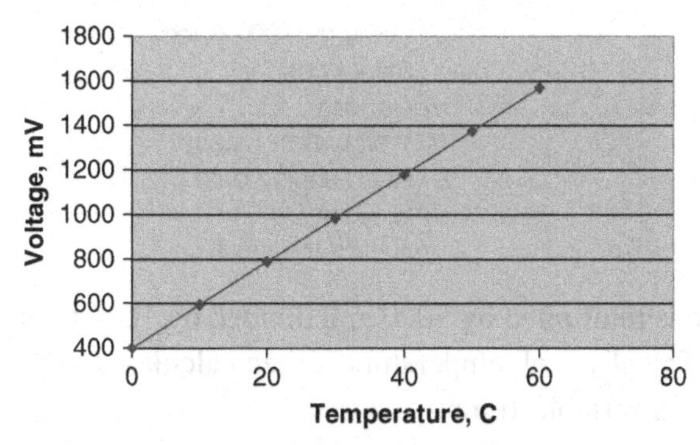

Figure 9.75: Change of Output Voltage with Ambient Temperature

The steps to calculate the temperature are as follows:

- Read the output voltage of the sensor using the A/D converter
- Convert the voltage to mV
- Use the above formula to calculate the temperature in °C

At the beginning of the program (see Figure 9.74), analog port pin RA1 and digital port pin RB0 are configured as inputs. The program then waits until the **Start/Stop** switch is pressed, and after the switch is released, the MDD library is initialized and new file ANALOG.TXT is created on the SD card. The main program loop starts with a **while** loop where the following code is executed every second inside the loop.

- Temperature data is read from analog port RA1 by calling the function Read_ADC_Chan. The port number is passed as an argument to the function. The function opens the ADC port using a clock frequency Fosc/64, the data is right-justified, interrupts are disabled, and port RA1 is configured as analog port. The C18 function SetChanADC selects the channel CH1 and then the function ConvertADC is called to start the A/D conversion, and the code waits until the conversion is ready. Once the data is converted, it is read from the A/D module by calling the function ReadADC, and the result is returned by the function to the calling program:

```
OpenADC(ADC_FOSC_64           &
        ADC_RIGHT_JUST        &
        ADC_0_TAD,
        ADC_CH1               &
        ADC_INT_OFF           &
        ADC_VREFPLUS_VDD      &
        ADC_VREFMINUS_VSS,
        13);
```

```
Delay10TCYx(10);                      // Delay for 100 cycles
SetChanADC(chan);                     // Select channel
ConvertADC();                         // Convert
while(BusyADC());                     // Wait until complete
Res = ReadADC();                      // Read analog data
CloseADC();                           // Close A/D
return Res;                           // Return the result
```

- The converted data is multiplied by 5000 and divided by 1024 to obtain the temperature data in millivolts. The physical temperature is then calculated in °C using the derived formula and stored in variable Temp:

```
result = Read_ADC_Chan(ADC_CH1);      // Get the data
mV = result * 5000.0 / 1024.0;        // Convert to mV
Temp = (mV – 400.0) / 19.5;           // Calculate temperature
```

- The decimal part of the temperature is stored in the variable i and the fractional part is stored in the variable j. In this program, only one digit is used after the decimal point:

```
i = (int)Temp;                        // Integer part
j = (int)((Temp–i)*10);               // Fractional part
```

- The temperature data is then written to the file using the MDD function FSfprintf:

```
FSfprintf(pntr, "%d.%d\n", i, j);
```

- The program then checks whether or not the Start/Stop button is pressed and if so closes the file and stops the program; otherwise, the above steps are repeated after a 1-s delay.

Figure 9.76 shows contents of file ANALOG.TXT for a typical run of the program.

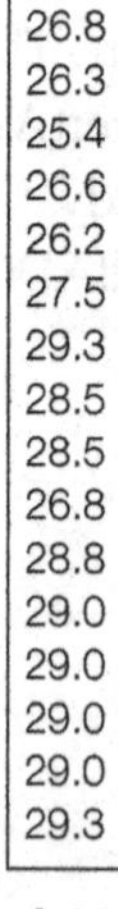
26.8
26.3
25.4
26.6
26.2
27.5
29.3
28.5
28.5
26.8
28.8
29.0
29.0
29.0
29.0
29.3

Figure 9.76: Typical ANALOG.TXT File

9.18.8 Suggestions for Future Work

Write a program similar to the program given in this project, but use two temperature sensors and save the data from both the sensors in the same file every minute.

9.19 PROJECT 18 – Temperature and Pressure Data Logging with Real-Time Clock

9.19.1 Description

This project is about data logging of temperature and pressure data on the SD card every second with real-time clock data. As in the previous project, an analog temperature sensor IC (MCP9701A) is used to sense the ambient temperature and is connected to port RA1 of the PIC18F8722 microcontroller. In addition, a pressure sensor IC (MPX4115A), connected to port RA2, is used to measure the ambient pressure. The temperature and the pressure data are stored in files TEMP.DAT and PRESS.DAT, respectively.

A real-time clock is made up in software using timer TMR0 of the microcontroller. The clock software is updated every second and keeps the absolute date and time information as long as the program is running. The clock is loaded from the PC by entering the current date and time from the keyboard.

An active low push-button **Start/Stop/Configure** switch is provided (connected to RB0) to control the operation of the program. The program operates in three modes: Configuration mode, Run mode, and Stop mode. The Configuration mode is entered when the **Start/Stop/Configure** button is pressed after the microcontroller is turned on or after the master reset is applied. During this mode, the user enters the real-time date and clock data. Pressing the **Start/Stop/Configure** button again places the program in the Run mode. In this mode, data is collected from the temperature and pressure sensors and stored in the files on the SD card every second. Pressing the **Start/Stop/Configure** button for a few seconds places the program into Stop mode where the files on the SD card are closed and the card can be removed safely from its holder.

9.19.2 Aim

The aim of this project is to show how multiple analog data can be collected and stored on the SD card continuously. In addition, the project shows how a real-time clock can be implemented in software.

9.19.3 Block Diagram

The block diagram of this project is shown in Figure 9.77.

9.19.4 Circuit Diagram

The circuit diagram of this project is shown in Figure 9.78. The temperature sensor and the pressure sensor ICs are connected to analog ports RA1 and RA2 of the microcontroller, respectively. The project is based on a PIC18F8722 microcontroller, and a 10-MHz crystal is used to provide clock pulses to the microcontroller. The serial port is connected to port pins TXD and RXD of the microcontroller via a MAX232-type RS232 level converter chip.

9.19.5 Operation of the Project

The operation of the project is shown in Figure 9.79.

9.19.6 Program Code

The program code (LOG.C) is shown in Figure 9.80.

9.19.7 Description of the Program Code

Before looking at the code in detail, it is worthwhile to see how the MPX4115A pressure sensor IC operates.

MPX4115A is a 6- or 8-pin pressure sensor IC that provides an output voltage proportional to the ambient atmospheric pressure. The pin configurations are as follows:

Pins	Descriptions
6-pin sensor	
1	Output voltage
2	Ground
3	+5-V supply
4–6	Not used
8-pin sensor	
1	Not used
2	+5-V supply
3	Ground
4	Output voltage
5–8	Not used

Figure 9.81 shows the 6-pin MPX4115A pressure sensor chip.

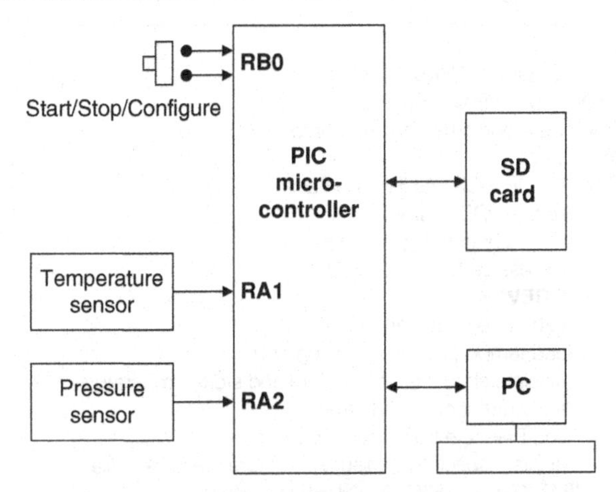

Figure 9.77: Block Diagram of the Project

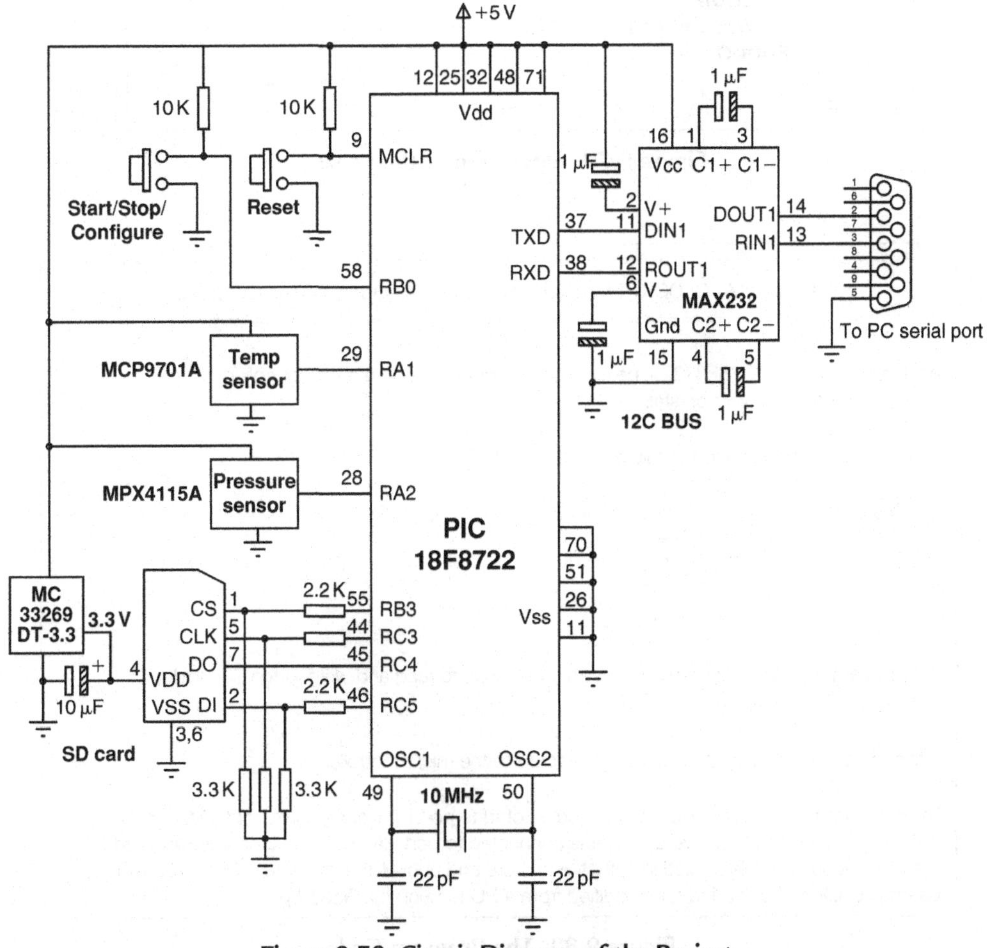

Figure 9.78: Circuit Diagram of the Project

```
BEGIN
    IF START/STOP/CONFIGURE is pressed
        Enter into configuration mode
        Read Date and time from keyboard
    ENDIF
    IF START/STOP/CONFIGURE is pressed
        Initialise the MDD library
        Create Temperature file on SD card
        Create Pressure file on SD card
        DO FOREVER
            Read current date and time
            Read temperature from analog port
            Calculate absolute temperature and store temperature in file
            Read current date and time
            Read Pressure from analog port
            Calculate absolute pressure and store pressure in file
            IF START/STOP/CONFIGURE is pressed
                Close files
                Wait here forever
            ENDIF
            Wait a second
        ENDDO
    ENDIF
END
```

Figure 9.79: Operation of the Project

```
/*****************************************************************************************************************
            REAL-TIME CLOCK TEMPERATURE AND PRESSURE DATA LOGGING
            =====================================================

In these projects, a PIC18F8722-type microcontroller is used. The microcontroller
is operated with a 10-MHz crystal.

An SD card is connected to the microcontroller as follows:

SD card                     microcontroller
CS                          RB3
CLK                         RC3
DO                          RC4
DI                          RC5

The program uses the Microchip MDD library functions to read and write to the SD card.


In this project two analog sensors are connected to the microcontroller:

An MCP9701-type analog temperature is connected to port RA1 of the microcontroller. Similarly, an
MPX4115A-type analog pressure sensor is connected to port RA2 of the microcontroller. In addition,
a real-time clock is implemented in software to store and update the absolute date and time. The
real-time clock data is loaded and modified from a PC through the RS232 port.
```

Figure 9.80: The Program Code

```
Two files are created on the SD card: TEMP.DAT and PRESS.DAT store the temperature and
pressure data, respectively.

MCP9701A chip is a 3-pin device where pin 1 is the supply (+5V), pin 2 is the Vout and pin 3 is the
ground.

The output voltage of the sensor is proportional to the temperature and is given by:

                              C = (Vout – 400) / 19.5

where, Vout is the sensor output voltage in mV

MPX4115 chip is a 6- or 8-pin chip. The 6-pin version is used in this project with the following pin
configuration:

          Pin 1 Output voltage
          Pin 2 Ground
          Pin 3 +5V supply
          Pins 4-6 not used

The output voltage of MPX4115 is proportional to the atmospheric pressure and is given by:

                              mb = (2.0V + 0.95)/0.009

where mb is the pressure in millibars and V is the sensor output voltage.

The system has three modes of operation: the Configuration Mode, Run mode, and the STOP mode.
The Configuration Mode is entered when button START/STOP/CONFIGURE is pressed. Current date
and time are entered from the PC keyboard and stored and updated every second in the program.

When the START/STOP/CONFIGURATION button is pressed the program enters the RUN mode where
data is collected every second from the two sensors and stored in files on the SD card every second.

When the START/STOP/CONFIGURE button is held down for a few seconds the program enters the
STOP mode where the files are closed on the SD card and thus the SD card can be removed from
its holder.



Author:   Dogan Ibrahim
Date:     August 2009
File:     LOG.C
*********************************************************************************************************/
#include <p18f8722.h>
#include <stdlib.h>
#include <stdio.h>
#include <string.h>
#include <delays.h>
#include <adc.h>
#include <usart.h>
#include <i2c.h>
#include <FSIO.h>
```

Figure 9.80: *Cont'd*

```
#pragma config WDT = OFF, OSC = HSPLL, LVP = OFF
#pragma config MCLRE = ON,MODE = MC, CCP2MX = PORTC

#define STRT PORTBbits.RB0
#define NUM(x,y) (10*(x - '0') + y - '0')
#define MSD(x) ((x / 10 + '0'))
#define LSD(x) ((x % 10) + '0')

unsigned char DateTime[] = "12/10/09 10:00:00 ";
unsigned char One_Sec_Flag;
unsigned char Days[]={0,31,28,31,30,31,30,31,31,30,31,30,31};

void One_Second_Delay(void);
void timer_ISR(void);

//
// Define the high interrupt vector to be at 0x08
//
#pragma code high_vector=0x08                    // Following code at void address 0x08
interrupt(void)

{
    _asm GOTO timer_ISR _endasm                  // Jump to ISR
}
#pragma code                                     // Return to default code section

//
// timer_ISR is an interrupt service routine (jumps here every 5ms)
//
#pragma interrupt timer_ISR
void timer_ISR()
{
    One_Sec_Flag = 1;                            // Set One_Sec_Flag
    TMR0L = 0x69;                                // Re-load TMR0
    TMR0H = 0x67;                                // Re-Load   TMR0H
    INTCON = 0x20;                               // Set T0IE and clear T0IF
}


//
// This function opens the A/D converter, reads analog data, converts into
// digital and then returns. The required channel number is passed as an
// argument to the function
//
int Read_ADC_Chan(unsigned char chan)
{
    int Res;

    OpenADC(ADC_FOSC_64          &               // Open the A/D
            ADC_RIGHT_JUST       &
            ADC_0_TAD,
            ADC_CH1              &
            ADC_INT_OFF          &
```

Figure 9.80: *Cont'd*

```
                ADC_VREFPLUS_VDD        &
                ADC_VREFMINUS_VSS,
                129);
    Delay10TCYx(10);                              // Delay for 100 cycles
    SetChanADC(chan);                             // Select channel
    ConvertADC();                                 // Convert
    while(BusyADC());                             // Wait until complete
    Res = ReadADC();                              // Read analog data
    CloseADC();                                   // Close A/D
    return Res;                                   // Return the result
}


//
// Configure Timer for 1 second interrupts. Load TIMER registers with
// 16 bit value 26473 = 0x6769
//
void Config_Timer(void)
{
   T0CON = 0x87;                                  // 16 bit,Prescaler = 256
   TMR0L = 0x69;                                  // Load TMR0L=0x69
   TMR0H = 0x67;                                  // Load TMR0H=0x67
   INTCON = 0xA0;                                 // Enable TMR0 interrupt
}



//
// This function initializes the USART
//
void Init_USART(void)
{
    Open1USART(USART_TX_INT_OFF  &
                 USART_RX_INT_OFF  &
                 USART_ASYNCH_MODE &
                 USART_EIGHT_BIT   &
                 USART_CONT_RX     &
                 USART_BRGH_LOW,
                 129);
}




//
// This function is used to read the real-time date and time from the PC
// keyboard and then load, store, and update the date and time in software
//
void Configure(void)
{
    char itm, RTCLen;
//
// Initialise USART to 4800 baud, 8 bits, no parity
//
    Init_USART();
```

Figure 9.80: *Cont'd*

```
//
// Read the date and time from the keyboard
//
    while(Busy1USART());
    putrs1USART(" Enter Date and Time (12/10/09 10:00:00): ");
//
// Read the date and time (until the Enter key is pressed)
//
    itm = 0;
    RTCLen = 0;
    while(itm != 0x0D)
    {
        while(!DataRdy1USART());
        itm = getc1USART();
        putc1USART(itm);
        DateTime[RTCLen] = itm;
        RTCLen++;
    }
    RTCLen--;
    DateTime[RTCLen] = '\0';
//
// send a message
//
    putrs1USART("\n\rDate and time set. Press START/STOP to start data logging...\n\r");
}



//
// This function updates the clock fields.
// 02/10/09 12:00:00
// 01234567890123456
//
void Update_Clock(void)
{
    unsigned char day,month,year,hour,minute,second;

    day = NUM(DateTime[0], DateTime[1]);
    month=NUM(DateTime[3], DateTime[4]);
    year = NUM(DateTime[6], DateTime[7]);
    hour = NUM(DateTime[9], DateTime[10]);
    minute = NUM(DateTime[12], DateTime[13]);
    second = NUM(DateTime[15], DateTime[16]);

    second++;
    if(second == 60)
    {
        second = 0; minute++;
        if(minute == 60)
        {
            minute = 0; hour++;
            if(hour == 24)
            {
                hour = 0; day++;
```

Figure 9.80: *Cont'd*

```
                  if(day == Days[month]+1)
                  {
                      day = 1; month++;
                      if(month == 13)
                      {
                          month = 1; year++;
                      }
                  }
              }
          }
      }
      DateTime[15] = MSD(second);
      DateTime[16] = LSD(second);
      DateTime[12] = MSD(minute);
      DateTime[13] = LSD(minute);
      DateTime[9]  = MSD(hour);
      DateTime[10] = LSD(hour);
      DateTime[6]  = MSD(year);
      DateTime[7]  = LSD(year);
      DateTime[3]  = MSD(month);
      DateTime[4]  = LSD(month);
      DateTime[0]  = MSD(day);
      DateTime[1]  = LSD(day);
}



void main(void)
{
      FSFILE *pntrTemp, *pntrPress;
      int result, i, j;
      float mV, Temp, Press;
//
// Configure PORT RB0 is inout (START/STOP switch, RA1, and RA2 are the
// analog inputs where the MCP9701A and MPX4115A are connected to
//
      TRISAbits.TRISA1 = 1;                          // RA1 is input
      TRISAbits.TRISA2 = 1;                          // RA2 is input
      TRISBbits.TRISB0 = 1;                          // RB0 is input

//
// Check if START/STOP button is pressed and if so enter the Configuration mode,
//  read the clock data and wait until START/STOP button is pressed and then
//  continue to data logging
//
      One_Sec_Flag = 0;

      while(STRT);                                // Wait until button pressed
      while(!STRT);
      //
      // Enter Configuration mode. Wait until the button is released
      // and call function Configure to read the real time date and time.
      //
      Configure();
```

Figure 9.80: *Cont'd*

```
//
// Configure Timer. Start timer interrupts
//
    Config_Timer();
//
// Look for leap year and adjust for February
//
    result = 10*(DateTime[6] - '0') + DateTime[7] - '0';
    result = result % 4;
    if(result == 0)
        Days[2] = 29;
    else
        Days[2] = 28;
//
// Wait until the START/STOP button is pressed and if so enter the Run mode.
// In this mode data is collected from the sensors and stored in two files
// on the SD card
//
    while(STRT);
    while(!STRT);

//
// Initialize MDD library
//
    while(!FSInit());
//
// Create files TEMP.DAT and PRESS.DAT (in write mode)
//
    pntrTemp  = FSfopenpgm("TEMP.DAT", "w+");
    pntrPress = FSfopenpgm("PRESS.DAT", "w+");

//
// ========================= START OF LOOP ==========================
//
// Read the temperature and store in the file
//

while(1)
{
    result = Read_ADC_Chan(ADC_CH1);                    // Get the temperature
    mV = result * 5000.0 / 1024.0;                      // Convert to mV
    Temp = (mV - 400.0) / 19.5;                         // Calculate temperature
    i = (int)Temp;                                      // Integer part
    j = (int)((Temp-i)*10);                             // Fractional part
//
// Write to file TEMP.DAT on the SD card. First get the date and time
//
    FSfprintf(pntrTemp, "%s %d.%d\n", DateTime, i, j);
//
// Read the pressure and store in the file
//
    result = Read_ADC_Chan(ADC_CH2);                    // Get the Pressure
    mV = result * 5000.0 / 1024.0;                      // Convert to mV
```

Figure 9.80: *Cont'd*

```
    Press = (2.0*mV + 950.0) / 9.0;                    // Calculate pressure
    i = (int)Press;                                    // Integer part
    j = (int)((Press-i)*10);                           // Fractional part
//
// Write to file PRESS.DAT on the SD card. First get the date and time
//
    FSfprintf(pntrPress, "%s %d.%d\n", DateTime, i, j);
//
// Check if START/STOP is pressed and if so, enter the Stop mode and close
// the files and stop the program
//
    if(STRT == 0)
    {
        FSfclose(pntrTemp);
        FSfclose(pntrPress);
        while(1);
    }
//
// Wait for a second. Variable "flag" is set whevenever an interrupt occurs
//

    while(!One_Sec_Flag);
    One_Sec_Flag = 0;
    Update_Clock();
}

}
```

Figure 9.80: *Cont'd*

Figure 9.81: MPX4115A Pressure Sensor

The output voltage V of MPX4115A is given by

$$V = 5.0 \times (0.009 \times kPa - 0.095)$$

or

$$kPa = \frac{\frac{V}{5.0} + 0.095}{0.009},$$

where kPa = Atmospheric pressure (kilopascals) and V = Output voltage (Volts).

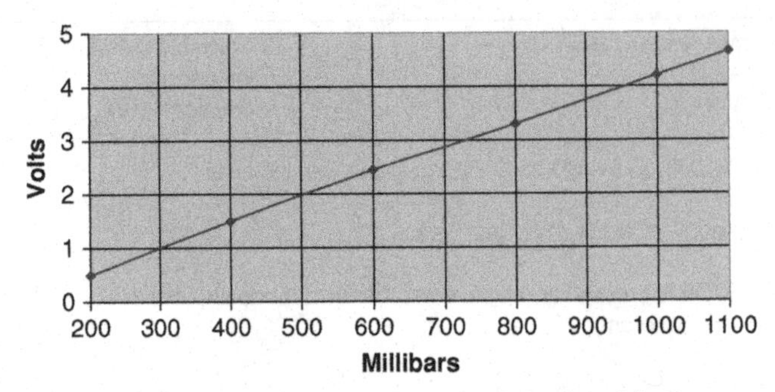

Figure 9.82: Variation of the Output Voltage with Pressure

In atmospheric pressure measurements, millibar is the most frequently used unit. The atmospheric pressure at sea level and at 15°C temperature is 1013.3 millibars. We can express the pressure in millibars if the above equation is multiplied by 10:

$$\text{mb} = 10 \times \frac{\frac{V}{5.0} + 0.095}{0.009}$$

or

$$\text{mb} = \frac{2.0V + 0.95}{0.009}$$

Figure 9.82 shows variation of the output voltage with the ambient pressure. Here, the area of interest is in the region 800–1000 millibars. To calculate the ambient pressure, we have to read the output voltage of the pressure sensor and then use the above formula to calculate the actual pressure in millibars.

The program (see Figure 9.80) consists of the following functions:

Timer_ISR: This is the interrupt service routine. The timer registers TMR0L and TMR0H are loaded and the timer interrupt is reenabled.

Read_ADC: This function opens the A/D converter and reads the analog data. The channel number is passed as an argument to the function. The A/D is opened with the following parameters:

- A/D clock is set to Fosc/64
- A/D data is right-justified
- A/D interrupts are disabled
- VDD is used as the A/D reference voltage
- VSS is used as the A/D ground voltage

The required A/D channel is selected using the SetChanADC statement. After this, the ConvertADC statement is used to start the conversion process. The program uses the

BusyADC function to wait until the conversion is complete. The converted data is then read using the ReadADC function of the C18 compiler. The function returns the converted digital data to the main calling program.

Config_Timer: This function configures the timer TMR0 so that interrupts are generated every second. The timer is configured with the following settings:

- TMR0 in 16-bit mode
- Prescaler = 256
- TMR0L = 0 × 69
- TMR0H = 0 × 67
- INTCON = 0 × A0 to enable timer interrupts and global interrupts

The timer TMR0 interrupts are generated at the intervals given by the formula:

$$\text{Interval} = 4 \times \text{ Tosc} \times \text{Prescaler} \times \text{ (65536} - \text{TMR0)},$$

where Tosc is the microcontroller clock period, Prescaler is the TMR0 prescaler value selected, and TMR0 is the value to be loaded into the 16-bit timer registers.

Here, the following values are used:

$$\text{Tosc} = 0.025\,\mu\text{s}\,(\text{Fosc} = 40\,\text{MHz})$$

$$\text{Prescaler} = 256$$

$$\text{TMR0} = 0 \times 6769\,\text{or decimal}\,26{,}473$$

The timer interrupt interval is then given by

$$\text{Interval} = 4 \times 0.1 \times 256 \times (65\,536 - 26\,473) = 1\,000\,012\,\mu\text{s}$$

or by 1.000012 s.

Init_USART: This function opens the USART and prepares it for serial communication. The following parameters are used:

- USART transmit and receive interrupts are disabled.
- USART is in asynchronous mode.
- USART is in 8-bit mode.
- Continuous receive is used.
- Baud rate is set to 4800.

Configure: This function reads the current date and time as entered from the PC keyboard. The function calls Init_USART to initialize the USART. Then the message

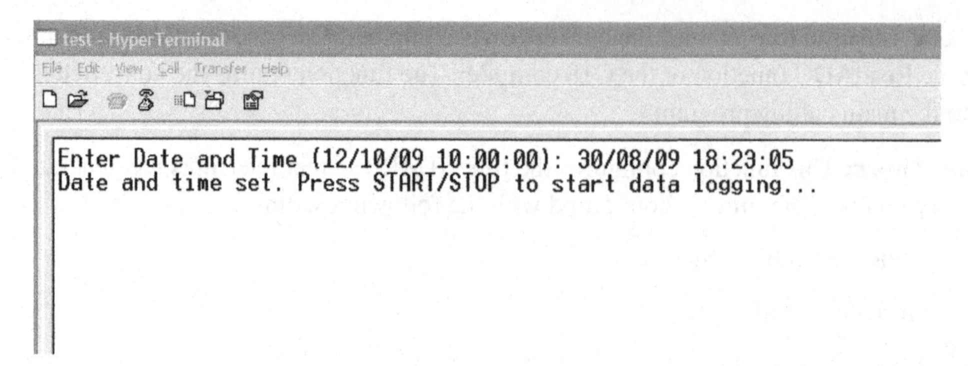

Figure 9.83: Prompting for Date and Time

"Enter Date and Time (12/10/09 10:00:00):"

is displayed and the user is expected to enter the current data and time (see Figure 9.83). The message

"Date and time set. Press **Start/Stop** to start data logging"

is displayed after reading the current date and time.

Update_Clock: This function updates the real-time clock software. Initially, Macro NUM is used to convert the date and time fields into decimal values. Then, the seconds field is incremented, and depending on the value of seconds, all other fields are updated accordingly. For example, if the seconds field is 60, it is set to 0 and the minutes field is incremented. If the minutes field is 60, it is set to 0 and the hours field is incremented, and so on. Before exiting the function, Macros MSD and LSD are used to convert the date and time fields back into character variables.

At the beginning of the main program (see Figure 9.80), analog ports RA1 and RA2 and digital port RB0 are all configured as inputs. The program then checks and enters the Configuration mode if the **Start/Stop/Configure** button is pressed. In this mode, the function Configure is called to read the real-time date and time from the user and also the timer TMR0 is configured by calling to the function Config_Timer. The program then checks for leap years and adjusts the number of days in February. The number of days in each month is stored in array Days.

The Run mode is entered when the button **Start/Stop/Configure** is pressed. The program initializes the MDD library, creates files TEMP.DAT and PRESS.DAT on the SD card, and then the main program loop is entered. Inside this loop, the temperature and pressure sensor data are read every second and are stored in files.

```
30/12/08  21:59:00  26.5
31/12/08  23:59:01  26.5
31/12/08  23:59:02  23.5
31/12/08  23:59:03  27.8
31/12/08  23:59:04  27.0
31/12/08  23:59:05  28.3
31/12/08  23:59:06  28.3
31/12/08  23:59:07  28.3
31/12/08  23:59:08  28.5
31/12/08  23:59:09  28.8
31/12/08  23:59:10  29.2
31/12/08  23:58:11  28.8
31/12/08  23:59:12  27.8
31/12/08  23:59:13  27.3
31/12/08  23:59:14  26.5
31/12/08  23:59:15  26.0
```

Figure 9.84: Contents of TEMP.DAT for a Typical Run

The temperature data is read by setting the channel to ADC_CH1. The value read is converted into millivolts, and then the absolute temperature is obtained in Celsius by subtracting 400 from the data and dividing it by 19.5. Variables *i* and *j* are set to store the integer and fractional parts of the temperature data. This data is written to file TEMP.DAT using the MDD function FSfprintf.

The pressure data is read by setting the channel number to ADC_CH2. The value read is converted into millivolts, and then the absolute pressure is obtained in millibars by multiplying the data by 2, adding 950, and then dividing by 9.

The variable **One_Sec_Flag** is set whenever an interrupt occurs, and this causes the real-time clock to be updated and the program to wait for a second before repeating the data collection process.

The Stop mode is entered when **Start/Stop/Configure** is pressed for a few seconds. In this mode, the two files are closed and the program waits forever so that the SD card can be removed from its holder safely.

Figure 9.84 shows the contents of file TEMP.DAT for a typical run.

9.19.8 Suggestions for Future Work

Modify the program given in Figure 9.80 so that the data collection interval can be read from the keyboard. In addition, use a DS1307 or a PCF8583-type RTC chip to store the real-time clock data. Store the data collection interval in the electrically erasable programmable read only memory (EEPROM) of the microcontroller.

Appendix A–MC33269 Data Sheet

Order this document by MC33269/D

MOTOROLA

MC33269

Advance Information

Low Dropout Positive Fixed and Adjustable Voltage Regulators

800 mA LOW DROPOUT THREE–TERMINAL VOLTAGE REGULATORS

The MC33269 series are low dropout, medium current, fixed and adjustable, positive voltage regulators specifically designed for use in low input voltage applications. These devices offer the circuit designer an economical solution for precision voltage regulation, while keeping power losses to a minimum.

The regulator consists of a 1.0 V dropout composite PNP–NPN pass transistor, current limiting, and thermal shutdown.

- 3.3 V, 5.0 V, 12 V and Adjustable Versions
- Space Saving DPAK, SOP–8 and SOT–223 Power Packages
- 1.0 V Dropout
- Output Current in Excess of 800 mA
- Thermal Protection
- Short Circuit Protection
- Output Trimmed to 1.0% Tolerance
- No Minimum Load Requirement for Fixed Voltage Output Devices

ORDERING INFORMATION

Device	Operating Temperature Range	Package
MC33269D	$T_J = -40°$ to $+125°C$	SOP–8
MC33269DT		DPAK
MC33269ST		SOT–223
MC33269T		Insertion Mount
MC33269D–3.3		SOP–8
MC33269DT–3.3		DPAK
MC33269ST–3.3		SOT–223
MC33269T–3.3		Insertion Mount
MC33269D–5.0		SOP–8
MC33269DT–5.0		DPAK
MC33269T–5.0		Insertion Mount
MC33269ST–5.0		SOT–223
MC33269D–12		SOP–8
MC33269DT–12		DPAK
MC33269T–12		Insertion Mount
MC33269ST–12		SOT–223

DEVICE TYPE/NOMINAL OUTPUT VOLTAGE

Device	Voltage	Device	Voltage
MC33269D	Adj	MC33269D–5.0	5.0 V
MC33269DT	Adj	MC33269DT–5.0	5.0 V
MC33269T	Adj	MC33269T–5.0	5.0 V
MC33269ST	Adj	MC33269ST–5.0	5.0 V
MC33269D–3.3	3.3 V	MC33269D–12	12 V
MC33269DT–3.3	3.3 V	MC33269DT–12	12 V
MC33269T–3.3	3.3 V	MC33269T–12	12 V
MC33269ST–3.3	3.3 V	MC33269ST–12	12 V

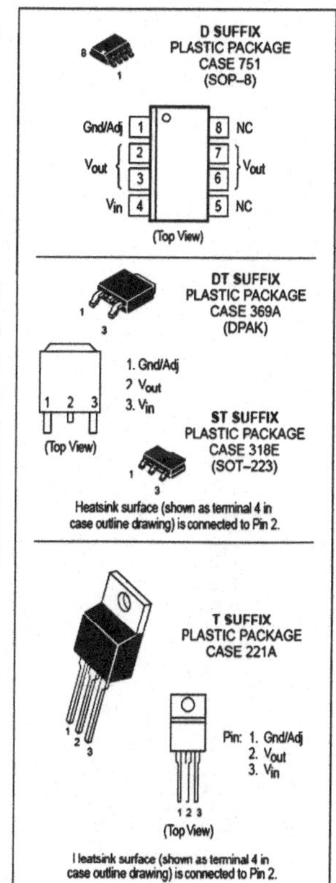

This document contains information on a new product. Specifications and information herein are subject to change without notice.

Rev 4

Appendix B–MAX232 Data Sheet

- **Meet or Exceed TIA/EIA-232-F and ITU Recommendation V.28**
- **Operate With Single 5-V Power Supply**
- **Operate Up to 120 kbit/s**
- **Two Drivers and Two Receivers**
- **±30-V Input Levels**
- **Low Supply Current . . . 8 mA Typical**
- **Designed to be Interchangeable With Maxim MAX232**
- **ESD Protection Exceeds JESD 22**
  - **2000-V Human-Body Model (A114-A)**
- **Applications**
  - **TIA/EIA-232-F**
  - **Battery-Powered Systems**
  - **Terminals**
  - **Modems**
  - **Computers**

MAX232 . . . D, DW, N, OR NS PACKAGE
MAX232I . . . D, DW, OR N PACKAGE
(TOP VIEW)

Pin	Name	Pin	Name
1	C1+	16	V_{CC}
2	V_{S+}	15	GND
3	C1–	14	T1OUT
4	C2+	13	R1IN
5	C2–	12	R1OUT
6	V_{S-}	11	T1IN
7	T2OUT	10	T2IN
8	R2IN	9	R2OUT

description/ordering information

The MAX232 is a dual driver/receiver that includes a capacitive voltage generator to supply EIA-232 voltage levels from a single 5-V supply. Each receiver converts EIA-232 inputs to 5-V TTL/CMOS levels. These receivers have a typical threshold of 1.3 V and a typical hysteresis of 0.5 V, and can accept ±30-V inputs. Each driver converts TTL/CMOS input levels into EIA-232 levels. The driver, receiver, and voltage-generator functions are available as cells in the Texas Instruments LinASIC™ library.

ORDERING INFORMATION

T_A	PACKAGE†		ORDERABLE PART NUMBER	TOP-SIDE MARKING
0°C to 70°C	PDIP (N)	Tube	MAX232N	MAX232N
	SOIC (D)	Tube	MAX232D	MAX232
		Tape and reel	MAX232DR	
	SOIC (DW)	Tube	MAX232DW	MAX232
		Tape and reel	MAX232DWR	
	SOP (NS)	Tape and reel	MAX232NSR	MAX232
–40°C to 85°C	PDIP (N)	Tube	MAX232IN	MAX232IN
	SOIC (D)	Tube	MAX232ID	MAX232I
		Tape and reel	MAX232IDR	
	SOIC (DW)	Tube	MAX232IDW	MAX232I
		Tape and reel	MAX232IDWR	

† Package drawings, standard packing quantities, thermal data, symbolization, and PCB design guidelines are available at www.ti.com/sc/package.

Appendix C—LM35 Data Sheet

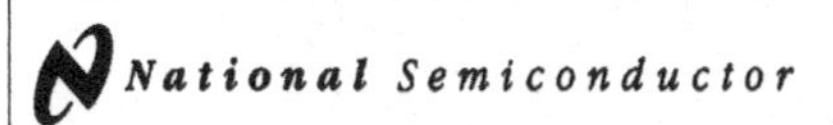

November 2000

LM35
Precision Centigrade Temperature Sensors

General Description

The LM35 series are precision integrated-circuit temperature sensors, whose output voltage is linearly proportional to the Celsius (Centigrade) temperature. The LM35 thus has an advantage over linear temperature sensors calibrated in ° Kelvin, as the user is not required to subtract a large constant voltage from its output to obtain convenient Centigrade scaling. The LM35 does not require any external calibration or trimming to provide typical accuracies of ±¼°C at room temperature and ±¾°C over a full −55 to +150°C temperature range. Low cost is assured by trimming and calibration at the wafer level. The LM35's low output impedance, linear output, and precise inherent calibration make interfacing to readout or control circuitry especially easy. It can be used with single power supplies, or with plus and minus supplies. As it draws only 60 μA from its supply, it has very low self-heating, less than 0.1°C in still air. The LM35 is rated to operate over a −55° to +150°C temperature range, while the LM35C is rated for a −40° to +110°C range (−10° with improved accuracy). The LM35 series is available packaged in hermetic TO-46 transistor packages, while the LM35C, LM35CA, and LM35D are also available in the plastic TO-92 transistor package. The LM35D is also available in an 8-lead surface mount small outline package and a plastic TO-220 package.

Features

- Calibrated directly in ° Celsius (Centigrade)
- Linear + 10.0 mV/°C scale factor
- 0.5°C accuracy guaranteeable (at +25°C)
- Rated for full −55° to +150°C range
- Suitable for remote applications
- Low cost due to wafer-level trimming
- Operates from 4 to 30 volts
- Less than 60 μA current drain
- Low self-heating, 0.08°C in still air
- Nonlinearity only ±¼°C typical
- Low impedance output, 0.1 Ω for 1 mA load

Typical Applications

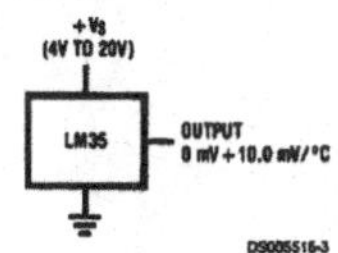

FIGURE 1. Basic Centigrade Temperature Sensor (+2°C to +150°C)

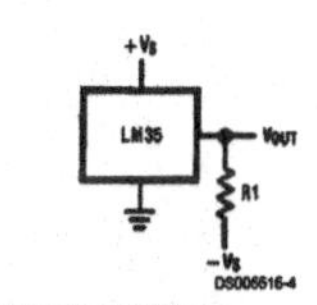

Choose $R_1 = -V_S/50\ \mu A$
V_{OUT} = +1,500 mV at +150°C
= +250 mV at +25°C
= −550 mV at −55°C

FIGURE 2. Full-Range Centigrade Temperature Sensor

 DS005516 www.national.com

Appendix D–MPX4115A Data Sheet

MOTOROLA
SEMICONDUCTOR TECHNICAL DATA

Order this document
by MPX4115A/D

Integrated Silicon Pressure Sensor for Manifold Absolute Pressure, Altimeter or Barometer Applications On-Chip Signal Conditioned, Temperature Compensated and Calibrated

MPX4115A
MPXA4115A
SERIES

INTEGRATED PRESSURE SENSOR
15 to 115 kPa (2.2 to 16.7 psi)
0.2 to 4.8 Volts Output

Motorola's MPX4115A/MPXA4115A series sensor integrates on-chip, bipolar op amp circuitry and thin film resistor networks to provide a high output signal and temperature compensation. The small form factor and high reliability of on-chip integration make the Motorola pressure sensor a logical and economical choice for the system designer.

The MPX4115A/MPXA4115A series piezoresistive transducer is a state-of-the-art, monolithic, signal conditioned, silicon pressure sensor. This sensor combines advanced micromachining techniques, thin film metallization, and bipolar semiconductor processing to provide an accurate, high level analog output signal that is proportional to applied pressure.

Figure 1 shows a block diagram of the internal circuitry integrated on a pressure sensor chip.

Features

- 1.5% Maximum Error over 0° to 85°C
- Ideally suited for Microprocessor or Microcontroller–Based Systems
- Temperature Compensated from –40° to +125°C
- Durable Epoxy Unibody Element or Thermoplastic (PPS) Surface Mount Package

Application Examples

- Aviation Altimeters
- Industrial Controls
- Engine Control
- Weather Stations and Weather Reporting Devices

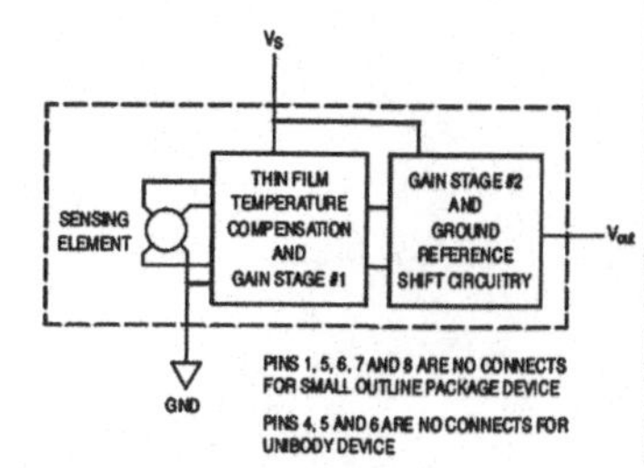

Figure 1. Fully Integrated Pressure Sensor Schematic

SMALL OUTLINE PACKAGE

MPXA4115A6U
CASE 482

MPXA4115AC6U
CASE 482A

PIN NUMBER			
1	N/C	5	N/C
2	V_S	6	N/C
3	Gnd	7	N/C
4	V_{out}	8	N/C

NOTE: Pins 1, 5, 6, 7, and 8 are internal device connections. Do not connect to external circuitry or ground. Pin 1 is noted by the notch in the lead.

UNIBODY PACKAGE

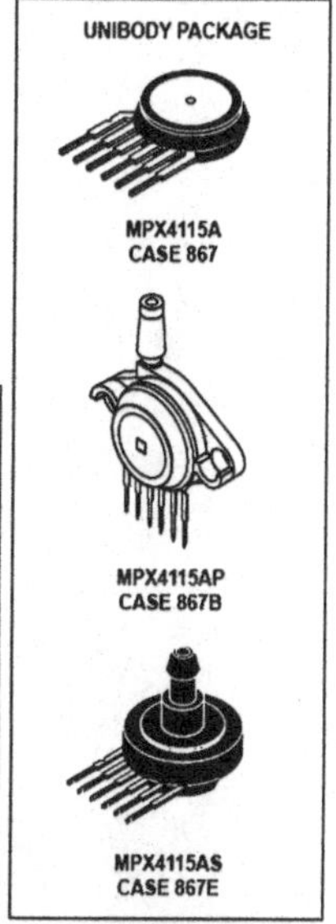

MPX4115A
CASE 867

MPX4115AP
CASE 867B

MPX4115AS
CASE 867E

PIN NUMBER			
1	V_{out}	4	N/C
2	Gnd	5	N/C
3	V_S	6	N/C

NOTE: Pins 4, 5, and 6 are internal device connections. Do not connect to external circuitry or ground. Pin 1 is noted by the notch in the lead.

REV 4

MOTOROLA

Index

G

H

I

L

Printed in the United States
By Bookmasters